Examples of Groups

Examples of Groups

Michael Weinstein

Polygonal Publishing House
80 Passaic Avenue
Passaic NJ, 07055

Paste-up and mechanicals by Jose Garcia
Artwork by Dorothy Brosterman
Cover design by Janet Wallach

Library of Congress Cataloging in Publication Data

Weinstein, Michael 1945–
Examples of groups.

Bibliography: p.
Includes indexes.
1. Groups, Theory of. I. Title.
QA171.W39 512'.22 76-51379

Printed in the United States of America

Dedicated to Belle, Abe, and Jerry who helped a lot

Contents

Logical Dependence of Chapters and Where to Find Things

It is recommended that this book be read "grasshopper" style, rather than from cover to cover. The reader, depending on his tastes, can begin reading in Chapters 4, 5, or 6 and turn to the earlier chapters and to the appendixes when he needs to. As can be seen from the chart, the appendixes provide background material necessary for the other chapters. It is suggested, therefore, that the reader scan the appendixes to see what topics are covered there before he begins reading elsewhere.

Many terms are used in the book that are not assumed to be known to the reader. Their definitions can be found in the glossary. The notation index performs a similar function for notation.

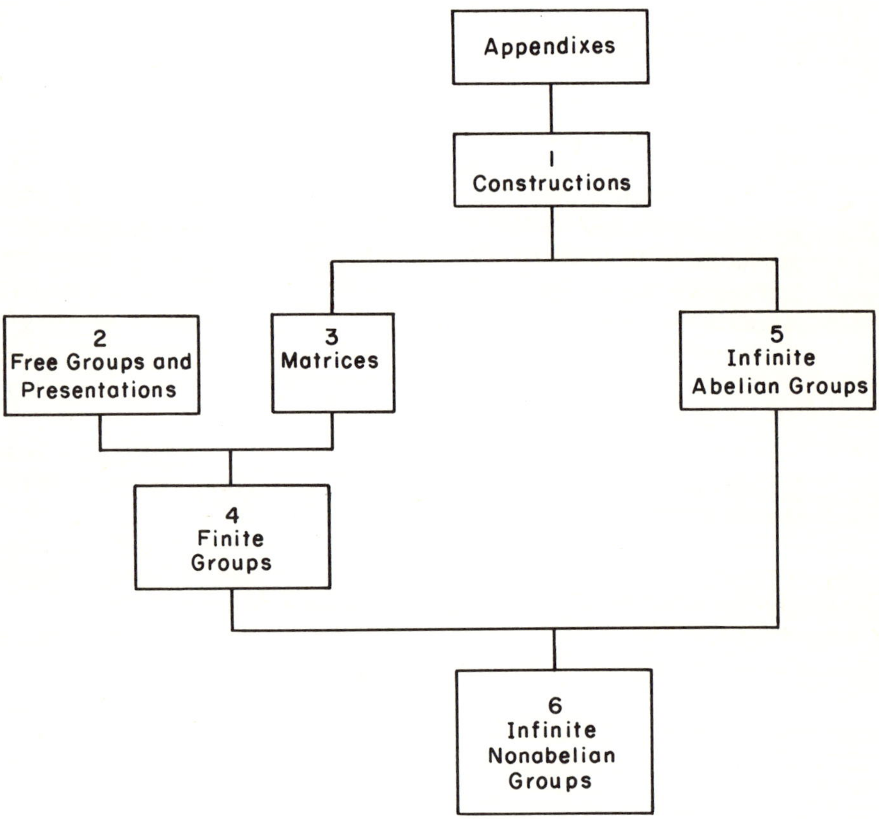

Preface

"Theories change but the groups remain."
Marshall Hall Jr. and James K. Senior [15]

Why study examples?

Several reasons come to mind. For the mathematician doing research, examples are all but indispensable to his work. To begin with, the direction of his research is guided by a thorough examination of all the pertinent examples he can get his hands on. Only after these examples are analyzed does he attempt to formulate their common properties into some sort of a theorem, and then attempt to prove the theorem.

For the student of group theory, regardless of his level, examples help to clarify and justify the definitions and theorems of the subject he is studying. Take, for instance, the definition of solvable group. As soon as the student sees this he will want to know the answers to at least two questions

(a) Are there such things?
(b) Are there groups that are nonsolvable?

These questions are more than just reasonable, they seem necessary to an understanding of the concept. Both questions must be answered by examples.

A third use of examples, related but not identical to the previous two, is the idea of a counterexample. Counterexamples are given, and rightfully so, a much more honored position in mathematices than the two kinds of examples previously discussed. Where the examples a research mathematician uses to formulate a theorem he later proves take a back seat to the theorem itself, and where the examples a student (or his instructor) proffers to answer questions such as (a) and (b) are of a rather personal and short-lived nature, a counterexample is as much a result in the theory of groups as a theorem is, and occupies a similarly high position. Theorems and counterexamples are simply results in opposite directions. Theorems assert that certain propositions about groups are true while counterexamples assert that other propositions are false.

This book was written with the dual purposes in mind of providing examples that serve to illustrate various group-theoretical concepts, and providing counterexamples.

One further topic requires some explanation. If certain methods and materials can be used to supply an example with interesting properties, it is not unreasonable to hope that a subtle change in these methods or materials will result in a group with equally interesting but somewhat different properties. It is not unreasonable to hope that the methods can be abstracted into a theory which yields not one but a number of different examples. When such a procedure is effected, it is called a construction. Constructions will be described abstractly, that is without regard to what material groups go into the construction, in Chapter 1, then particularized when they are needed in the later chapters.

Throughout this book, the ideas of group properties and classes of groups will play an extremely important role. Although logically distinct, these two concepts will be used almost interchangeably. A **class** of groups, $\mathcal{H}$, is a collection of groups such that $A \in \mathcal{H}$ and $A \cong B$ imply $B \in \mathcal{H}$ (classes are closed under isomorphisms). Based on this, a group **property** can be defined as a proposition about groups such that the collection of groups satisfying the proposition is a class of groups. Alternately, *property* could have been defined first, then *class* defined in terms of property. For all practical purposes, the two concepts are identical.

This book is intended for anyone who has taken an introductory course in group theory, or even a more general introductory course in abstract algebra. It is, in fact, ideally suited as a supplement to an introductory group theory textbook.

In more detail, the requisite group-theoretical concepts assumed on the part of the reader are the following: groups and homomorphisms; subgroups and quotient groups; normal, characteristic, and fully-invariant subgroups; p-Sylow subgroups; the three isomorphism theorems and the correspondence theorem; commutators and commutator subgroups; permutation groups; cyclic groups; centralizers and normalizers; the direct product of two groups; automorphisms and automorphism groups; and Lagrange's theorem. All of this material can be found in any of the references Baumslag and Chandler [2], Hall [14], Rotman [31], or Scott [33].

The mathematical prerequisites outside of group theory necessary to read this book will be slight. Knowledge of such basic topics as functions, unions and intersections, subsets, and cardinality will be assumed, as will a nodding acquaintance with rings and fields. All other necessary concepts will be explained.

The exercises are an important part of the book insofar as most of them continue the presentation of material pertinent to the subject of examples and constructions. As all of the proofs given in the text are complete, none of these proofs will be made to depend on previously given exercises. Note, however, that many of the exercises themselves will rely on previous ones. Hints to the solution of some of the more difficult exercises are given at the end of the book.

The author would like to express his sincere thanks and appreciation to Franklin Haimo, Mark P. Hale Jr., Leo Katzenstein, and Fred Richman who read various parts of the first draft of this book and made many valuable suggestions. Special thanks also go to Rich Perline for general encouragement while the book was in progress.

Michael Weinstein

PART ONE
MATERIALS AND CONSTRUCTIONS

1

Constructions

As the name implies, a construction is a manufacturing process used to supply examples of groups. As in any such process, certain basic materials are required to begin. It should come as no surprise that groups themselves are the most common basic material used in the construction of new groups. Other common materials include ordered n-tuples, functions, and homomorphisms. It will be explained at the beginning of each construction described in this chapter exactly what materials are required. The collection of these materials will usually be called a **system**.

After each of the various constructions is described, certain problems and questions naturally suggest themselves. The most important of these is the question of which properties of groups are preserved by the construction. If all the groups used as basic materials in a construction satisfy a given property, one wants to know whether or not this fact is sufficient to ensure that the constructed group will satisfy the property as well. When it is sufficient, the property is said to be **preserved** by the construction. Such considerations will occupy a considerable amount of this chapter and its exercises.

A second major consideration will be to investigate subgroups of the constructed groups. First of all there is the problem of determining in some way subgroups such as $Z(G)$, G', $H(G)$, etcetera, for a constructed group G. Secondly there is the question of whether or not subgroups of the material groups induce in some fashion subgroups of the constructed group. And thirdly there is the possibility that the construction or something about it will suggest ways of defining certain subgroups of the constructed groups that are not induced by subgroups of the material groups.

A third area of investigation will be a consideration of homomorphisms among constructed groups. The fundamental question here is: Can a homomorphism between two material groups, one used to construct group A and the other used to construct group B, be made to induce in some fashion a homomorphism between A and B?

Category theory will make brief appearances from time to time in this chapter, usually only in the exercises. Roughly speaking, a categorical statement about groups is one that can be expressed solely in terms of the groups, possible homomorphisms among them, and the concept of composition of homomorphisms. A categorical statement avoids any mention of the various elements of the groups and the group multiplications. Categorical statements can be made about the constructed groups for many of the constructions described in this chapter. Moreover most of these statements have true converses. That is, if the categorical statement given in this chapter for a certain constructed group A also applies to a group B, then $B \cong A$. These converses will not usually be given as exercises because of the length and difficulty involved in stating them. Once stated, however, their proofs are not difficult.

1.1 Central Pushouts

The construction about to be described is variously called by different authors the pushout, the central product, the direct product with amalgamated center, or the direct product with amalgamated subgroup. It is particularly useful for constructing nilpotent groups.

Materials for the construction are three groups, one of which must be abelian, and two injective homomorphisms.

DEFINITION (A,H,K,h,k) is a **central pushout system** if and only if A is an abelian group, H and K are groups, $h{:}A \to Z(H)$ is an injective homomorphism, and $k{:}A \to Z(K)$ is an injective homomorphism.

It is helpful to think of a central pushout system as a diagram

$$\begin{array}{ccc} A & \xrightarrow{\;k\;} & K \\ {\scriptstyle h}\downarrow & & \\ H & & \end{array}$$

We define the central pushout of the system (A,H,K,h,k) to be the group $(H \times K)/D$ where $D = \{(h(x), k(x^{-1})) \mid x \in A\}$. In order for this definition to make sense, it must be shown that D is a normal subgroup.

THEOREM 1.1.1 *If (A,H,K,h,k) is a central pushout system, then D as defined above is a normal subgroup of $H \times K$.*

Proof: Let $s = (h(x), k(x^{-1}))$ and $t = (h(y), k(y^{-1}))$ belong to D. Then $st^{-1} =$

$(h(x)h(y^{-1}), k(x^{-1})k(y)) = (h(x)h(y^{-1}), k(y)k(x^{-1}))$ since $k(y) \in Z(K)$ by definition of central pushout system. Therefore $st^{-1} = (h(xy^{-1}), k(yx^{-1}))$ which equals $(h(xy^{-1}), k((xy^{-1})^{-1}))$. Thus $st^{-1} \in D$ showing that D is a subgroup. To show normality, let $(h(x), k(x^{-1}))$ belong to D and let (u,v) belong to $H \times K$. Then $(u,v)^{-1}(h(x),k(x^{-1}))(u,v) = (u^{-1}h(x)u, v^{-1}k(x^{-1})v)$. But $h(x) \in Z(H)$ and $k(x^{-1}) \in Z(K)$ so the last ordered pair becomes $(h(x)u^{-1}u, k(x^{-1})v^{-1}v)$ which is $(h(x),k(x^{-1})) \in D$.

For a central pushout system (A,H,K,h,k) and its central pushout G, we define two further functions, $h^*: H \to G$ and $k^*: K \to G$ as follows: $h^*(x) = D(x,1)$ for all $x \in H$, and $k^*(x) = D(1,x)$ for all $x \in K$. Since $h^*(x)h^*(y) = D(x,1)$ $D(y,1) = D(xy,1) = h^*(xy)$, h^* is a homomorphism, and similarly so is k^*. They will be referred to as the **associated homomorphisms.**

THEOREM 1.1.2 *The associated homomorphisms are injective.*
Proof: Suppose $h^*(x) = 1$. That is, $D(x,1) = 1$ so $(x,1) \in D$. Consequently there exists $a \in A$ such that $(x,1) = (h(a), k(a^{-1}))$. Then $1 = k(a^{-1})$ implies $a^{-1} = 1$ by injectivity of k. This further implies $a = 1$ so $x = h(a) = h(1) = 1$. Therefore h^* is injective. Similarly so is k^*.

THEOREM 1.1.3 *If (A,H,K,h,k) is a central pushout system and G is its central pushout, then the following diagram commutes.*

$$\begin{array}{ccc} A & \xrightarrow{k} & K \\ {\scriptstyle h}\downarrow & & \downarrow{\scriptstyle k^*} \\ H & \xrightarrow[h^*]{} & G \end{array}$$

Proof: Let $x \in A$. Then $h^* \circ h(x) = D(h(x),1)$ while $k^* \circ k(x) = D(1,k(x))$. But $(h(x),1)(1,k(x))^{-1} = (h(x), k(x^{-1})) \in D$. Therefore $D(h(x),1) = D(1,k(x))$ and so the diagram commutes.

THEOREM 1.1.4 *Let (A,H,K,h,k) be a central pushout system and let G be its central pushout. Then*
(a) *$N \lhd H$ implies $h^*[N] \lhd G$.*
(b) *$N \lhd K$ implies $k^*[N] \lhd G$.*
Proof: To show Part (a), let $x \in h^*[N]$ and let $g \in G$. Thus $x = h^*(w) = D(w,1)$ for some $w \in N$ and $g = D(p,q)$ for some $p \in H$ and $q \in K$. Therefore $g^{-1}xg = D(p,q)^{-1}D(w,1)D(p,q) = D(p^{-1}wp,1) = h^*(p^{-1}wp)$. But $w \in N$ implies that $p^{-1}wp \in N$ because N is normal in H. Thus $g^{-1}xg = h^*(p^{-1}wp) \in h^*[N]$ showing $h^*[N] \lhd G$, proving Part (a). The proof of Part (b) is similar.

COROLLARY *Let (A,H,K,h,k) be a central pushout system and let G be its central pushout. Then $h^*[H]$, $k^*[K]$, and $h^*[H] \cap k^*[K]$ are all normal subgroups of G.*

Proof: Since H is a normal subgroup of itself, the theorem shows $h^*[H] \lhd G$ and similarly $k^*[K] \lhd G$. Finally $h^*[H] \cap k^*[K]$ is the intersection of two normal subgroups so it too is normal.

The last normal subgroup mentioned in the above corollary plays a sufficiently important role in the theory of central pushouts to warrant its own notation.

NOTATION If (A,H,K,h,k) is a central pushout system and G is its central pushout, then $\mathrm{R}(G)$ will denote $h^*[H] \cap k^*[K]$.

THEOREM 1.1.5 *Let $\mathcal{H}$ be a quotient-closed class of groups such that $A,B \in \mathcal{H}$ implies $A \times B \in \mathcal{H}$. Suppose (A,H,K,h,k) is a central pushout system and G is its central pushout. Then $H,K \in \mathcal{H}$ implies $G \in \mathcal{H}$ (that is, $\mathcal{H}$ is preserved by central pushouts).*

Proof: The group $H \times K$ belongs to $\mathcal{H}$ since H and K do. This in turn implies $G = (H \times K)/D \in \mathcal{H}$.

THEOREM 1.1.6 *Suppose $\mathcal{H}$ is one of the following classes:* (*In* (c) *and* (d), *n is a fixed natural number. In* (b), *p is a fixed prime.*)

(a) *torsion groups,*
(b) *p-groups,*
(c) *groups of exponent n,*
(d) $\{G \mid G$ *is class-m nilpotent where* $m \leqslant n\}$,
(e) *nilpotent groups.*

Then $\mathcal{H}$ is preserved by central pushouts.

Proof: If A and B are torsion groups then clearly $A \times B$ is likewise. It is also clear that the class of torsion groups is quotient closed. Applying Theorem 1.1.5 proves Part (a). Parts (b) and (c) are similar.

To show Part (d), note that $\mathcal{H}$ is quotient closed by Theorem A.5.2. Next suppose $A,B \in \mathcal{H}$. Thus A is class-m nilpotent and B is class-m' nilpotent where $m,m' \leqslant n$. Then by Theorems A.2.4 and A.5.4, $A \times B$ is class-max$\{m,m'\}$ nilpotent. Since $\max\{m,m'\} \leqslant n$, $A \times B \in \mathcal{H}$ too. Consequently Theorem 1.1.5 may be applied to complete the proof of Part (d). Part (e) is proved similarly.

THEOREM 1.1.7 *Let (A,H,K,h,k) be a central pushout system and let G be its central pushout. Then*

(a) $\mathrm{R}(G) = h^* \circ h[A] = k^* \circ k[A]$.
(b) $G/\mathrm{R}(G) = (h^*[H]/\mathrm{R}(G)) \oplus (k^*[K]/\mathrm{R}(G))$.

Proof: First note that

$$h^* \circ h[A] = k^* \circ k[A] \tag{1}$$

is just a restatement of Theorem 1.1.3. Now $h[A] \subseteq H$ implies

$$h^* \circ h[A] \subseteq h^*[H] \tag{2}$$

and similarly

$$k^*\circ k[A] \subseteq k^*[K] \tag{3}$$

Equation 1 and Inclusion 3 imply $h^*\circ h[A] \subseteq k^*[K]$. This and Inclusion 2 yield $h^*\circ h[A] \subseteq \mathrm{R}(G)$.

Conversely let $D(x,y) \in \mathrm{R}(G)$. Thus there exist $p \in H$ and $q \in K$ such that $D(x,y) = h^*(p) = k^*(q)$. Hence $D(x,y) = D(p,1) = D(1,q)$ so $(x,y)(p,1)^{-1} \in D$ and $(x,y)(1,q)^{-1} \in D$. Since $(xp^{-1},y) \in D$, there exists some $a \in A$ such that $(xp^{-1},y) = (h(a),k(a^{-1}))$. Since $(x,yq^{-1}) \in D$, there exists some $b \in A$ such that $(x,yq^{-1}) = (h(b),k(b^{-1}))$. Therefore $y = k(a^{-1})$ and $x = h(b)$. Set $c = a^{-1}b$. Then $c \in A$ and $(x,y)(h(c),1)^{-1} = (h(b),k(a^{-1}))(h(c^{-1}),1) = (h(b),k(a^{-1}))\cdot(h(b^{-1}a),1) = (h(a),k(a^{-1})) \in D$. Thus $D(x,y) = D(h(c),1) = h^*\circ h(c) \in h^*\circ h[A]$ completing the proof of Part (a).

By the Corollary to Theorem 1.1.4, $h^*[H]$ and $k^*[K]$ are normal subgroups of G so $h^*[H]/\mathrm{R}(G)$ and $k^*[K]/\mathrm{R}(G)$ are normal subgroups of $G/\mathrm{R}(G)$. Note that as a consequence of the definition of $\mathrm{R}(G)$, $(h^*[H]/\mathrm{R}(G)) \cap (k^*[K]/\mathrm{R}(G))$ is trivial. Finally note that any $D(x,y) \in G$ can be expressed as $D(x,1)D(1,y) = h^*(x)k^*(y)$ so $G = h^*[H]k^*[K]$. Thus $G/\mathrm{R}(G) = (h^*[H]/\mathrm{R}(G))(k^*[K]/\mathrm{R}(G))$ completing the proof of Part (b).

THEOREM 1.1.8 *Let (A,H,K,h,k) be a central pushout system and let G be its central pushout. Then* $\mathrm{R}(G) \subseteq \mathrm{Z}(G)$.

Proof: Let $D(x,y) \in \mathrm{R}(G)$, let $D(p,q) \in G$, and seek to show these two elements commute. Since $\mathrm{R}(G) = h^*\circ h[A]$ by Theorem 1.1.7, $D(x,y) \in \mathrm{R}(G)$ implies $D(x,y) = D(h(a),1)$ for some a. Therefore $D(x,y)D(p,q) = D(h(a),1)D(p,q) = D(h(a)p,q) = D(ph(a),q)$ (since $h(a) \in \mathrm{Z}(H)$) $= D(p,q)D(h(a),1) = D(p,q)D(x,y)$.

In the remainder of this section an additional assumption will be made on the two homomorphisms h and k in a central pushout system, namely that these be onto $\mathrm{Z}(H)$ and $\mathrm{Z}(K)$ respectively. This requires $\mathrm{Z}(H) \cong A \cong \mathrm{Z}(K)$. When this assumption is made, considerably more can be said about the central pushout.

THEOREM 1.1.9 *Let (A,H,K,h,k) be a central pushout system where $h{:}A \to \mathrm{Z}(H)$ and $k{:}A \to \mathrm{Z}(K)$ are isomorphisms and let G be its central pushout. Then*

(a) $\mathrm{R}(G) = \mathrm{Z}(G)$.

(b) $\mathrm{Z}(G) \cong A$.

(c) $G/\mathrm{Z}(G) = (h^*[H]/\mathrm{Z}(G)) \oplus (k^*[K]/\mathrm{Z}(G))$.

Proof: By Theorem 1.1.8, $\mathrm{R}(G) \subseteq \mathrm{Z}(G)$, so to prove Part (a) only the reverse inclusion need be shown. Let $D(p,q) \in \mathrm{Z}(G)$. First it is claimed that $p \in \mathrm{Z}(H)$. For all $x \in H$, $D(x,q^{-1}) \in G$ so $D(p,q)D(x,q^{-1}) = D(x,q^{-1})D(p,q)$. Hence $D(px,1) = D(xp,1)$. In other words, $h^*(px) = h^*(xp)$. But this implies $px = xp$ by injectivity of h^* (Theorem 1.1.2), and establishes the claim. Similarly $q \in \mathrm{Z}(K)$. Now by hypothesis h is surjective. Hence there exists $a \in A$ such that $p = h(a)$. Thus $h^*(p) = h^*(h(a))$ which belongs to $\mathrm{R}(G)$ by Theorem 1.1.7, Part (a). Similarly

$k^*(q) \in \mathrm{R}(G)$. From $D(p,1) \in \mathrm{R}(G)$ and $D(1,q) \in \mathrm{R}(G)$ conclude the product $D(p,q) \in \mathrm{R}(G)$ as required. This proves Part (a). Part (b) follows immediately since $\mathrm{R}(G) \cong A$ (by virtue of the function $h^* \circ h : A \to \mathrm{R}(G)$). Finally substituting $\mathrm{Z}(G)$ for $\mathrm{R}(G)$ in Theorem 1.1.7 yields Part (c).

THEOREM 1.1.10 *Let (A,H,K,h,k) be a central pushout system where both $h:A \to \mathrm{Z}(H)$ and $k:A \to \mathrm{Z}(K)$ are isomorphisms, and let G be its central pushout. Then if A, H, and K are finite groups,*

$$\mathrm{card}(G) = \frac{\mathrm{card}(H)\mathrm{card}(K)}{\mathrm{card}(A)}.$$

Proof: By Theorem 1.1.9, Part (c), $\mathrm{card}(G/\mathrm{Z}(G)) = \mathrm{card}(h^*[H]/\mathrm{Z}(G))$ times $\mathrm{card}(k^*[K]/\mathrm{Z}(G))$. Now $h^*[H] \cong H$ because h^* is injective (Theorem 1.1.2) and $\mathrm{Z}(G) \cong A$ by Theorem 1.1.9 again. Thus

$$\mathrm{card}(h^*[H]/\mathrm{Z}(G)) = \frac{\mathrm{card}(H)}{\mathrm{card}(A)}$$

and similarly

$$\mathrm{card}(k^*[K]/\mathrm{Z}(G)) = \frac{\mathrm{card}(K)}{\mathrm{card}(A)}.$$

Thus

$$\begin{aligned}\mathrm{card}(G) &= \mathrm{card}(G/\mathrm{Z}(G))\mathrm{card}(\mathrm{Z}(G)) \\ &= \frac{\mathrm{card}(H)}{\mathrm{card}(A)}\ \frac{\mathrm{card}(K)}{\mathrm{card}(A)}\ \mathrm{card}(A) \\ &= \frac{\mathrm{card}(H)\mathrm{card}(K)}{\mathrm{card}(A)}.\end{aligned}$$

THEOREM 1.1.11 *Let (A,H,K,h,k) be a central pushout system where both $h:A \to \mathrm{Z}(H)$ and $k:A \to \mathrm{Z}(K)$ are isomorphisms, and let G be its central pushout. Then for all natural numbers n, $h^*[\mathrm{Z}_n(H)] \subseteq \mathrm{Z}_n(G)$ and $k^*[\mathrm{Z}_n(K)] \subseteq \mathrm{Z}_n(G)$.*

Proof: The inclusion for h^* will be shown; the corresponding proof for k^* is similar. Clearly the result is true when $n = 0$ so it may be assumed that $1 \leqslant n$.

First note that for any $x \in \mathrm{Z}(H)$, $x = h(a)$ for some $a \in A$ since h is an isomorphism. Therefore $h^*(x) = h^* \circ h(a) \in h^* \circ h[A] = \mathrm{Z}(G)$ by Parts (a) of Theorems 1.1.7 and 1.1.9. Thus $h^*[\mathrm{Z}(H)] \subseteq \mathrm{Z}(G)$. Theorem A.10.1 can now be applied to produce a surjective homomorphism $f:H/\mathrm{Z}(H) \to h^*[H]/\mathrm{Z}(G)$ defined by $f(\mathrm{Z}(H)x) = \mathrm{Z}(G)h^*(x)$. Then by the Lemma to Theorem A.5.2,

$$f[\mathrm{Z}_{n-1}(H/\mathrm{Z}(H))] \subseteq \mathrm{Z}_{n-1}(h^*[H]/\mathrm{Z}(G)). \tag{1}$$

Let $x \in \mathrm{Z}_n(H)$. Then $\mathrm{Z}(H)x \in \mathrm{Z}_n(H)/\mathrm{Z}(H) = \mathrm{Z}_{n-1}(H/\mathrm{Z}(H))$ by Theorem A.5.1, Part (c). Consequently $f(\mathrm{Z}(H)x) \in \mathrm{Z}_{n-1}(h^*[H]/\mathrm{Z}(G))$ by Inclusion (1). But $G/\mathrm{Z}(G) = (h^*[H]/\mathrm{Z}(G)) \oplus (k^*[K]/\mathrm{Z}(G))$ (Theorem 1.1.9) which implies that $Z_{n-1}(G/Z(G)) = Z_{n-1}(h^*[H]/Z(G)) \oplus Z_{n-1}(k^*[K]/Z(G))$ by the Lemma to Theorem A.5.4. Thus $f(Z(H)x) \in Z_{n-1}(G/Z(G))$, and so $Z(G)h^*(x)$ belongs to $Z_n(G)/Z(G)$ by the definition of f and by a second application of Theorem A.5.1, Part (c). Conclude $h^*(x) \in Z_n(G)$.

THEOREM 1.1.12 *Let (A,H,K,h,k) be a central pushout system where both $h:A \to Z(H)$ and $k:A \to Z(K)$ are isomorphisms, and let G be its central pushout. Then $h^*[H]/Z(G) \cong H/Z(H)$ and $k^*[K]/Z(G) \cong K/Z(K)$.*

Proof: First note that $h^*[Z(H)] \subseteq Z(G)$ by Theorem 1.1.11. Applying Theorem A.10.1 yields a surjective homomorphism $f:H/Z(H) \to h^*[H]/Z(G)$ defined by $f(Z(H)x) = Z(G)h^*(x)$. To show f is also injective suppose $f(Z(H)x) = 1$. Then $Z(G)h^*(x) = 1$ so $h^*(x) \in Z(G) = h^* \circ h[A]$ by Theorems 1.1.7, Part (a) and 1.1.9, Part (a). Hence $h^*(x) = h^*(h(a))$ for some $a \in A$. But then $x = h(a)$ by injectivity of h^* (Theorem 1.1.2). Consequently $x \in Z(H)$, so $Z(H)x = 1$ proving f is injective. Conclude $h^*[H]/Z(G) \cong H/Z(H)$. The proof that $k^*[K]/Z(G) \cong K/Z(K)$ is similar.

EXERCISE

1. Show the following classes are preserved by central pushouts.
   (a) finitely generated,
   (b) solvable,
   (c) divisible,
   (d) semidivisible,
   (e) perfect.

1.2 Semidirect Products

This construction is probably the single most important in group theory. It can be used to produce a wide number and variety of both finite and infinite nonabelian groups. Moreover several other constructions depend on semidirect products. (One of these, the wreath product, will be examined in Section 1.7.)

DEFINITION (A,g,B) is a **semidirect system** if and only if A and B are groups and $g:A \to \text{Aut}(B)$ is a homomorphism.

NOTE: Semidirect systems and direct systems (see Section 1.5) are similar in name only. The two concepts are otherwise unrelated.

NOTATION Let (A,g,B) be a semidirect system and let $a \in A$. Then g_a (rather than $g(a)$) will denote the value of g at a. Since $g_a \in \text{Aut}(B)$, g_a is itself a function (from B into B). If $b \in B$, the value of g_a at b will be denoted $g_a(b)$.

DEFINITION AND NOTATION If (A,g,B) is a semidirect system, then $B \times_g A$ is the **semidirect product** of (A,g,B) if and only if $B \times_g A$ is the group (this will be proved below) with underlying set $B \times A$ and with multiplication defined by $(b,a)(b',a') = (bg_a(b'),aa')$.

LEMMA *Let S be a set with an associative multiplication, a right identity element (that is, $e \in S$ such that $xe = x$ for all $x \in S$), and right inverses (that is, for every $x \in S$, there exists some y such that $xy = e$). Then S is a group.*
Proof: What must be shown is that e also acts as a left identity element and that the right inverses are also left inverses. Let $x \in S$ and seek to show $ex = x$. By hypothesis, there exists $y \in S$ such that $xy = e$ and there exists $z \in S$ such that $yz = e$. Then $exy = ee = e = xy$ so $exyz = xyz$ implying $exe = xe$. However $xe = x$ so $ex = x$ as required.

Again let $x \in S$, so there exists $y \in S$ such that $xy = e$. Seek to show $yx = e$ as well. Since $y \in S$, there exists $z \in S$ such that $yz = e$. Therefore $z = ez = xyz = xe = x$. This and $yz = e$ imply $yx = e$.

THEOREM 1.2.1 *If (A,g,B) is a semidirect system, then $B \times_g A$ is a group. In addition,* $(1,1)$ *is the identity element of this group and inverses are described by* $(b,a)^{-1} = (g_{a^{-1}}(b^{-1}),a^{-1})$.
Proof: To show the group operation is associative, let (b,a), (x,y), and (p,q) belong to $B \times A$. Then $((b,a)(x,y))(p,q) = (bg_a(x),ay)(p,q) = (w,ayq)$ where $w = bg_a(x)g_{ay}(p)$. On the other hand, $(b,a)((x,y)(p,q)) = (b,a)(xg_y(p),yq) = (w^*,ayq)$ where $w^* = bg_a(xg_y(p))$. But g_a is an automorphism of B so $w^* = bg_a(x)g_a(g_y(p))$ which equals $bg_a(x)(g_a \circ g_y)(p)$. But $g{:}A \to \text{Aut}(B)$ is a homomorphism. Therefore $g_{ay} = g_a \circ g_y$. Thus $w^* = bg_a(x)g_{ay}(p) = w$ establishing associativity.

To complete the proof it suffices, by the lemma, to show the existence of a right identity element and of right inverses.

Consider any element (b,a). Then $(b,a)(1,1) = (bg_a(1),a)$. But since g_a is an automorphism of B, $g_a(1) = 1$. Therefore $(b,a)(1,1) = (b,a)$. Hence $(1,1)$ is the required right identity element.

Again let $(b,a) \in B \times A$. Then $(b,a)(g_{a^{-1}}(b^{-1}),a^{-1}) = (bg_a(g_{a^{-1}}(b^{-1})),aa^{-1}) = (bg_{aa^{-1}}(b^{-1}),aa^{-1})$. But $g_{aa^{-1}} = g_1$ which is the identity automorphism of B since $g{:}A \to \text{Aut}(B)$ is a homomorphism. Therefore $g_{aa^{-1}}(b^{-1}) = b^{-1}$. Consequently $(b,a)(g_{a^{-1}}(b^{-1}),a^{-1}) = (bb^{-1},aa^{-1}) = (1,1)$ as required.

NOTATION Let (A,g,B) be a semidirect system and let G be its semidirect product. Suppose S is a subgroup of A and T is a subgroup of B. Then $^*S$ will denote $\{(1,s) \mid s \in S\}$ and T^* will denote $\{(t,1) \mid t \in T\}$.

THEOREM 1.2.2 *Let (A,g,B) be a semidirect system and let G be its semidirect product. Then*
(a) *$^*S$ is a subgroup of G whenever S is a subgroup of A.*
(b) *T^* is a subgroup of G whenever T is a subgroup of B.*

(c) $B^* \lhd G$.
(d) $G = (B^*)(^*A)$.
(e) $B^* \cap {}^*A = \{1\}$.
(f) $B^* \cong B$ *and* $G/B^* \cong A$.

Proof: Let $(1,s)$, $(1,s') \in {}^*S$. Then $(1,s)(1,s') = (1g_s(1), ss') = (1,ss') \in {}^*S$. Also if $(1,s) \in {}^*S$, then $(1,s)^{-1} = (g_{s^{-1}}(1^{-1}),s^{-1}) = (1,s^{-1}) \in {}^*S$. Thus ${}^*S$ is a subgroup of G, proving Part (a).

Let $(t,1)$ and $(t',1)$ belong to T^*. Then $(t,1)(t',1) = (tg_1(t'),1) = (tt',1) \in T^*$. Also if $(t,1) \in T^*$, then $(t,1)^{-1} = (g_{1^{-1}}(t^{-1}),1^{-1}) = (t^{-1},1) \in T^*$. Thus T^* is a subgroup of G, proving Part (b).

To show B^* is normal in G, let $(b,1)$ belong to B^*, let (x,a) belong to G, and consider $(x,a)^{-1}(b,1)(x,a)$. Ignoring the first coordinate of the product of these three ordered pairs, note that the second coordinate is $a^{-1}\cdot 1 \cdot a = 1$. Therefore $(x,a)^{-1}(b,1)(x,a) \in B^*$ establishing Part (c).

That $G = (B^*)(^*A)$ follows immediately after noting that $(b,a) = (b,1)(1,a)$ for all $(b,a) \in G$. This proves Part (d).

Next suppose $(b,a) \in B^* \cap {}^*A$. Then $b = 1$ and $a = 1$ so $(b,a) = (1,1)$, the identity element of G, proving Part (e).

Finally by Parts (d) and (e), $G/B^* = (B^*)(^*A)/B^* \cong {}^*A/(^*A \cap B^*) \cong {}^*A$. But clearly the function $f{:}A \to {}^*A$ defined by $f(a) = (1,a)$ is an isomorphism, as is the function $g{:}B \to B^*$ defined by $g(b) = (b,1)$. Thus $B^* \cong B$ and $G/B^* \cong A$, proving Part (f).

The following theorem provides examples of semidirect systems.

THEOREM 1.2.3 *Suppose B is a group and $\phi \in$ Aut(B). If $\phi^m = 1$, define $g{:}Z(m) \to$ Aut(B) by $g_k = \phi^k$. If ϕ does not have finite order in* Aut(B), *define $g{:}Z \to$ Aut(B) by $g_k = \phi^k$. In either case, g is a homomorphism.*

Proof: Suppose $\phi^m = 1$. To show $g{:}Z(m) \to$ Aut(B) is a function suppose $p = q$ in $Z(m)$. Thus $p \equiv q \bmod m$ so $mc = p - q$ for some integer c. Thus $\phi^{p-q} = \phi^{mc} = 1$, hence $\phi^p = \phi^q$ showing g is a function. Moreover g is a homomorphism because $g_{p+q} = \phi^{p+q} = \phi^p \circ \phi^q = g_p \circ g_q$ for all $p,q \in Z(m)$.

The proof that $g{:}Z \to$ Aut(B) is a homomorphism when ϕ does not have finite order is straightforward.

In the remainder of this section we consider three kinds of semidirect products. In a general semidirect product $B \times_g A$ there are, of course, two arbitrary groups, B and A. In the three particularizations about to be described only the group B will be arbitrary. The second group, A, will either be fixed or will depend on B. These particular semidirect products should not be thought of as examples. They are bona fide constructions which use only one group, B, as a basic material, rather than two. Examples are obtained by particularizing B.

NOTATION If B is an abelian group, then $\theta{:}B \to B$ will denote the function

$\theta(x) = -x$.

THEOREM 1.2.4 *If B is an abelian group, then* $\theta \in \mathrm{Aut}(B)$. *Furthermore the order of* θ *in* $\mathrm{Aut}(B)$ *is* 1 *if* $\exp(B) = 2$, *and* 2 *otherwise.*

Proof: For $x,y \in B$ $\theta(x+y) = -(x+y) = -x-y = \theta(x)+\theta(y)$. So θ is a homomorphism. If $\theta(x) = 0$, then $-x = 0$ which implies $x = 0$, hence θ is injective. Finally $\theta(-x) = x$ for all $x \in B$, so θ is surjective too, and is therefore an automorphism of B.

If $\exp(B) = 2$, then $2x = 0$ for all $x \in B$ which implies $\theta(x) = -x = x$ for all $x \in B$, hence $\theta = i$ where i denotes the identity element of $\mathrm{Aut}(B)$. On the other hand if $\exp(B) \neq 2$, then there exists some $b \in B$ such that $2b \neq 0$. Hence $\theta(b) = -b \neq b$ so $\theta \neq i$. But $\theta^2(x) = -(-x) = x$ for all $x \in B$, so $\theta^2 = i$ and hence θ has order 2.

By Theorems 1.2.3 and 1.2.4, $(Z(2),g,B)$ is a semidirect system whenever B is an abelian group such that $\exp(B) \neq 2$ and $g{:}Z(2) \to \mathrm{Aut}(B)$ is defined by $g_0 = i$ (the identity element of $\mathrm{Aut}(B)$) and $g_1 = \theta$.

DEFINITION AND NOTATION If B is an abelian group such that $\exp(B) \neq 2$ and g is defined as above, then $Dih(B)$ is the **generalized dihedral group** determined by B if and only if $Dih(B) = B \times_g Z(2)$.

NOTE: As per Sections 4.1 and 6.5 respectively, the so-called **dihedral groups** are obtained by taking $Dih(B)$ for B finite and cyclic, and the **infinite dihedral group** is obtained by letting $B = Z$.

THEOREM 1.2.5 *If B is an abelian group, such that* $\exp(B) \neq 2$, *then the following rule for commutators holds in* $Dih(B)$.

$$(b,a)^{-1}(b',a')^{-1}(b,a)(b',a') = \begin{cases} (0,0) & \text{if } a = 0 \text{ and } a' = 0, \\ (-2b,0) & \text{if } a = 0 \text{ and } a' = 1, \\ (2b',0) & \text{if } a = 1 \text{ and } a' = 0, \\ (2b-2b',0) & \text{if } a = 1 \text{ and } a' = 1. \end{cases}$$

Proof: From the rule for inversion in a semidirect product,

$$(b,a)^{-1} = (g_{-a}(-b),-a) = \begin{cases} (-b,0) & \text{if } a = 0, \\ (b,1) & \text{if } a = 1. \end{cases}$$

Similarly

$$(b',a')^{-1} = \begin{cases} (-b',0) & \text{if } a' = 0, \\ (b',1) & \text{if } a' = 1. \end{cases}$$

Thus

$$(b,a)^{-1}(b',a')^{-1} = \begin{cases} (-b,0)(-b',0) = (-b-b',0) & \text{if } a=0 \text{ and } a'=0, \\ (-b,0)(b',1) \quad = (-b+b',1) & \text{if } a=0 \text{ and } a'=1, \\ (b,1)(-b',0) \quad = (b+b',1) & \text{if } a=1 \text{ and } a'=0, \\ (b,1)(b',1) \quad = (b-b',0) & \text{if } a=1 \text{ and } a'=1. \end{cases}$$

while

$$(b,a)(b',a') = \begin{cases} (b+b',0) & \text{if } a=0 \text{ and } a'=0, \\ (b+b',1) & \text{if } a=0 \text{ and } a'=1, \\ (b-b',1) & \text{if } a=1 \text{ and } a'=0, \\ (b-b',0) & \text{if } a=1 \text{ and } a'=1. \end{cases}$$

Combining yields

$$(b,a)^{-1}(b',a')^{-1}(b,a)(b',a') = \begin{cases} (-b-b',0)(b+b',0) = (0,0) \\ (-b+b',1)(b+b',1) = (-2b,0) \\ (b+b',1)(b-b',1) \quad = (2b',0) \\ (b-b',0)(b-b',0) \quad = (2b-2b',0). \end{cases}$$

COROLLARY 1 *If B is an abelian group and* $\exp(B) \neq 2$, *then* $Dih(B)' = (2B)^*$.

Proof: Let $x = (b,a)$ and $y = (b',a')$ be any two elements of $Dih(B)$. By Theorem 1.2.5, $x^{-1}y^{-1}xy = (0,0)$, $(-2b,0)$, $(2b',0)$, or $(2(b-b'),0)$ which (in any case) belongs to $(2B)^*$, proving $Dih(B)' \subseteq (2B)^*$. Conversely let $(x,0) \in (2B)^*$. Then $x \in 2B$ implies $-x = 2b$ for some $b \in B$. Then by Theorem 1.2.5 again, $(b,0)^{-1}(0,1)^{-1}(b,0)(0,1) = (-2b,0) = (x,0)$. Thus $(x,0) \in Dih(B)'$.

COROLLARY 2 *If B is an abelian group and* $\exp(B) \neq 2$ *and B is 2-divisible, then* $Dih(B)' = B^*$.

Proof: If B is 2-divisible, then $B = 2B$ so $Dih(B)' = B^*$ by Corollary 1.

COROLLARY 3 *If B is an abelian group and* $\exp(B) \neq 2$, *then* $Z(Dih(B)) = B[2]^*$ *and hence* $Z(Dih(B)) \cong B[2]$.

Proof: Let $b \in B[2]$. Then by the theorem, for all $(b',a') \in Dih(B)$,

$$(b,0)^{-1}(b',a')^{-1}(b,0)(b',a') = \begin{cases} (0,0) & \text{if } a'=0, \\ (-2b,0) & \text{if } a'=1. \end{cases}$$

But $b \in B[2]$ implies $-2b = 0$ so the above commutator is trivial in either case. Thus $(b,0)$ belongs to the center showing $B[2]^* \subseteq Z(Dih(B))$.

Conversely let (b,a) belong to the center. First it must be shown that $a = 0$. Suppose to the contrary that $a = 1$. Since $\exp(B) \neq 2$, there exists $x \in B$ such that $2x \neq 0$. Then by the theorem, $(b,1)^{-1}(x,0)^{-1}(b,1)(x,0) = (2x,0)$. But since $(b,1)$ belongs to the center, the above commutator is trivial implying $2x = 0$ which is false. Conclude $a = 0$. Finally it must be shown that $b \in B[2]$. Once more by the theorem $(b,0)^{-1}(0,1)^{-1}(b,0)(0,1) = (-2b,0)$. This and $(b,0) \in Z(Dih(B))$ imply $(-2b,0) = (0,0)$ so $2b = 0$ and hence $b \in B[2]$.

THEOREM 1.2.6 *Let B be an abelian group such that* $\exp(B) \neq 2$. *Then every subgroup of B^* is normal in $Dih(B)$.*

Proof: Let C be a subgroup of B^*. Let $c \in C$, let $x \in Dih(B)$, and seek to show $x^{-1}cx \in C$. Since $c \in C \subseteq B^*$, $c = (b,0)$ for some b. Let $x = (b',a')$. Then

$$x^{-1} = \begin{cases} (-b',0) & \text{if } a' = 0, \\ (b',1) & \text{if } a' = 1. \end{cases}$$

Consequently

$$x^{-1}cx = \begin{cases} (-b',0)(b,0)(b',0) = (-b'+b,0)(b',0) = (b,0) & \text{if } a' = 0, \\ (b',1)(b,0)(b',1) \quad = (b'-b,1)(b',1) \quad = (-b,0) & \text{if } a' = 1. \end{cases}$$

Note, however, that $c^{-1} = (b,0)^{-1} = (-b,0)$ so

$$x^{-1}cx = \begin{cases} c & \text{if } a' = 0, \\ c^{-1} & \text{if } a' = 1. \end{cases}$$

Conclude $x^{-1}cx \in C$.

If B happens to be 2-divisible, then Theorem 1.2.6 has the following converse.

THEOREM 1.2.7 *Let B be an abelian group such that* $\exp(B) \neq 2$ *and B is 2-divisible. Then every proper normal subgroup of $Dih(B)$ is contained in B^*.*

Proof: For convenience denote $Dih(B)$ by D. Let $C \lhd D$ such that $C \neq D$. Then $B^* \subseteq B^*C \subseteq D$. This and $D/B^* \cong Z(2)$ force B^*C to equal B^* or D. The first case implies $C \subseteq B^*$ as was to be shown. To complete the proof then it suffices to show $B^*C = D$ leads to a contradiction.

Note that $B^* \cong B$ so B^* is abelian and hence so too is $B^*/(B^* \cap C)$. But since $B^*C = D$, $B^*/(B^* \cap C) \cong B^*C/C = D/C$. Thus D/C is abelian, implying $D' \subseteq C$ by Theorem A.6.4. However since B is 2-divisible, $D' = B^*$ by Corollary 2 to Theorem 1.2.5. Thus $B^* \subseteq C$, implying $D = B^*C = C$, the desired contradiction.

NOTATION Suppose B and C are abelian groups, neither having exponent 2, and $f{:}B \to C$ is a homomorphism. Let $f^*{:}Dih(B) \to Dih(C)$ denote the function defined by $f^*(b,a) = (f(b),a)$.

THEOREM 1.2.8 *Suppose B and C are abelian groups, neither having exponent 2, and $f{:}B \to C$ is a homomorphism. Then*

(a) *f^* is a homomorphism.*

(b) *If f is injective, so is f^*.*

(c) *If f is surjective, so is f^*.*

Proof: For $(b,a), (b',a') \in Dih(B)$,

$$f^*((b,a)(b',a')) = \begin{cases} f^*(b+b',a') \quad = (f(b)+f(b'),a') & \text{if } a = 0, \\ f^*(b-b',1+a') = (f(b)-f(b'),1+a') & \text{if } a = 1. \end{cases}$$

On the other hand,

$$f^*(b,a)f^*(b',a') = (f(b),a)(f(b'),a') = \begin{cases} (f(b)+f(b'),a') & \text{if } a=0, \\ (f(b)-f(b'),1+a') & \text{if } a=1. \end{cases}$$

Conclude $f^*((b,a)(b',a')) = f^*(b,a)f^*(b',a')$ proving Part (a).

Now suppose f is injective. Then $f^*(b,a) = (0,0)$ implies $(f(b),a) = (0,0)$ so $f(b) = 0$ and $a = 0$. Injectivity of f then implies $b = 0$ so $(b,a) = (0,0)$ showing f^* is injective proving Part (b).

Next suppose f is surjective. Then given $(c,a) \in Dih(C)$, $c = f(b)$ for some $b \in B$ so $f^*(b,a) = (c,a)$ showing f^* is surjective, proving Part (c).

THEOREM 1.2.9 *Let B be an abelian group and suppose T is a subgroup of B. Then $Dih(B)/T^* \cong Dih(B/T)$.* (It is assumed that $\exp(B/T) \neq 2$.)

Proof: The natural map $e{:}B \to B/T$ induces a surjective homomorphism e^* from $Dih(B)$ onto $Dih(B/T)$ by Theorem 1.2.8. It suffices to show $T^* = \ker e^*$.

Let $(t,0) \in T^*$. Then $t \in T$ so $e^*(t,0) = (e(t),0) = 1$ and hence $(t,0) \in \ker e^*$. Conversely if $(x,y) \in \ker e^*$, then $1 = e^*(x,y) = (T + x,y)$ so $T + x = 0$ and $y = 0$. Thus $x \in T$ and $y = 0$, hence $(x,y) = (x,0) \in T^*$.

THEOREM 1.2.10 *If B is an abelian group such that* $\exp(B) \neq 2$, *then $Dih(B)$ is solvable.*

Proof: By Theorem 1.2.2, $Dih(B)$ is an extension of B^* by $Z(2)$. But B^* and $Z(2)$ are both abelian and hence solvable (B^* is abelian because $B^* \cong B$). Conclude $Dih(B)$ is solvable by Theorem A.6.6.

THEOREM 1.2.11 *If B is an abelian group such that* $\exp(B) \neq 2$, *then $Dih(B)$ is hypercyclic.*

Proof: Seek to show a nontrivial quotient group $Dih(B)/K$ has a nontrivial cyclic normal subgroup. Either $B^* \cap K \neq B^*$ or $B^* \cap K = B^*$. In the first case, since B^* is abelian so is $B^*/(B^* \cap K)$. Let $C/(B^* \cap K)$ be any nontrivial cyclic subgroup of $B^*/(B^* \cap K)$. Then $C \subseteq B^*$ implies that $C \lhd Dih(B)$ by Theorem 1.2.6. Consequently $CK \lhd Dih(B)$ which further implies normality of CK/K in $Dih(B)/K$. Now $CK/K \cong C/(C \cap K)$, but $C \cap K = B^* \cap K$ because $C \subseteq B^*$. Thus $CK/K \cong C/(B^* \cap K)$ so CK/K is the required nontrivial cyclic normal subgroup of $Dih(B)/K$.

In the second case ($B^* \cap K = B^*$), $B^* \subseteq K$. This, $K \neq Dih(B)$, and $Z(2) \cong Dih(B)/B^*$ imply $K = B^*$. Thus $Dih(B)/K$ is the required nontrivial cyclic normal subgroup of itself.

THEOREM 1.2.12 *If B is an abelian group such that* $\exp(B) \neq 2$ *and B is 2-divisible, then* $(b,1)$ *and* $(b',1)$ *are conjugate in $Dih(B)$ for all $b, b' \in B$.*

Proof: By 2-divisibility, there exists $x \in B$ such that $2x = b - b'$. Consequently $(x,0)^{-1}(b,1)(x,0) = (-x,0)(b,1)(x,0) = (-x+b,1)(x,0) = (b-2x,1) = (b',1)$.

LEMMA *Divisible abelian groups have no maximal subgroups.*
Proof: Let A be a divisible abelian group and suppose M is a maximal subgroup. Then because A is abelian, A/M is a cyclic group of cardinality p for some prime p. But by divisibility, A has no proper subgroups of finite index (Theorem A.8.2), the desired contradiction.

THEOREM 1.2.13 *If B is an abelian group such that* $\exp(B) \neq 2$ *and B is divisible, then $Dih(B)$ has a unique maximal subgroup, namely B^*.*
Proof: Since $Dih(B)/B^* \cong Z(2)$, B^* is a maximal subgroup. Now let M be any maximal subgroup. Then $B^* \cap M$ is either equal to or properly contained in B^*. In the first case $B^* \subseteq M$ so $B^* = M$ as required, by maximality of B^*. It suffices then to show that the second case cannot happen. If $B^* \cap M \subset B^*$, then by the lemma $B^* \cap M$ is not a maximal subgroup of B^* so there is some subgroup N such that

$$B^* \cap M \subset N \subset B^*. \tag{1}$$

Since $N \subseteq B^*$, N is normal by Theorem 1.2.6. Therefore NM is a subgroup of $Dih(B)$, and furthermore $M \subseteq NM \subseteq Dih(B)$. Thus by maximality of M, $NM = M$ or $Dih(B)$. If $NM = M$, then $N \subseteq M$. This and $N \subseteq B^*$ imply $N \subseteq B^* \cap M$ which contradicts Expression (1). If $NM = Dih(B)$, then for all $x \in B^*$, $x = nm$ where $n \in N$ and $m \in M$. Consequently $n^{-1}x = m$ so $n^{-1}x \in B^* \cap M \subseteq N$ which further implies $x \in N$. Thus $B^* \subseteq N$ again in contradiction to Expression (1).

We now consider a second particularization of semidirect products. Again we begin with an abelian group B which we will presently require to be unbounded. Define $e: B \times B \to B \times B$ by $e(x,y) = (x + y, y)$. Then e is a homomorphism because for (x,y) and (x',y') belonging to $B \times B$, $e((x,y) + (x',y')) = e(x + x', y + y') = (x + x' + y + y', y + y') = (x + y, y) + (x' + y', y') = e(x,y) + e(x',y')$. Furthermore e is injective for if $e(x,y) = e(x',y')$, then $(x + y, y) = (x' + y', y')$ so $x + y = x' + y'$ and $y = y'$. Thus $x = x'$ too, showing injectivity. Furthermore e is surjective, for if $(x,y) \in B \times B$, then $(x - y, y) \in B \times B$ and $e(x - y, y) = (x,y)$. Conclude $e \in \mathrm{Aut}(B \times B)$.

THEOREM 1.2.14 *If B is an unbounded abelian group and e is defined as above, then* $o(e)$ *is infinite in* $\mathrm{Aut}(B \times B)$.
Proof: Suppose to the contrary that $o(e) = n$ for some n. Thus $e^n = 1$. But there exists some $a \in B$ such that $na \neq 0$ (if $nb = 0$ for all $b \in B$, then B is bounded). Hence $e^n = 1$ implies $e^n(0,a) = (0,a)$ so $(na, a) = (0,a)$ implying $na = 0$. This contradiction completes the proof.

Theorems 1.2.3 and 1.2.14 now show that the function $g: Z \to \mathrm{Aut}(B \times B)$ defined by $g_k = e^k$ is a homomorphism. That is, $(Z, g, B \times B)$ is a semidirect system.

NOTATION If B is an unbounded abelian group, then $M(B)$ will denote the semidirect product of $(Z, g, B \times B)$.

Thus the underlying set of the group $M(B)$ is $B \times B \times Z$, multiplication is described by $(x,y,i)(x',y',j) = (x+x'+iy',y+y',i+j)$, and inversion is described by $(x,y,i)^{-1} = (-x+iy,-y,-i)$.

NOTES:

(1) Alternatively to requiring that B be unbounded, it can be assumed that $\exp(B) = n$ for some natural number n. In this case a modification of Theorem 1.2.14 shows that e has order n in $\mathrm{Aut}(B \times B)$. Theorem 1.2.3 then shows that $(Z(n),g,B \times B)$ is a semidirect system where $g{:}Z(n) \to \mathrm{Aut}(B \times B)$ is defined by $g_k = e^k$. Thus semidirect products with underlying sets $B \times B \times Z(n)$ can be constructed.

(2) A generalization of the construction $M(B)$ is given in Miscellaneous Exercise I.6 at the end of Chapter 6.

THEOREM 1.2.15 *Let B be an unbounded abelian group. Then for all $x,x',y \in B$, $(x,y,1)$ and $(x',y,1)$ are conjugate in $M(B)$.*
Proof: Conjugation by $(0,x'-x,0)$ yields $(0,x'-x,0)^{-1}(x,y,1)(0,x'-x,0) = (0,x-x',0)(x,y,1)(0,x'-x,0) = (x,x-x'+y,1)(0,x'-x,0) = (x',y,1)$.

THEOREM 1.2.16 *Let B be an unbounded abelian group. Then $Z(M(B)) = \{(x,0,0) \mid x \in B\}$ and hence $Z(M(B)) \cong B$.*
Proof: Let $x \in B$. To show $(x,0,0)$ belongs to the center, let $(x',y',k) \in M(B)$. Then $(x,0,0)(x',y',k) = (x+x',y',k)$ while $(x',y',k)(x,0,0) = (x'+x,y',k)$. Conclude $\{(x,0,0) \mid x \in B\} \subseteq Z(M(B))$. Conversely let $(x,y,i) \in Z(M(B))$ and seek to show $y = 0$ and $i = 0$. First observe that $(x,y,i)(0,0,1) = (0,0,1)(x,y,i)$ hence $(x,y,i+1) = (x+y,y,1+i)$ which implies $x = x+y$ and hence $y = 0$. Next note that for all $a \in B$, $(x,0,i)(0,a,0) = (0,a,0)(x,0,i)$. Thus $(x+ia,a,i) = (x,a,i)$ showing $x+ia = x$ and hence $ia = 0$ for all $a \in B$. This and B unbounded imply $i = 0$.

This proves the first claim. As for the second, the function $h{:}B \to Z(M(B))$ defined by $h(x) = (x,0,0)$ is easily seen to be an isomorphism.

THEOREM 1.2.17 *If B is an unbounded abelian group, then $M(B)$ is class-2 nilpotent.*
Proof: Let $(x,y,i) \in M(B)$ and consider, for all $(x',y',k) \in M(B)$, the commutator $(x,y,i)^{-1}(x',y',k)^{-1}(x,y,i)(x',y',k)$. Ignoring first coordinates, this commutator is $(\ ,-y,-i)(\ ,-y',-k)(\ ,y,i)(\ ,y',k) = (\ ,0,0)$ which belongs to $Z(M(B))$ by Theorem 1.2.16. This shows $(x,y,i) \in Z_2(M(B))$ for all $(x,y,i) \in M(B)$ and hence that $M(B) = Z_2(M(B))$.

Theorem 1.2.18 *If B is an unbounded abelian group, then $M(B)/Z(M(B)) \cong B \times Z$.*
Proof: Define $d{:}M(B) \to B \times Z$ by $d(x,y,i) = (y,i)$. It is clear that d is a homomorphism and is furthermore surjective. Moreover, $(x,y,i) \in \ker d$ if and only if $y = 0$ and $i = 0$, so that $\ker d = Z(M(B))$ by Theorem 1.2.16. Consequently

$M(B)/Z(M(B)) \cong B \times Z$ by the first isomorphism theorem.

LEMMA *Extensions of torsion-free groups by torsion-free groups are torsion-free.*

Proof: Suppose G/H and H are torsion-free. Suppose $x \in G$ such that $x^n = 1$ for some positive integer n. Since G/H is torsion-free, $(Hx)^n = Hx^n = 1$ implies $Hx = 1$. Thus $x \in H$. This and $x^n = 1$ imply $x = 1$ since H is torsion-free.

THEOREM 1.2.19 *If B is a nontrivial torsion-free abelian group, then $M(B)$ is torsion-free.*

Proof: Since both B and Z are torsion-free, so is $B \times Z$ as is easily seen. Hence by Theorem 1.2.18, $M(B)/Z(M(B))$ is torsion-free. Next note that $Z(M(B))$ is also torsion-free since it is isomorphic to B by Theorem 1.2.16. Application of the lemma completes the proof.

NOTATION Suppose B and C are unbounded abelian groups and $f:B \to C$ is a homomorphism. Let $f^*:M(B) \to M(C)$ denote the function defined by $f^*(x,y,i) = (f(x),f(y),i)$.

THEOREM 1.2.20 *Suppose B and C are unbounded abelian groups and $f:B \to C$ is a homomorphism. Then*

(a) *f^* is a homomorphism.*

(b) *If f is injective, so is f^*.*

(c) *If f is surjective, so is f^*.*

Proof: Let (x,y,i), $(x',y',j) \in M(B)$. Then $f^*((x,y,i)(x',y',j)) = f^*(x+x'+iy', y+y', i+j) = (f(x+x'+iy'), f(y+y'), i+j) = (f(x)+f(x')+if(y'), f(y)+f(y'), i+j)$ whereas $f^*(x,y,i)f^*(x',y',j) = (f(x),f(y),i)(f(x'),f(y'),j) = (f(x)+f(x')+if(y'), f(y)+f(y'), i+j)$. This shows f^* is a homomorphism.

Now assume f is injective and suppose $f^*(x,y,i) = 1$. Thus $(f(x),f(y),i) = (0,0,0)$ hence $f(x) = 0$, $f(y) = 0$, and $i = 0$. Consequently $x = 0$ and $y = 0$ by injectivity of f, so $(x,y,i) = (0,0,0) = 1$ proving f^* is injective.

Finally assume f is surjective and suppose $(x,y,i) \in M(C)$. Then there exist $p,q \in B$ such that $x = f(p)$ and $y = f(q)$ by surjectivity of f. Consequently $f^*(p,q,i) = (x,y,i)$ proving f^* is surjective.

We are now ready for the third particularization of semidirect products. As this construction will only be used once in the book, its exposition will be extremely brief.

Let B be any group and let $i:\mathrm{Aut}(B) \to \mathrm{Aut}(B)$ be the identity map. Then $(\mathrm{Aut}(B),i,B)$ is a semidirect system.

DEFINITION AND NOTATION If B is a group, then $\mathrm{Hol}(B)$ is the **holomorph** of B if and only if $\mathrm{Hol}(B) = B \times_i \mathrm{Aut}(B)$.

THEOREM 1.2.21 *Let B be a group and let $k \in \mathrm{Aut}(B)$. Then there exists an inner automorphism ϕ of $\mathrm{Hol}(B)$ such that $\phi(x,1) = (k(x),1)$ for all $(x,1) \in B^*$.*
Proof: Let ϕ be the inner automorphism determined by $(1,k^{-1})$. Thus for $(x,1) \in B^*$, $\phi(x,1) = (1,k^{-1})^{-1}(x,1)(1,k^{-1}) = (1,k)(x,1)(1,k^{-1}) = (k(x),k)(1, k^{-1}) = (k(x)k(1),1) = (k(x),1)$.

EXERCISES

1. Let $G = B \times_g A$. Suppose C is a characteristic subgroup of B. Show $C^* \lhd G$.
2. State and prove the converses of Theorem 1.2.8, Parts (b) and (c).
3. Suppose B, C, and D are abelian groups none of which have exponent 2 and $f{:}B \to C$ and $g{:}C \to D$ are homomorphisms. Then $(g \circ f)^*$ and $g^* \circ f^*$ are both homomorphisms from $Dih(B)$ into $Dih(D)$. Show $(g \circ f)^* = g^* \circ f^*$.
4. Let B be an abelian group such that $\exp(B) \neq 2$ and B is 2-divisible. Show all subnormal subgroups of $Dih(B)$ are normal.
5. Let A be an unbounded abelian group. Show that for positive integers n, $(x,y,i)^n = (nx + (\frac{1}{2})(n^2 - n)iy, ny, ni)$.
6. Suppose $M(B) \cong M(C)$ for unbounded abelian groups B, C. Show $B \cong C$.
7. Let B be an unbounded abelian group, and let C be a subgroup of B. Show $(C \times C)^* \lhd M(B)$.
8. Let B be an unbounded abelian group, and let C be a subgroup of B such that B/C is also unbounded. Show $M(B)/(C \times C)^* \cong M(B/C)$.
9. State and prove the converses of Theorem 1.2.20, Parts (b) and (c).
10. Suppose B, C, and D are unbounded abelian groups and $f{:}B \to C$ and $g{:}C \to D$ are homomorphisms. Then $(g \circ f)^*$ and $g^* \circ f^*$ are both homomorphisms from $M(B)$ into $M(D)$. Show $(g \circ f)^* = g^* \circ f^*$.
11. Let B be an unbounded abelian group and let n be a positive integer. Define $H_n \subseteq M(B)$ as $H_n = \{(x,y,m) \mid x,y \in nB \text{ and } n \text{ divides } m\}$. Show
    (a) $H_n \lhd M(B)$.
    (b) $M(B)/H_n \cong (B/nB) \times (B/nB) \times Z(n)$.
    (c) If $\cap \{nB \mid n \in N, n \neq 0\}$ is trivial, then so is $\cap \{H_n \mid n \in N, n \neq 0\}$.
12. Show $M(B)' = Z(M(B))$ for all unbounded abelian groups B.

1.3 Direct Products

The material for the direct product construction is known as an indexed set of groups. The need for indexed sets can be explained as follows. If the groups A and B are different, then the group $A \times B$ can be thought of as some sort of an output of the set of groups $\{A,B\}$. A problem arises, however, if $A = B$. Now the set of groups $\{A,B\} = \{A\}$ has only a single element in it. Can $A \times B$ (or $A \times A$, which is the same thing) still be considered the output of the set $\{A\}$?

If it can, then so too can the group $A \times A \times A$ because $\{A,A,A\} = \{A\}$. But the groups $A \times A$ and $A \times A \times A$ are not generally isomorphic so it seems wrong to consider them both as the output of the set $\{A\}$.

In order to resolve this problem we must find some method of indexing the various occurences of the group A so that they may be distinguished from each other. No difficulty arises if we speak of $A \times A$ as the output of the set of groups $\{A_1,A_2\}$ and $A \times A \times A$ as the output of the set $\{A_1,A_2,A_3\}$. Here the indexes 1,2, and 3 help to keep track of how many times A occurs.

Informally, an indexed set of groups may be thought of as a set of groups each of which has a tag on it. The tags serve to distinguish various occurences of the same group. (That is, if such multiple occurences happen. They need not. It is quite permissible for the groups in an indexed set to be pairwise nonisomorphic.) These tags may be anything at all. A common procedure is to use natural numbers as tags, assuming, of course, that the set of groups to be indexed is countable.

Formally an indexed set of groups consists of a set of groups S; a set of indices I (no requirements here $-$ I can be any set); and a surjection $g:I \to S$. The reason for requiring surjectivity is to ensure that every group in S gets indexed. If $A = g(i)$, we denote the group A by A_i.

NOTATION A set of groups indexed by the set I will be denoted $\{A_i\}_{i \in I}$.

NOTE: If the letter A in $\{A_i\}_{i \in I}$ is replaced by the symbol for a particular group, the resulting expression will be used to denote the indexed set of groups, indexed by I, each of which is equal to the particular group in question. For instance $\{Z(7)_i\}_{i \in N}$ denotes the indexed set of groups, indexed by the natural numbers, each of which is equal to the cyclic group $Z(7)$.

We now proceed to the definition of direct product. What we are calling direct products here are also known as unrestricted direct products, complete direct products, strong direct products, and cartesian products.

DEFINITION AND NOTATION If $\{A_i\}_{i \in I}$ is an indexed set of groups, let K be the set of elements that belong to A_i for some $i \in I$ (i.e., $K = \cup\{A_i \mid i \in I\}$). Then $\times\{A_i\}_{i \in I}$ is the **direct product** of $\{A_i\}_{i \in I}$ if and only if $\times\{A_i\}_{i \in I} = \{f \mid f:I \to K \text{ and } f(i) \in A_i \text{ for all } i \in I\}$.

The group operation on $\times\{A_i\}_{i \in I}$ is defined "componentwise." That is, by the equation $(fg)(i) = f(i)g(i)$. It is easily seen that $\times\{A_i\}_{i \in I}$ is a group under this operation. Its identity element is the function sending each i into the identity element of A_i (in other words $1(i) = 1$ for all i), and inverses in the direct product are described by $f^{-1}(i) = f(i)^{-1}$.

Improtant functions that are associated with a direct product are the coordinate projections. These are defined as follows.

DEFINITION Let $\{A_i\}_{i \in I}$ be an indexed set of groups and let $k \in I$. Then $\pi_k : \mathsf{X}\{A_i\}_{i \in I} \to A_k$ is the kth **coordinate projection** if and only if $\pi_k(f) = f(k)$ for all f.

THEOREM 1.3.1 *Let $\{A_i\}_{i \in I}$ be an indexed set of groups, and let $k \in I$. Then π_k is a surjective homomorphism.*

Proof: Let $f, g \in \mathsf{X}\{A_i\}_{i \in I}$. Then $\pi_k(fg) = (fg)(k) = f(k)g(k) = \pi_k(f)\pi_k(g)$.

To show surjectivity, let $a \in A_k$. Define f by

$$f(i) = \begin{cases} a & \text{if } i = k, \\ 1 & \text{if } i \neq k. \end{cases}$$

Then f belongs to the direct product and $\pi_k(f) = a$.

The next four theorems give some miscellaneous properties of groups that are preserved by direct products.

THEOREM 1.3.2 *If $\{A_i\}_{i \in I}$ is an indexed set of abelian groups, then the direct product $\mathsf{X}\{A_i\}_{i \in I}$ is abelian.*

Proof: Let $f, g \in \mathsf{X}\{A_i\}_{i \in I}$ and seek to show f and g commute. For all $i \in I$, $f(i)$ and $g(i)$ belong to A_i so $f(i)g(i) = g(i)f(i)$. Consequently $(fg)(i) = (gf)(i)$ for all $i \in I$, hence $fg = gf$.

THEOREM 1.3.3 *If $\{A_i\}_{i \in I}$ is an indexed set of reduced abelian groups, then the direct product $\mathsf{X}\{A_i\}_{i \in I}$ is reduced.*

Proof: Let D be a divisible subgroup of $\mathsf{X}\{A_i\}_{i \in I}$ and seek to show that D is trivial. Let $d \in D$. By Theorem A.8.1, $\pi_i[D]$ is divisible for all $i \in I$. But $\pi_i[D]$ is a subgroup of the reduced group A_i so $\pi_i[D]$ is trivial for all $i \in I$. Therefore $d(i) = \pi_i(d) = 0$ for all i, hence $d = 0$ as required.

THEOREM 1.3.4 *If $\{A_i\}_{i \in I}$ is an indexed set of divisible groups, then the direct product $\mathsf{X}\{A_i\}_{i \in I}$ is divisible.*

Proof: Let $f \in \mathsf{X}\{A_i\}_{i \in I}$ and let n be a positive integer. Then for all $i \in I$, $f(i) \in A_i$ so the set $M_i = \{y \mid y \in A_i \text{ and } y^n = f(i)\}$ is nonempty by divisibility of A_i. Apply the axiom of choice to yield a function g such that $g(i) \in M_i$ for all i. Thus g belongs to the direct product and $g^n = f$.

THEOREM 1.3.5 *If $\{A_i\}_{i \in I}$ is a collection of torsion-free groups, then the direct product $\mathsf{X}\{A_i\}_{i \in I}$ is torsion-free.*

Proof: Let $g \in \mathsf{X}\{A_i\}_{i \in I}$ such that $g^n = 1$. Then for all $i \in I$, $(g^n)(i) = 1(i) = 1$. Thus $g(i)^n = 1$ which implies $g(i) = 1$ for all i since A_i is torsion-free. Conclude $g = 1$.

THEOREM 1.3.6 *Let $\{A_i\}_{i \in I}$ be an indexed set of groups and let n be a natural number. Then $(\mathsf{X}\{A_i\}_{i \in I})^{(n)} \subseteq \mathsf{X}\{A_i^{(n)}\}_{i \in I}$.*

Proof: For convenience let X denote $\times\{A_i\}_{i\in I}$ and let W denote $\times\{A_i^{(n)}\}_{i\in I}$. The proof will be by induction on n. When $n = 0, A_i^{(n)} = A_i$ for all i and so $W = X$. But also $X^{(n)} = X$ and hence $X^{(n)} \subseteq W$ so induction begins. Assume now the induction hypothesis $X^{(n-1)} \subseteq \times\{A_i^{(n-1)}\}_{i\in I}$. Let $f \in X^{(n)}$. Thus $f = g_1 g_2 \ldots g_s$ where for each t, g_t is a commutator $g_t = h_t^{-1}k_t^{-1}h_t k_t$ where $h_t, k_t \in X^{(n-1)}$. Applying the induction hypothesis yields $h_t, k_t \in \times\{A_i^{(n-1)}\}_{i\in I}$. Thus $h_t(i)$ and $k_t(i)$ belong to $A_i^{(n-1)}$ for all i. Hence $h_t(i)^{-1}k_t(i)^{-1}h_t(i)k_t(i) \in A_i^{(n)}$ for all i. That is, $g_t(i) \in A_i^{(n)}$ for all i and so each $g_t \in W$. taking the product of the functions g_t yields $f \in W$ completing the proof.

We next investigate some special subgroups of a direct product.

NOTATION Let $G = \times\{A_i\}_{i\in I}$ and let $J \subseteq I$. Then $\bar{D}_G(J)$ will denote $\{f \mid f \in G$ and $f(i) = 1$ for all $i \in I\backslash J\}$.

THEOREM 1.3.7 *If $G = \times\{A_i\}_{i\in I}$ and $J \subseteq I$, then*
(a) $\bar{D}_G(J) \lhd G$.
(b) $\bar{D}_G(J)$ *is isomorphic to* $\times\{A_j\}_{j\in J}$.
(c) $G = \bar{D}_G(J) \oplus \bar{D}_G(I\backslash J)$.

Proof: Note that $f \in \bar{D}_G(J)$ if and only if $\pi_i(f) = 1$ for all $i \in I\backslash J$. Thus $\bar{D}_G(J) = \cap\{\ker \pi_i \mid i \in I\backslash J\}$. Hence $\bar{D}_G(J)$ is the intersection of normal subgroups of G and is therefore a normal subgroup itself. This proves Part (a).

Let H denote $\times\{A_j\}_{j\in J}$. For all $f \in H$ let f' be defined by

$$f'(i) = \begin{cases} f(i) & \text{if } i \in J, \\ 1 & \text{if } i \in I\backslash J. \end{cases}$$

Thus $f' \in G$ and more particularly $f' \in \bar{D}_G(J)$. Hence the function ϕ defined by $\phi(f) = f'$ is on H into $\bar{D}_G(J)$. For $f,g \in H$,

$$(f'g')(i) = f'(i)g'(i) = \begin{cases} f(i)g(i) = (fg)(i) & \text{if } i \in J \\ 1 & \text{if } i \in I\backslash J \end{cases}$$

$$= (fg)'(i).$$

Thus $f'g' = (fg)'$ so ϕ is a homomorphism. This homomorphism is injective because if $\phi(f)$ is the identity, then $f'(i) = 1$ for all $i \in I$. Hence $f(i) = 1$ for all i, so f must be the identity of H. To see that ϕ is also onto $\bar{D}_G(J)$, let $d \in \bar{D}_G(J)$ and let f denote the restriction of d to J. Then $f \in H$ and $\phi(f) = d$. This establishes surjectivity and completes the proof of Part (b).

To prove Part (c), first note that both $\bar{D}_G(J)$ and $\bar{D}_G(I\backslash J)$ are normal subgroups by Part (a). Now for any $g \in G$, define g' as in the previous paragraph, and define g'' by

$$g''(i) = \begin{cases} g(i) & \text{if } i \in I\backslash J, \\ 1 & \text{if } i \in J. \end{cases}$$

Thus $g' \in \bar{D}_G(J)$, $g'' \in \bar{D}_G(I \setminus J)$, and clearly $g = g'g''$. This shows that $G = \bar{D}_G(J)\bar{D}_G(I \setminus J)$. All that remains to be shown is that $\bar{D}_G(J)$ and $\bar{D}_G(I \setminus J)$ intersect trivially. Let f belong to their intersection. Then $f(i) = 1$ for all $i \in I \setminus J$ since $f \in \bar{D}_G(J)$ and $f(i) = 1$ for all $i \in [I \setminus (I \setminus J)] = J$ since $f \in \bar{D}_G(I \setminus J)$. Hence $f(i) = 1$ for all i and so f is the identity element of G as required.

COROLLARY *Let $\{A_i\}_{i \in I}$ be an indexed set of groups, let $G = \times\{A_i\}_{i \in I}$, and let $j \in I$. Then*

(a) $\bar{D}_G(I \setminus \{j\}) = \ker \pi_j$.

(b) $A_j \cong \bar{D}_G(\{j\})$.

(c) *A_j is isomorphic to a direct factor of G.*

Proof: Observe that $f \in \bar{D}_G(I \setminus \{j\})$ if and only if $\pi_i(f) = f(i) = 1$ for all $i \in I \setminus (I \setminus \{j\}) = \{j\}$; that is, if and only if $f \in \ker \pi_j$. This proves Part (a).

By Part (a) and Theorem 1.3.7, Part (c),

$$G = \bar{D}_G(\{j\}) \oplus \ker \pi_j. \tag{1}$$

Therefore $\bar{D}_G(\{j\}) \cong G/\ker \pi_j \cong \operatorname{Im} \pi_j$ which equals A_j by Theorem 1.3.1, thus proving Part (b).

Part (c) follows from Part (b) and Equation (1).

THEOREM 1.3.8 *If G is a group and if $\mathcal{E}$ is a collection of normal subgroups of G, then $G/\cap\mathcal{E}$ can be embedded in $\times\{G/E\}_{E \in \mathcal{E}}$.*

Proof: For convenience let W denote $\cap\mathcal{E}$. Define $d: G/W \to \times\{G/E\}_{E \in \mathcal{E}}$ by $d(Wx) = \{(E, Ex) \mid E \in \mathcal{E}\}$. That is, $d(Wx)$ is the function sending E into Ex. To show d is a function suppose $Wx = Wy$. Then for all $E \in \mathcal{E}$, $xy^{-1} \in W \subseteq E$ implies $Ex = Ey$, so $d(Wx)(E) = d(Wy)(E)$. Thus $d(Wx) = d(Wy)$ so d is a function. To show d is a homomorphism, suppose $Wx, Wy \in G/W$ and denote $d(Wx)$, $d(Wy)$, and $d(WxWy) = d(Wxy)$ by f, g, and k respectively. Then for all $E \in \mathcal{E}$, $k(E) = E_{xy} = E_xE_y = f(E)g(E) = (fg)(E)$ which means that $k = fg$. Thus d is a homomorphism. Finally to show injectivity, suppose $d(Wx) = 1$ where 1 denotes the function sending each E into the identity element of G/E, which is E. On the other hand, $d(Wx)$ sends E into E_x. Thus $E_x = E$ for all $E \in \mathcal{E}$ and so $x \in \cap\mathcal{E} = W$ implying $Wx = 1$. Conclude d is injective.

The remainder of this section will be devoted to a brief discussion of residually $\mathcal{H}$ groups where $\mathcal{H}$ is a class of groups. As can be seen from the following definition, this concept arises naturally when the direct product construction is considered.

DEFINITION Let $\mathcal{H}$ be a class of groups and let A be a group. Then A is **residually** $\mathcal{H}$ if and only if A can be embedded in a direct product $\times\{B_i\}_{i \in I}$ where each $B_i \in \mathcal{H}$.

If the class of groups $\mathcal{H}$ happens to be subgroup closed, then the following characterization or alternate definition of residually $\mathcal{H}$ groups is available.

THEOREM 1.3.9 *If $\mathcal{H}$ is a subgroup-closed class of groups and G is a group, then the following are equivalent:*

(a) *G is residually $\mathcal{H}$.*

(b) *For all $x \in G \setminus \{1\}$, there exists B such that $x \notin B \lhd G$ and $G/B \in \mathcal{H}$.*

(c) *G has a collection of normal subgroups $\mathcal{E}$ such that $\cap\mathcal{E} = \{1\}$ and $G/E \in \mathcal{H}$ for all $E \in \mathcal{E}$.*

Proof: First it is claimed that if G satisfies Condition (b), then so does each of its subgroups. Let H be a subgroup. Then $x \in H \setminus \{1\}$ implies $x \in G \setminus \{1\}$ so there exists a normal subgroup B of G such that $x \notin B$ and $G/B \in \mathcal{H}$. Let $F = H \cap B$. Then $F \lhd H$ and $x \notin F$ (because $x \notin B$). Finally $H/F = H/(H \cap B) \cong HB/B$. But HB/B is a subgroup of G/B and hence belongs to $\mathcal{H}$ since $\mathcal{H}$ is subgroup closed. Thus $H/F \in \mathcal{H}$: Conclude H satisfies Condition (b), proving the claim.

Next it will be shown that Condition (a) implies Condition (b). By the above claim it suffices to show that Condition (b) holds for direct products of groups belonging to $\mathcal{H}$. So let $C = \times\{A_i\}_{i\in I}$ where each $A_i \in \mathcal{H}$. To show C satisfies Condition (b), let $f \in C$, $f \neq 1$. Then there exists some $i \in I$ such that $f(i) \neq 1$ hence $f \notin B$ where $B = \bar{\mathrm{D}}_C(I \setminus \{i\})$. But $C = \bar{\mathrm{D}}_C(\{i\}) \oplus B$ by Theorem 1.3.7 so $C/B \cong \bar{\mathrm{D}}_C(\{i\}) \cong A_i$ (by the Corollary to Theorem 1.3.7) which belongs to $\mathcal{H}$. Hence C satisfies Condition (b).

Next it will be shown that Condition (b) implies Condition (c). Suppose for all $x \neq 1$ there exists $B(x)$ such that $x \notin B(x) \lhd G$ and $G/B(x) \in \mathcal{H}$. Let $\mathcal{E}$ denote $\{B(x) \mid x \neq 1\}$. Then $\cap\mathcal{E} = \{1\}$ because if $t \in \cap\mathcal{E}$ and $t \neq 1$, then $t \in B(t)$ which is false. Thus Condition (c) holds.

Finally it must be shown that Condition (c) implies Condition (a). Let G satisfy Condition (c). Then triviality of $\cap\mathcal{E}$ implies $G/\cap\mathcal{E} \cong G$ so by Theorem 1.3.8, G can be embedded in $\times\{G/E\}_{E\in\mathcal{E}}$. Thus G is residually $\mathcal{H}$ since each $G/E \in \mathcal{H}$.

THEOREM 1.3.10 *Let $\mathcal{H}$ be a subgroup-closed class of groups. If G is a subdirectly irreducible, residually $\mathcal{H}$ group, then $G \in \mathcal{H}$.*

Proof: By definition of subdirect irreducibility there exists some $h \in \mathrm{H}(G)$ such that $h \neq 1$. Then by Theorem 1.3.9, there is a normal subgroup B of G such that $h \notin B$ and $G/B \in \mathcal{H}$. If B is nontrivial, then $\mathrm{H}(G) \subseteq B$ and hence $h \in B$, which is false, so B must be trivial. Then $G \cong G/B$ and so $G \in \mathcal{H}$ as required.

The following theorem about residually finite abelian groups will be used a number of times in Chapter 5.

THEOREM 1.3.11 *Let A be a residually finite abelian group. Then*

(a) $\mathrm{U}(A) = \{0\}$.

(b) *A is reduced.*

Proof: Let $x \in \mathrm{U}(A)$, suppose $x \neq 0$ and seek a contradiction. By Theorem 1.3.9 there exists B such that $x \notin B$ and A/B is finite, say of cardinality n. Then $x \in nA$ since $x \in \mathrm{U}(A)$ and so $x = ny$ for some y. Thus $B + x = B + ny = n(B + y) = 0$ because $\mathrm{card}(A/B) = n$. Thus $x \in B$. This contradiction proves Part (a).

To show Part (b), suppose D is a divisible subgroup of A. As it is easily seen that $D \subseteq \mathrm{U}(A)$, triviality of D follows from Part (a).

EXERCISES

In Exercises 1–4 let $X = \times\{A_i\}_{i \in I}$.

1. If each A_i is abelian, show $nX = \times\{nA_i\}_{i \in I}$ for all natural numbers n.
2. If each A_i is abelian, show $X[n] = \times\{A_i[n]\}_{i \in I}$ for all natural numbers n.
3. If each A_i is abelian show $\mathrm{U}(X) = \times\{\mathrm{U}(A_i)\}_{i \in I}$.
4. Show $\mathrm{Z}_n(X) = \times\{\mathrm{Z}_n(A_i)\}_{i \in I}$ for all natural numbers n.

In Exercises 5–9 show the given classes are preserved by direct products. (In Exercises 7–9 n denotes a fixed natural number.)

5. R-groups.
6. $\{G \mid$ elements of relatively prime order in G commute$\}$.
7. $\{G \mid \exp(G) = n\}$.
8. class-n nilpotent.
9. length-n solvable.
10. Suppose $\{A_i\}_{i \in I}$ and $\{B_i\}_{i \in I}$ are indexed sets of groups and suppose for all $i \in I$ that $\phi_i : A_i \to B_i$ is a homomorphism. Define ϕ on $\times\{A_i\}_{i \in I}$ into $\times\{B_i\}_{i \in I}$ by $\phi(f)(i) = \phi_i(f(i))$. Show
    (a) ϕ is a homomorphism.
    (b) ϕ is injective if and only if each ϕ_i is injective.
    (c) ϕ is surjective if and only if each ϕ_i is surjective.
11. Suppose $I = \{i_1, i_2, \ldots, i_n\}$. Show
    $$\times\{A_i\}_{i \in I} \cong A_{i_1} \times A_{i_2} \times \ldots \times A_{i_n}.$$
12. Let $\mathcal{H}$ be a subgroup-closed class of groups such that $A, B \in \mathcal{H}$ implies that $A \times B \in \mathcal{H}$. Show that finite, residually $\mathcal{H}$ groups belong to $\mathcal{H}$.
13. Let $\mathcal{H}$ be a subgroup-closed class of groups. Show that residually residually $\mathcal{H}$ groups are residually $\mathcal{H}$ groups.
14. Let $\{A_i\}_{i \in I}$ be an indexed set of groups. Suppose G is a group and $g_i : G \to A_i$ is a homomorphism for all $i \in I$. Show there exists a unique homomorphism $k : G \to \times\{A_i\}_{i \in I}$ such that $\pi_i \circ k = g_i$ for all $i \in I$.
15. Let I and K be equivalent sets. Let A be a group, let $A_i = A$ for all $i \in I$, and let $A_k = A$ for all $k \in K$. Show $\times\{A_i\}_{i \in I} \cong \times\{A_k\}_{k \in K}$.
16. Let A be a nontrivial group and let $A_i = A$ for all natural numbers i. Show the group $\times\{A_i\}_{i \in N}$ is both nonhopfian and noncohopfian.

1.4 Restricted Direct Products

In Section 1.3 we examined the direct product of an indexed set of groups. The direct product has a most interesting subgroup called the restricted direct product. It will be seen that the direct and restricted direct product constructions lead, in general, to radically different groups.

DEFINITION AND NOTATION If $f \in \mathsf{X}\{A_i\}_{i \in I}$, then $S(f)$ is the **support** of f if and only if $S(f) = \{i \mid i \in I \text{ and } f(i) \neq 1\}$.

DEFINITION AND NOTATION If $\{A_i\}_{i \in I}$ is an indexed set of groups, then $\Sigma\{A_i\}_{i \in I}$ is the **restricted direct product** of $\{A_i\}_{i \in I}$ if and only if $\Sigma\{A_i\}_{i \in I} = \{f \mid f \in \mathsf{X}\{A_i\}_{i \in I} \text{ and } S(f) \text{ is finite}\}$.

Restricted direct products are also known as incomplete direct products, weak direct products, and direct sums. Note that if the index set I is finite, then $\Sigma\{A_i\}_{i \in I}$ and $\mathsf{X}\{A_i\}_{i \in I}$ are the same group. The difference appears only when I is infinite.

THEOREM 1.4.1 *Let $\{A_i\}_{i \in I}$ be an indexed set of groups. Then $\Sigma\{A_i\}_{i \in I} \lhd \mathsf{X}\{A_i\}_{i \in I}$.*

Proof: Let $f,g \in \Sigma\{A_i\}_{i \in I}$. Then since $S(f)$ and $S(g)$ are both finite, so is their union. Thus to show $fg^{-1} \in \Sigma\{A_i\}_{i \in I}$, it suffices to show $S(fg^{-1}) \subseteq S(f) \cup S(g)$. Suppose $i \notin S(f) \cup S(g)$. Then $f(i) = 1$ and $g(i) = 1$. Therefore $(fg^{-1})(i) = f(i)g^{-1}(i) = 1 \cdot 1^{-1} = 1$, hence $i \notin S(fg^{-1})$. Thus $S(fg^{-1}) \subseteq S(f) \cup S(g)$ as required. This proves $\Sigma\{A_i\}_{i \in I}$ is a subgroup of $\mathsf{X}\{A_i\}_{i \in I}$.

To show normality, let $f \in \Sigma\{A_i\}_{i \in I}$ and let $g \in \mathsf{X}\{A_i\}_{i \in I}$. It suffices to show $S(g^{-1}fg) \subseteq S(f)$. So suppose $i \notin S(f)$. Then $f(i) = 1$, hence $g(i)^{-1}f(i)g(i) = 1$ implying $(g^{-1}fg)(i) = 1$ so that $i \notin S(g^{-1}fg)$ as required.

Important functions that are associated with a restricted direct product are the coordinate injections. These are defined as follows.

DEFINITION AND NOTATION Let $\{A_i\}_{i \in I}$ be an indexed set of groups, and let $k \in I$. Then $\lambda_k : A_k \to \Sigma\{A_i\}_{i \in I}$ is the kth **coordinate injection** if and only if, for all $x \in A_k$, $\lambda_k(x)$ is defined by

$$\lambda_k(x)(i) = \begin{cases} x & \text{if } i = k, \\ 1 & \text{otherwise.} \end{cases}$$

THEOREM 1.4.2 *Let $\{A_i\}_{i \in I}$ be an indexed set of groups and let $k \in I$. Then λ_k is an injective homomorphism.*

Proof: Let $x,y \in A_k$ and denote $\lambda_k(x)$, $\lambda_k(y)$, and $\lambda_k(xy)$ by f, g, and h respectively. Then for all $i \in I$,

$$(fg)(i) = f(i)g(i) = \begin{cases} xy & \text{if } i = k, \\ 1 & \text{otherwise,} \end{cases}$$
$$= h(i).$$

Thus $fg = h$ and so λ_k is a homomorphism. To show injectivity, suppose $\lambda_k(x) = 1$ where 1 is the identity element of $\Sigma\{A_i\}_{i\in I}$. Then $\lambda_k(x)(k) = 1$ so $x = 1$ and hence λ_k is injective.

THEOREM 1.4.3 *Let $\{A_i\}_{i\in I}$ be an indexed set of groups, let $f \in \Sigma\{A_i\}_{i\in I}$, and let* $S(f) = \{i_1, i_2, \ldots, i_n\}$. *Then*

$$f = \lambda_{i_1}(f(i_1))\lambda_{i_2}(f(i_2))\ldots\lambda_{i_n}(f(i_n)).$$

Proof: Let the right-hand side of the above equation be denoted by g and seek to show $g = f$. Suppose $i \notin S(f)$. Then for $1 \leqslant t \leqslant n$, $i \neq i_t$ so

$$\lambda_{i_t}(f(i_t))(i) = 1$$

hence $g(i) = 1 \cdot 1 \cdot \ldots \cdot 1 = 1$. Also $f(i) = 1$ since $i \notin S(f)$. Thus $f(i) = g(i)$ for all $i \notin S(f)$. On the other hand if $i \in S(f)$, then $i = i_t$ for some t. In this case

$$\lambda_{i_t}(f(i_t))(i) = f(i_t)$$

while

$$\lambda_{i_k}(f(i_k))(i) = 1 \quad \text{for } 1 \leqslant k \leqslant n \text{ such that } k \neq t.$$

Hence $g(i) = 1 \cdot 1 \cdot \ldots \cdot f(i_t) \cdot \ldots \cdot 1 = f(i_t) = f(i)$. Conclude $f(i) = g(i)$ for all $i \in I$. Therefore $f = g$.

Some miscellaneous properties of groups preserved by restricted direct products are examined in the next four theorems.

THEOREM 1.4.4 *If $\{A_i\}_{i\in I}$ is an indexed set of abelian groups, then the restricted direct product $\Sigma\{A_i\}_{i\in I}$ is abelian.*
Proof: Since the restricted direct product is a subgroup of the direct product, the result follows from Theorem 1.3.2.

THEOREM 1.4.5 *Let $\{A_i\}_{i\in I}$ be an indexed set of groups and let G denote $\Sigma\{A_i\}_{i\in I}$.*
(a) *If each A_i is semidivisible, then G is semidivisible.*
(b) *If each A_i is divisible, then G is divisible.*
Proof: Suppose each A_i is semidivisible and let n be a positive integer. To show every $f \in G$ can be expressed as a product of nth powers it suffices, by the representation for f given in Theorem 1.4.3, to show that every $\lambda_i(x)$ can be so ex-

pressed (all $i \in I$, all $x \in A_i$). But this is easily done; since $x \in A_i$ there exist $y_1, y_2, \ldots, y_k \in A_i$ such that $x = y_1{}^n y_2{}^n \ldots y_k{}^n$. Thus $\lambda_i(x) = \lambda_i(y_1)^n \lambda_i(y_2)^n \ldots \lambda_i(y_k)^n$. This proves part (a).

As for Part (b), suppose each A_i is divisible, let n be a positive integer, and let $f \in G$. Then f has finite support, say $\{i_1, i_2, \ldots, i_k\}$. For $1 \leqslant t \leqslant k$, there exists y_t such that $f(i_t) = y_t{}^n$. Define g by

$$g(i) = \begin{cases} y_t & \text{if } i = i_t \text{ for some } t, \\ 1 & \text{otherwise.} \end{cases}$$

Then g has finite support so belongs to G and clearly $g^n = f$.

THEOREM 1.4.6 *Let* $\{A_i\}_{i \in I}$ *be an indexed set of groups and let* G *denote* $\Sigma\{A_i\}_{i \in I}$.
(a) *If each* A_i *is torsion, then* G *is torsion.*
(b) *If each* A_i *is a p-group, then* G *is a p-group.*
(c) *If* $\exp(A_i) = n$ *for all* $i \in I$, *then* $\exp(G) = n$.
Proof: Suppose each A_i is torsion. Let $f \in G$ and let $S(f) = \{i_1, i_2, \ldots, i_k\}$. Then, for $1 \leqslant t \leqslant k$,

$$f(i_t) \in A_{i_t}$$

so $f(i_t)$ has finite order, say n_t. Let $m = n_1 n_2 \ldots n_k$. Then $f(i_t)^m = 1$ for each $i_t \in S(f)$. Also, clearly, $f(i)^m = 1$ for all $i \notin S(f)$. Thus $f(i)^m = 1$ for all $i \in I$, so $f^m = 1$ showing f has finite order and completing the proof of Part (a). Parts (b) and (c) are similar.

THEOREM 1.4.7 *If* $\{A_i\}_{i \in I}$ *is an indexed set of torsion-free groups, then the restricted direct product* $\Sigma\{A_i\}_{i \in I}$ *is torsion-free.*
Proof: Since the restricted direct product is a subgroup of the direct product, the result follows from Theorem 1.3.5.

The next theorem supplies information on the derived series of a restricted direct product. The situation here is actually somewhat more satisfying then the direct-product analog (Theorem 1.3.6) since the inclusion in Theorem 1.4.8 can be shown to be an equality (see Exercise 19).

THEOREM 1.4.8 *Let* $\{A_i\}_{i \in I}$ *be an indexed set of groups and let n be a natural number. Then* $(\Sigma\{A_i\}_{i \in I})^{(n)} \subseteq \Sigma\{A_i{}^{(n)}\}_{i \in I}$.
Proof: Let $f \in (\Sigma\{A_i\}_{i \in I})^{(n)}$. Then $f \in (\mathsf{X}\{A_i\}_{i \in I})^{(n)}$ by Theorem A.6.1 and hence $f \in \mathsf{X}\{A_i{}^{(n)}\}_{i \in I}$ by Theorem 1.3.6. This and the fact that f has finite support (since $f \in (\Sigma\{A_i\}_{i \in I})^{(n)} \subseteq \Sigma\{A_i\}_{i \in I}$) imply $f \in \Sigma\{A_i{}^{(n)}\}_{i \in I}$.

COROLLARY *If* $\{A_i\}_{i \in I}$ *is an indexed set of length-n solvable groups, then* $\Sigma\{A_i\}_{i \in I}$ *is length-n solvable.*
Proof: By the theorem, $(\Sigma\{A_i\}_{i \in I})^{(n)}$ is trivial so $\Sigma\{A_i\}_{i \in I}$ is length-m sol-

vable for some $m \leqslant n$. But any A_i can be embedded in $\Sigma\{A_i\}_{i\in I}$ (Theorem 1.4.2), hence $n \leqslant m$ by Theorem A.6.5. Conclude the result.

As we did for direct products in Section 1.3, we now look at some special subgroups of the restricted direct product.

NOTATION Let $G = \Sigma\{A_i\}_{i\in I}$, and let $J \subseteq I$. Then $\mathrm{D}_G(J)$ will denote $\{f \mid f \in G$ and $f(i) = 1$ for all $i \in I\backslash J\}$. Note that $\mathrm{D}_G(J) = \bar{\mathrm{D}}_P(J) \cap G$ where P denotes $\times\{A_i\}_{i\in I}$.

THEOREM 1.4.9 *If $G = \Sigma\{A_i\}_{i\in I}$ and $J \subseteq I$, then*
(a) $\mathrm{D}_G(J) \lhd G$.
(b) $\mathrm{D}_G(J) \cong \Sigma\{A_j\}_{j\in J}$.
(c) $G = \mathrm{D}_G(J) \oplus \mathrm{D}_G(I \setminus J)$.

Proof: Let $P = \times\{A_i\}_{i\in I}$. Then, as noted above, $\mathrm{D}_G(J) = \bar{\mathrm{D}}_P(J) \cap G$. This and $\bar{\mathrm{D}}_P(J) \lhd P$ (Theorem 1.3.7) imply $\mathrm{D}_G(J) \lhd G$ proving Part (a).

The proofs for Parts (b) and (c) are exactly the same as the proofs for the corresponding parts of Theorem 1.3.7 with $\bar{\mathrm{D}}_G(J)$ and $\bar{\mathrm{D}}_G(I \setminus J)$ replaced by $\mathrm{D}_G(J)$ and $\mathrm{D}_G(I \setminus J)$ respectively and with $H = \times\{A_j\}_{j\in J}$ replaced by $H = \Sigma\{A_j\}_{j\in J}$.

COROLLARY *Let $\{A_i\}_{i\in I}$ be an indexed set of groups, let $G = \Sigma\{A_i\}_{i\in I}$, and let $j \in I$. Then*
(a) $\mathrm{D}_G(\{j\}) = \mathrm{Im}\, \lambda_j$.
(b) $A_j \cong \mathrm{D}_G(\{j\})$.
(c) *A_j is isomorphic to a direct factor of G.*

Proof: That $\mathrm{Im}\, \lambda_j \subseteq \mathrm{D}_G(\{j\})$ is easily seen. Conversely suppose $f \in \mathrm{D}_G(\{j\})$. Let $x = f(j)$. Then for all $i \neq j$, $f(i) = 1$ because $f \in \mathrm{D}_G(\{j\})$. Hence f and $\lambda_j(x)$ agree on all $i \neq j$. But also $\lambda_j(x)(j) = x = f(j)$. Conclude $f = \lambda_j(x) \in \mathrm{Im}\, \lambda_j$. This proves Part (a). Part (b) follows Part (a) because λ_j is injective by Theorem 1.4.2. Finally apply Part (c) of the theorem to get $G = \mathrm{D}_G(\{j\}) \oplus \mathrm{D}_G(I \setminus \{j\})$. This and Part (b) imply Part (c).

THEOREM 1.4.10 *Suppose $\{A_i\}_{i\in I}$ and $\{B_i\}_{i\in I}$ are indexed sets of groups and suppose for all $i \in I$ that $\phi_i: A_i \to B_i$ is a homomorphism. Define the function $\phi: \Sigma\{A_i\}_{i\in I} \to \Sigma\{B_i\}_{i\in I}$ by $\phi(f)(i) = \phi_i(f(i))$ for all $i \in I$. Then*
(a) *ϕ is a homomorphism.*
(b) *If each ϕ_i is injective, so is ϕ.*
(c) *If each ϕ_i is surjective, so is ϕ.*

Proof: First note that if $i \notin \mathrm{S}(f)$, then $f(i) = 1$ hence $\phi_i(f(i)) = 1$ so $\phi(f)(i) = 1$ implying $i \notin \mathrm{S}(\phi(f))$. Thus $\mathrm{S}(\phi(f)) \subseteq \mathrm{S}(f)$ and so $\phi(f) \in \Sigma\{B_i\}_{i\in I}$ whenever $f \in \Sigma\{A_i\}_{i\in I}$.

To complete the proof of Part (a), let $g,h \in \Sigma\{A_i\}_{i\in I}$. Then $\phi(gh)(i) = \phi_i((gh)(i)) = \phi_i(g(i)h(i)) = \phi_i(g(i))\phi_i(h(i)) = \phi(g)(i)\phi(h)(i) = (\phi(g)\phi(h))(i)$ for all $i \in I$. Thus $\phi(gh) = \phi(g)\phi(h)$ as required.

To show Part (b), suppose each ϕ_i is injective and suppose $\phi(g) = 1$ for some g. Thus $\phi(g)(i) = 1$ for all $i \in I$, and hence $\phi_i(g(i)) = 1$ for all such i. Injectivity of each ϕ_i then implies $g(i) = 1$ for all i, hence $g = 1$ showing ϕ is injective.

To show Part (c), suppose each ϕ_i is surjective, let $h \in \Sigma\{B_i\}_{i\in I}$, and suppose $S(h) = \{i_1, i_2, \ldots, i_n\}$. Then for $1 \leqslant k \leqslant n$, there exists

$$w_k \in A_{i_k} \quad \text{such that } \phi_{i_k}(w_k) = h(i_k)$$

by the surjectivity assumption. Define f by

$$f(i) = \begin{cases} w_k & \text{if } i = i_k \text{ for some } k, \\ 1 & \text{otherwise.} \end{cases}$$

Clearly f has finite support so belongs to $\Sigma\{A_i\}_{i\in I}$. Then $\phi(f)(i) =$

$$\phi_i(f(i)) = \begin{cases} \phi_{i_k}(w_k) = h(i_k) & \text{if } i = i_k \text{ for some } k, \\ \phi_i(1) = 1 & \text{otherwise.} \end{cases}$$

Thus $\phi(f)$ and h agree on all $i \in I$, hence are equal completing the proof.

THEOREM 1.4.11 *Let I and K be equivalent sets. Let A be a group, and let $A_i = A$ for all $i \in I$ and $A_k = A$ for all $k \in K$. Then $\Sigma\{A_i\}_{i\in I} \cong \Sigma\{A_k\}_{k\in K}$.*

Proof: Let $\alpha: K \to I$ denote an equivalence (a bijective function). Define ϕ on $\Sigma\{A_i\}_{i\in I}$ to $\Sigma\{A_k\}_{k\in K}$ by $\phi(g)(k) = g(\alpha(k))$. It must be checked that $S(\phi(g))$ is finite when $S(g)$ is. Suppose $S(g) = \{i_1, i_2, \ldots, i_n\}$. Then there exist $k_1, k_2, \ldots, k_n \in K$ such that $i_t = \alpha(k_t)$ for $1 \leqslant t \leqslant n$. It suffices to show $S(\phi(g)) \subseteq \{k_1, k_2, \ldots, k_n\}$. Let $k \in S(\phi(g))$. Then $\phi(g)(k) \neq 1$, which implies $g(\alpha(k)) \neq 1$. Therefore $\alpha(k) \in S(g)$ so $\alpha(k) = i_t = \alpha(k_t)$ for some t. Hence $k = k_t$ by injectivity of α.

To show that ϕ is a homomorphism, let $f, g \in \Sigma\{A_i\}_{i\in I}$. Then for all $k \in K$, $\phi(fg)(k) = (fg)(\alpha(k)) = f(\alpha(k))g(\alpha(k)) = \phi(f)(k)\phi(g)(k) = (\phi(f)\phi(g))(k)$. Thus $\phi(fg) = \phi(f)\phi(g)$, hence ϕ is a homomorphism.

To see that ϕ is injective, suppose $\phi(f) = 1$. Then for all $k \in K$, $1 = \phi(f)(k) = f(\alpha(k))$. But for all $i \in I$, $i = \alpha(k)$ for some k so $f(i) = 1$ for all $i \in I$. Thus $f = 1$ proving ϕ is injective.

Finally to see surjectivity of ϕ, let $h \in \Sigma\{A_k\}_{k\in K}$. Define $f \in \Sigma\{A_i\}_{i\in I}$ by $f(i) = h(\alpha^{-1}(i))$. The proof that f has finite support is similar to the finite-support proof in the first paragraph. Therefore for all $k \in K$, $\phi(f)(k) = f(\alpha(k)) = h(\alpha^{-1} \circ \alpha(k)) = h(k)$, hence $\phi(f) = h$ so ϕ is surjective.

THEOREM 1.4.12 *Let A be a group, let $A_i = A$ for all natural numbers i, and let $G = \Sigma\{A_i\}_{i\in N}$. Then*

(a) $A \times G \cong G$.

(b) $G \times G \cong G$.

Proof: Let P denote the set of positive integers. Then by Theorems 1.4.9 and A.2.5 $G \cong D_G(\{0\}) \times D_G(P)$. But the Corollary to Theorem 1.4.9 yields $D_G(\{0\}) \cong A_0 \cong A$. Also by Theorem 1.4.9, $D_G(P) \cong \Sigma\{A_p\}_{p\in P}$ which in turn is isomorphic to G by Theorem 1.4.11 because P and N are equivalent (in particular $\alpha{:}N \to P$ defined by $\alpha(i) = i + 1$ is a bijection). Conclude $G \cong A \times G$ proving Part (a).

For Part (b), let E and O denote the sets of even and odd natural numbers respectively. Then $G \cong \Sigma\{A_e\}_{e\in E} \times \Sigma\{A_o\}_{o\in O}$ by Theorems 1.4.9 and A.2.5. But E and N are equivalent ($\alpha{:}N \to E$ defined by $\alpha(n) = 2n$ is a bijection) as are O and N ($\alpha{:}N \to O$ defined by $\alpha(n) = 2n + 1$). Conclude $G \cong G \times G$ by Theorem 1.4.11.

We turn our attention now to four theorems concerning restricted direct products of abelian groups. The first of these is category-theoretic in nature. In the language of category theory, it says that for the category of abelian groups, restricted direct products are coproducts.

THEOREM 1.4.13 *Suppose $\{A_i\}_{i\in I}$ is an indexed set of abelian groups, B is an abelian group, and $d_i{:}A_i \to B$ is a homomorphism for all $i \in I$. Then there exists a unique homomorphism $\beta{:}\Sigma\{A_i\}_{i\in I} \to B$ such that $\beta\circ\lambda_i = d_i$ for all $i \in I$.*

Proof: The usual notation for sums (Σ) will be used in this proof. In order to avoid confusion between this and the restricted direct product notation, let $\Sigma\{A_i\}_{i\in I}$ be denoted by G. Then Σ will refer only to sums in this proof.

For all $f \in G$ let $\beta(f)$ be the sum (in the group B) of all terms $d_i(f(i))$. That is, $\beta(f) = \Sigma_{i\in I} d_i(f(i))$. It makes sense to talk about this sum because f has finite support. Thus $\beta{:}G \to B$. To see β is a homomorphism, let $f,g \in G$. Then $\beta(f+g) = \Sigma_{i\in I} d_i((f+g)(i)) = \Sigma_{i\in I} d_i(f(i) + g(i)) = \Sigma_{i\in I}(d_i(f(i)) + d_i(g(i)))$ because d_i is a homomorphism. Since B is abelian, this becomes $\beta(f+g) = [\Sigma_{i\in I} d_i(f(i))] + [\Sigma_{i\in I} d_i(g(i))] = \beta(f) + \beta(g)$.

Next it will be shown that $\beta\circ\lambda_k = d_k$ for all $k \in I$. Let x belong to A_k. Then $\beta(\lambda_k(x)) = \Sigma_{i\in I} d_i(\lambda_k(x)(i))$. But $\lambda_k(x)(i)$ is nonzero only when $i = k$ and so $\beta(\lambda_k(x)) = d_k(\lambda_k(x)(k)) = d_k(x)$ as required.

To show that β is unique, suppose $\beta'{:}G \to B$ is a homomorphism such that $\beta'\circ\lambda_i = d_i$ for all $i \in I$. Thus $\beta'(\lambda_i(x)) = d_i(x) = \beta(\lambda_i(x))$ for all $i \in I$ and all $x \in A_i$. Then since β' and β agree on every $\lambda_i(x)$ (and since they are both homomorphisms) they must agree on every $f \in G$ because of the representation for an arbitrary f given in Theorem 1.4.3. Conclude $\beta' = \beta$, hence β is unique.

THEOREM 1.4.14 *Let $\{A_i\}_{i\in I}$ be an indexed set of abelian groups. Then $\mathrm{t}(\Sigma\{A_i\}_{i\in I}) = \Sigma\{\mathrm{t}(A_i)\}_{i\in I}$.*

Proof: Let $g \in \mathrm{t}(\Sigma\{A_i\}_{i\in I})$. Thus $g \in \Sigma\{A_i\}_{i\in I}$ and g has finite order, call it n. Then $ng = 0$, hence for all $i \in I$, $(ng)(i) = 0$ implying $ng(i) = 0$ so $g(i) \in \mathrm{t}(A_i)$. This and finite support of g (since $g \in \Sigma\{A_i\}_{i\in I}$) imply $g \in \Sigma\{\mathrm{t}(A_i)\}_{i\in I}$.

Conversely $\Sigma\{\mathrm{t}(A_i)\}_{i\in I}$ is a torsion group by Theorem 1.4.6 and is clearly a subgroup of $\Sigma\{A_i\}_{i\in I}$. Hence $\Sigma\{\mathrm{t}(A_i)\}_{i\in I} \subseteq \mathrm{t}(\Sigma\{A_i\}_{i\in I})$ completing the proof.

THEOREM 1.4.15 *Let* $\{A_i\}_{i\in I}$ *be an indexed set of abelian groups. Then* $\Sigma\{A_i\}_{i\in I}/\Sigma\{\mathrm{t}(A_i)\}_{i\in I} \cong \Sigma\{A_i/\mathrm{t}(A_i)\}_{i\in I}$.

Proof: For all $i \in I$ let $e_i: A_i \to A_i/\mathrm{t}(A_i)$ be the natural map. Applying Theorem 1.4.10 yields a surjection $e: \Sigma\{A_i\}_{i\in I} \to \Sigma\{A_i/\mathrm{t}(A_i)\}_{i\in I}$ defined by $e(f)(i) = e_i(f(i))$. It suffices then to show $\ker e = \Sigma\{\mathrm{t}(A_i)\}_{i\in I}$.

Let $f \in \Sigma\{\mathrm{t}(A_i)\}_{i\in I}$. Then $f(i) \in \mathrm{t}(A_i)$ for all $i \in I$ so $e_i(f(i)) = 0$. Hence $e(f)(i) = 0$ for all $i \in I$ implying $e(f) = 0$ and thus that $f \in \ker e$. Reversing the argument shows $\ker e \subseteq \Sigma\{\mathrm{t}(A_i)\}_{i\in I}$, completing the proof.

THEOREM 1.4.16 *If* $\{A_i\}_{i\in I}$ *is an indexed set of abelian groups and n is a natural number, then* $n(\Sigma\{A_i\}_{i\in I}) = \Sigma\{nA_i\}_{i\in I}$.

Proof: Let $f \in n(\Sigma\{A_i\}_{i\in I})$. Thus $f = ng$ where $g \in \Sigma\{A_i\}_{i\in I}$. Then for all $i \in I$, $f(i) = ng(i) \in nA_i$. This and finite support of f (because $f \in \Sigma\{A_i\}_{i\in I}$) imply $f \in \Sigma\{nA_i\}_{i\in I}$. Conversely, if $f \in \Sigma\{nA_i\}_{i\in I}$, then for all $i \in \mathrm{S}(f)$ there exists $a_i \in A_i$ such that $f(i) = na_i$. Define g by

$$g(i) = \begin{cases} a_i & \text{if } i \in \mathrm{S}(f), \\ 0 & \text{otherwise.} \end{cases}$$

Then g has finite support so belongs to $\Sigma\{A_i\}_{i\in I}$ and $f = ng$. Conclude that $f \in n(\Sigma\{A_i\}_{i\in I})$.

By Theorem 1.4.1, $\Sigma\{A_i\}_{i\in I} \lhd \mathsf{X}\{A_i\}_{i\in I}$ for any indexed set of groups $\{A_i\}_{i\in I}$. Consequently $\mathsf{X}\{A_i\}_{i\in I}/\Sigma\{A_i\}_{i\in I}$ is itself a group and can thus be considered a new construction on the material $\{A_i\}_{i\in I}$. We close the section with an interesting theorem on this construction. Some properties preserved by this construction will be examined in the exercises.

THEOREM 1.4.17 *Suppose* $\{G_p\}_{p\in PR}$ *is an indexed set of groups such that* G_p *is a p-group for all primes p. Then* $\mathsf{X}\{G_p\}_{p\in PR}/\Sigma\{G_p\}_{p\in PR}$ *is divisible.*

Proof: For convenience denote the direct product by A and the restricted direct product by B. Let n be a positive integer and let $Bf \in A/B$. Then for all primes p, $f(p)$ belongs to G_p so its order is some power of p, say $p^{m(p)}$. Suppose p does not divide n. Then n and p are relatively prime, hence so too are n and $p^{m(p)}$. Thus there exist integers i and k such that $in + kp^{m(p)} = 1$. There may be more than one pair of integers (i, k) that satisfy the last equation so for definiteness choose the pair such that $|i|$ is smallest. Define $h \in A$ by

$$h(p) = \begin{cases} 1 & \text{if } p \mid n, \\ f(p)^i & \text{otherwise.} \end{cases}$$

It is claimed that the support of $h^n f^{-1}$ is finite and in particular is a subset of the finite set $\{p \mid p \text{ divides } n\}$. For all p not dividing n, $(h^n f^{-1})(p)$ is equal to $f(p)^{in} f(p)^{-1} = f(p)^{in-1} = f(p)^{km(p)} = 1$ and hence $p \notin \mathrm{S}(h^n f^{-1})$ establishing the claim. Thus $h^n f^{-1} \in B$. Conclude $Bf = Bh^n = (Bh)^n$.

EXERCISES

1. Prove a result analogous to Section 1.3, Exercise 2 for restricted direct products.
2. Prove a result analogous to Section 1.3, Exercise 3 for restricted direct products.
3. Prove a result analogous to Section 1.3, Exercise 4 for restricted direct products.

In Exercises 4–16 show the given classes are preserved by restricted direct products.

4. class-n nilpotent
5. reduced abelian
6. M_1
7. M_2
8. M_3
9. perfect
10. FC
11. locally $\mathcal{H}$ where $\mathcal{H}$ is any subgroup-closed class such that $A, B \in \mathcal{H}$ implies $A \times B \in \mathcal{H}$
12. R-groups
13. $\{G \mid$ elements of relatively prime order in G commute$\}$
14. $\{G \mid G$ has no nontrivial finite quotient groups$\}$
15. hypercentral
16. hyper $\mathcal{H}$ where $\mathcal{H}$ is any class
17. Let A be a nontrivial group and let $A_i = A$ for all natural numbers i. Show the group $\Sigma\{A_i\}_{i \in N}$ is both nonhopfian and noncohopfian.
18. State and prove the converses of Theorem 1.4.10, Parts (b) and (c).
19. If $\{A_i\}_{i \in I}$ is an indexed set of groups and n is a natural number show that $(\Sigma\{A_i\}_{i \in I})^{(n)} = \Sigma\{A_i^{(n)}\}_{i \in I}$.

In Exercises 20–23 show that $\mathsf{X}\{A_i\}_{i \in I}/\Sigma\{A_i\}_{i \in I}$ belongs to the given classes whenever each A_i does.

20. divisible
21. torsion-free
22. R-groups
23. $\{G \mid$ elements of relatively prime order in G commute$\}$
24. Suppose $\{G_p\}_{p \in PR}$ is an indexed set of groups such that G_p is a p-group for all primes p. Show $\mathsf{X}\{G_p\}_{p \in PR}/\Sigma\{G_p\}_{p \in PR}$ is an R-group.

1.5 Direct Limits

The word *limit* suggests the concept of convergence in analysis. Indeed, the idea of a direct limit of groups is somewhat similar to the idea of a convergent sequence. Very roughly, direct limits provide a construction for a group that a sequence of groups $\{G_n\}_{n\in N}$ can be said to converge to. As it turns out, however, the direct limit construction can handle more general indexed sets than sequences. There is no need to require only a countable number of groups in the indexed set.

DEFINITION $(D,\leqslant)$ is a **directed set** if and only if D is a nonempty set and $\leqslant$ is a relation on D such that
(a) $d \leqslant d$ for all $d \in D$ ($\leqslant$ is reflexive).
(b) $d \leqslant e$ and $e \leqslant f$ imply $d \leqslant f$ ($\leqslant$ is transitive).
(c) $d,e \in D$ implies there exists $f \in D$ such that $d \leqslant f$ and $e \leqslant f$. (This is known as the Moore–Smith property.)

DEFINITION If $(D,\leqslant)$ is a directed set, then $(\{A_d\}_{d\in D}, \{\pi_d^e \mid d \leqslant e\})$ is a **direct system** over $(D,\leqslant)$ if and only if
(a) A_d is a group for all $d \in D$.
(b) $\pi_d^e : A_d \to A_e$ is a homomorphism for all $d,e \in D$ such that $d \leqslant e$. (These homomorphisms are called **bonding maps.**)
(c) $\pi_d^d : A_d \to A_d$ is the identity homomorphism for all $d \in D$.
(d) If $d,e,i \in D$ such that $d \leqslant e$ and $e \leqslant i$, then $\pi_e^i \circ \pi_d^e = \pi_d^i$.

NOTE: Direct systems and semidirect systems (Section 1.2) are similar in name only. The two concepts are otherwise unrelated.

Suppose $(\{A_d\}_{d\in D}, \{\pi_d^e \mid d \leqslant e\})$ is a direct system. We define the **direct limit** of this system as follows: Let $M = \{(x,d) \mid d \in D \text{ and } x \in A_d\}$ and consider the following relation on M: $(x,d) \equiv (y,e)$ if and only if there exists $i \in D$ such that $d,e \leqslant i$ and $\pi_d^i(x) = \pi_e^i(y)$. We will show that $\equiv$ is an equivalence relation, and then define the direct limit to be the set of equivalence classes of M with a suitable multiplication.

THEOREM 1.5.1 *With assumptions and notation as in the previous paragraph,* $\equiv$ *is an equivalence relation on M.*

Proof: If $(x,d) \in M$, then $d \leqslant d$ and $\pi_d^d(x) = \pi_d^d(x)$ so $\equiv$ is reflexive. The proof that $\equiv$ is symmetric is similarly straightforward. Finally suppose $(x,d) \equiv (x^*,d^*)$ and $(x^*,d^*) \equiv (x^{**},d^{**})$. Then there exist $e,i \in D$ such that $d,d^* \leqslant e$; $\pi_d^e(x) = \pi_{d^*}^e(x^*)$; $d^*,d^{**} \leqslant i$; and $\pi_{d^*}^i(x^*) = \pi_{d^{**}}^i(x^{**})$. By the Moore–Smith property, there exists $h \in D$ such that $e,i \leqslant h$. Then $d,d^*,d^{**} \leqslant h$ by transitivity of $\leqslant$. Now by several applications of Part (d) in the definition of *direct system* we get $\pi_d^h(x) = \pi_e^h(\pi_d^e(x)) = \pi_e^h(\pi_{d^*}^e(x^*)) = \pi_{d^*}^h(x^*) = \pi_i^h(\pi_{d^*}^i(x^*)) = \pi_i^h(\pi_{d^{**}}^i(x^{**})) = \pi_{d^{**}}^h(x^{**})$. Conclude $(x,d) \equiv (x^{**},d^{**})$.

As was remarked before Theorem 1.5.1, we now define the direct limit L to be the set of equivalence classes. The class containing the pair (x,d) will be denoted by $[(x,d)]$. Hence $L = \{[(x,d)] \mid (x,d) \in M\}$.

The group operation on L is defined as follows: If $[(x,d)]$ and $[(y,e)]$ belong to L, then by the Moore–Smith property, there exists $i \in D$ such that $d,e \leqslant i$. Define the product $[(x,d)]\,[(y,e)]$ to be $[(\pi_d^i(x)\pi_e^i(y),i)]$.

THEOREM 1.5.2 *With assumptions and notation as above, L is a group. The identity element is* $[(1,d)]$ *where d is any element of D, and in fact* $[(1,d)] = [(1,e)]$ *for all* $d,e \in D$. *Inverses are described by* $[(x,d)]^{-1} = [(x^{-1},d)]$.

Proof: To begin with it must be shown that the operation defined above is a function. So suppose that

$$(x,d) \equiv (x^*,d^*) \quad \text{and} \quad (y,e) \equiv (y^*,e^*) \tag{1}$$

further suppose that $i,i^* \in D$ such that $d,e \leqslant i$ and $d^*,e^* \leqslant i^*$, and seek to show

$$(\pi_d^i(x)\pi_e^i(y),i) \equiv (\pi_{d^*}^{i^*}(x^*)\pi_{e^*}^{i^*}(y^*),i^*). \tag{2}$$

The equivalences in Line (1) yield existence of $k,m \in D$ such that $d,d^* \leqslant k$; $e,e^* \leqslant m$;

$$\pi_d^k(x) = \pi_{d^*}^k(x^*) \quad \text{and} \quad \pi_e^m(y) = \pi_{e^*}^m(y^*). \tag{3}$$

By several applications of the Moore–Smith property there exists $p \in D$ such that $k,m,i,i^* \leqslant p$. By the first equivalence in Line (3), $\pi_k^p(\pi_d^k(x)) = \pi_k^p(\pi_{d^*}^p(x^*))$ and hence $\pi_d^p(x) = \pi_{d^*}^p(x^*)$ which further implies $\pi_i^p(\pi_d^i(x)) = \pi_{i^*}^p(\pi_{d^*}^{i^*}(x^*))$. Similarly the second equivalence in Line (3) yields $\pi_i^p(\pi_e^i(y)) = \pi_{i^*}^p(\pi_{e^*}^{i^*}(y^*))$. Now multiply (in the group A_p) these last two equations and use the fact that π_i^p and $\pi_{i^*}^p$ are homomorphisms to get $\pi_i^p(\pi_d^i(x)\pi_e^i(y)) = \pi_{i^*}^p(\pi_{d^*}^{i^*}(x^*)\pi_{e^*}^{i^*}(y^*))$. This and the definition of $\equiv$ yields Line (2) as required. Thus multiplication is a function as was to be shown.

To show associativity, suppose $[(x,d)]$, $[(y,e)]$, and $[(z,c)]$ belong to L. By the Moore–Smith property (applied twice) there exists $h \in D$ such that $d,e,c \leqslant h$. Then $([(x,d)]\,[(y,e)])[(z,c)] = [(\pi_d^h(x)\pi_e^h(y),h)]\,[(z,c)]$ which is equal to $[(\pi_h^h(\pi_d^h(x)\pi_e^h(y))\pi_c^h(z),h)] = [(\pi_d^h(x)\pi_e^h(y)\pi_c^h(z),h)]$ because π_h^h is the identity homomorphism on A_h. A similar demonstration shows $[(x,d)]([(y,e)]\,[(z,c)])$ is the same thing, hence multiplication is associative.

Next it will be shown that $[(1,d)] = [(1,e)]$ for all $d,e \in D$. By the Moore–Smith property there exists $i \in D$ such that $d,e \leqslant i$. Then $\pi_d^i(1) = 1 = \pi_e^i(1)$. This proves $(1,d) \equiv (1,e)$, hence $[(1,d)] = [(1,e)]$.

To show $[(1,d)]$ is the identity element, let $[(x,e)] \in L$. Then by the above $[(1,d)]\,[(x,e)] = [(1,e)]\,[(x,e)] = [(\pi_e^e(1)\pi_e^e(x),e)] = [(x,e)]$. A similar demonstration shows $[(x,e)]\,[(1,d)] = [(x,e)]$.

Finally for any $[(x,d)] \in L$, $[(x,d)]\,[(x^{-1},d)] = [(\pi_d^d(x)\pi_d^d(x^{-1}),d)] = [(xx^{-1},d)] = [(1,d)]$ which is the identity of L. A similar demonstration shows $[(x^{-1},d)]\,[(x,d)] = [(1,d)]$. Thus $[(x^{-1},d)]$ is the inverse of $[(x,d)]$.

DEFINITION AND NOTATION If $(\{A_d\}_{d\in D}, \{\pi_d^e | d \leqslant e\})$ is a direct system over $(D,\leqslant)$ and L is its direct limit, then $\{h_d\}_{d\in D}$ is the set of **associated maps** if and only if for all $d \in D$, $h_d : A_d \to L$ is defined by $h_d(x) = [(x,d)]$.

THEOREM 1.5.3 *Let $(\{A_d\}_{d\in D}, \{\pi_d^e | d \leqslant e\})$ be a direct system over $(D,\leqslant)$ and let L be its direct limit. Then each h_d is a homomorphism.*
Proof: Let $x,y \in A_d$. Then since $d \leqslant d$, $[(x,d)]\,[(y,d)] = [(\pi_d^d(x)\pi_d^d(y),d)] = [(xy,d)]$ because π_d^d is the identity homomorphism on A_d. Hence $h_d(x)h_d(y) = h_d(xy)$.

THEOREM 1.5.4 *Let $(\{A_d\}_{d\in D}, \{\pi_d^e | d \leqslant e\})$ be a direct system over $(D,\leqslant)$ and let L be its direct limit. Then $h_d = h_e \circ \pi_d^e$ for all $d \leqslant e$.*
Proof: Let $x \in A_d$. Then $h_e(\pi_d^e(x)) = [(\pi_d^e(x),e)]$, while $h_d(x) = [(x,d)]$. But $d \leqslant e$ and $e \leqslant e$ and $\pi_e^e(\pi_d^e(x)) = \pi_d^e(x)$ so $(\pi_d^e(x),e) \equiv (x,d)$. Thus $[(\pi_d^e(x),e)] = [(x,d)]$ proving $h_e(\pi_d^e(x)) = h_d(x)$ as required.

THEOREM 1.5.5 *Let $(\{A_d\}_{d\in D}, \{\pi_d^e | d \leqslant e\})$ be a direct system over $(D,\leqslant)$ and let L be its direct limit. Suppose $x \in A_d$ such that $h_d(x) = 1$. Then there exists $e \in D$ such that $d \leqslant e$ and $\pi_d^e(x) = 1$.*
Proof: By Theorem 1.5.2, $[(1,d)]$ is the identity of L. Thus $[(x,d)] = h_d(x) = [(1,d)]$. Hence $(x,d) \equiv (1,d)$ so there exists $e \in D$ such that $d \leqslant e$ and $\pi_d^e(x) = \pi_d^e(1) = 1$.

THEOREM 1.5.6 *Let $(\{A_d\}_{d\in D}, \{\pi_d^e | d \leqslant e\})$ be a direct system over $(D,\leqslant)$ and let L be its direct limit. If all of the bonding maps are injective, then so are all of the associated maps.*
Proof: Suppose $h_d(x) = 1$. By Theorem 1.5.5 there exists $e \in D$ such that $d \leqslant e$ and $\pi_d^e(x) = 1$. Hence $x = 1$ by injectivity of π_d^e. Conclude h_d is injective.

THEOREM 1.5.7 *Let $(\{A_d\}_{d\in D}, \{\pi_d^e | d \leqslant e\})$ be a direct system over $(D,\leqslant)$ and let L be its direct limit. Suppose for some particular $d \in D$ that $B \lhd A_d$ and suppose further that $\pi_d^e[B] \lhd A_e$ for all e such that $d \leqslant e$. Then $h_d[B] \lhd L$.*
Proof: Let $x \in h_d[B]$, let $y \in L$, and seek to show $y^{-1}xy \in h_d[B]$. Since $x \in h_d[B]$, $x = h_d(b)$ where $b \in B$. Since $y \in L$, $y = [(t,c)]$ for some $c \in D$ and $t \in A_c$. By the Moore–Smith property, $d,c \leqslant e$ for some e. Therefore $x = h_d(b) = h_e(\pi_d^e(b))$ (by Theorem 1.5.4) and $y = h_c(t) = h_e(\pi_c^e(t))$ (again by Theorem 1.5.4). Consequently $y^{-1}xy = h_e(\pi_c^e(t)^{-1}\pi_d^e(b)\pi_c^e(t))$. But by hypothesis $\pi_d^e[B]$ is normal in A_e so $\pi_c^e(t)^{-1}\pi_d^e(b)\pi_c^e(t) = \pi_d^e(b^*)$ for some $b^* \in B$. Thus $y^{-1}xy = h_e(\pi_d^e(b^*)) = h_d(b^*) \in h_d[B]$ as required.

THEOREM 1.5.8 *Let $(\{A_d\}_{d\in D}, \{\pi_d^e | d \leqslant e\})$ be a direct system over $(D,\leqslant)$ and let L be its direct limit. Suppose $\pi_d^e[\mathrm{Z}_m(A_d)] \subseteq \mathrm{Z}_m(A_e)$ for all natural numbers m and for all $d \leqslant e$. Then $h_d[\mathrm{Z}_n(A_d)] \subseteq \mathrm{Z}_n(L)$ for all n and d.*
Proof: This will be shown by induction. When $n = 0$, $\mathrm{Z}_n(A_d)$ is trivial hence so

too is $h_d[Z_n(A_d)]$ so induction begins. Assume now the induction hypothesis $h_d[Z_{n-1}(A_d)] \subseteq Z_{n-1}(L)$ for all $d \in D$, let $x \in Z_n(A_d)$, and seek to show that $h_d(x) \in Z_n(L)$.

Thus it must be shown that

$$h_d(x)^{-1}y^{-1}h_d(x)y \in Z_{n-1}(L) \quad \text{for all } y \in L. \tag{1}$$

For such y, $y = [(p,c)] = h_c(p)$ for some $c \in D$ and some $p \in A_c$. By the Moore–Smith property there exists $e \in D$ such that $d,c \leqslant e$. Therefore $h_d(x) = h_e(\pi_d^e(x))$ (Theorem 1.5.4) and $y = h_e(\pi_c^e(p))$ (again by Theorem 1.5.4). Consequently

$$h_d(x)^{-1}y^{-1}h_d(x)y = h_e(\pi_d^e(x)^{-1}\pi_c^e(p)^{-1}\pi_d^e(x)\pi_c^e(p)). \tag{2}$$

But $x \in Z_n(A_d)$ implies $\pi_d^e(x) \in Z_n(A_e)$ by hypothesis which in turn implies $\pi_d^e(x)^{-1}\pi_c^e(p)^{-1}\pi_d^e(x)\pi_c^e(p) \in Z_{n-1}(A_e)$. Therefore by the induction hypothesis $h_e(\pi_d^e(x)^{-1}\pi_c^e(p)^{-1}\pi_d^e(x)\pi_c^e(p)) \in Z_{n-1}(L)$. This and Line (2) imply Line (1).

THEOREM 1.5.9 *Let $(\{A_d\}_{d\in D}, \{\pi_d^e \mid d \leqslant e\})$ be a direct system over $(D,\leqslant)$ and let L be its direct limit. Suppose $\pi_d^e[Z_m(A_d)] \subseteq Z_m(A_e)$ for all natural numbers m and for all $d \leqslant e$. Suppose further that each A_d is class-n nilpotent for some fixed n. Then L is also nilpotent, of nilpotency class less than or equal to n.*

Proof: It suffices to show $L = Z_n(L)$. Let $x \in L$. Then $x = [(p,d)] = h_d(p)$ for some $d \in D$ and some $p \in A_d$. But since A_d is class-n nilpotent, $p \in Z_n(A_d)$. Therefore $x \in Z_n(L)$ by Theorem 1.5.8 completing the proof.

THEOREM 1.5.10 *Let $(\{A_d\}_{d\in D}, \{\pi_d^e \mid d \leqslant e\})$ be a direct system over $(D,\leqslant)$ and let L be its direct limit.*

(a) *If each A_d is torsion, then L is torsion.*
(b) *If each A_d is a p-group, then L is a p-group.*
(c) *If* $\exp(A_d) = n$ *for all d, and in addition if all bonding maps are injective, then* $\exp(L) = n$.

Proof: To show Part (a), suppose $[(x,d)] \in L$. Then $x \in A_d$ so x has, say, order k. By the rule for multiplication in L, $[(x,d)]^k = [(\pi_d^d(x)^k,d)] = [(x^k,d)]$ (because π_d^d is the identity) $= [(1,d)] = 1$. Thus $[(x,d)]$ has finite order. This proves Part (a). Part (b) is similar.

Assume now the conditions of Part (c). Then similar to the previous parts, every element of L, when raised to the nth power, is trivial. Now suppose every element of L, when raised to the mth power, is trivial and seek to show $n \mid m$. Choose any $d \in D$. Then for all $x \in A_d$, $1 = [(x,d)]^m = h_d(x)^m = h_d(x^m)$, hence $1 = x^m$ for all $x \in A_d$ by injectivity of h_d (Theorem 1.5.6). This and $\exp(A_d) = n$ imply $n \mid m$ as required.

Theorem 1.5.11, to be considered after a preliminary lemma, addresses itself to showing that under suitable conditions direct limits are subdirectly irreducible. One of these conditions is a requirement on the directed set, namely that this directed set have a least element. A **least element** is an element $0 \in D$ such that $0 \leqslant d$ for all $d \in D$. An example of a directed set with least element is N, the set of natural numbers, with the usual ordering. There will be two applications of Theorem 1.5.11 in this book. Both will be to direct systems over $(N,\leqslant)$.

LEMMA *Let A and B be groups and let $g:A \to B$ be a homomorphism. Suppose $K \lhd B$, and let $g^{-1}[K] = \{x \mid x \in A$ and $g(x) \in K\}$. Then $g^{-1}[K] \lhd A$.*
Proof: It suffices to show that $g^{-1}[K]$ is the kernel of some homomorphism. Let $e:B \to B/K$ be the natural homomorphism. Then $x \in g^{-1}[K]$ if and only if $g(x) \in K$ which is true if and only if $Kg(x) = 1$. Thus $g^{-1}[K] = \ker e \circ g$.

THEOREM 1.5.11 *Let $(D,\leqslant)$ be a directed set with a least element 0. Let $(\{A_d\}_{d \in D}, \{\pi_d^e \mid d \leqslant e\})$ be a direct system over $(D,\leqslant)$, and let L be its direct limit. Suppose further that*
(a) *A_0 is subdirectly irreducible.*
(b) *All bonding maps are injective.*
(c) *$\pi_d^e[\mathrm{H}(A_d)] \subseteq \mathrm{H}(A_e)$ for all $d \leqslant e$.*
Then L is subdirectly irreducible.
Proof: Since A_0 is subdirectly irreducible, there exists $x \in \mathrm{H}(A_0)$, $x \neq 1$. By Theorem 1.5.6, h_0 is injective, therefore $h_0(x) \neq 1$. It suffices then to show $h_0(x) \in \mathrm{H}(L)$. So let K be a nontrivial normal subgroup of L and seek to show $h_0(x) \in K$. Since K is nontrivial, there exists $[(y,d)] \in K \setminus \{1\}$. Thus $h_d(y) \neq 1$, hence $y \neq 1$. By the lemma, $h_d^{-1}[K] \lhd A_d$. Also $y \in h_d^{-1}[K]$. This and $y \neq 1$ imply $h_d^{-1}[K]$ is nontrivial. Consequently $\mathrm{H}(A_d) \subseteq h_d^{-1}[K]$. Now $0 \leqslant d$ by definition of *least element* so $\pi_0^d[\mathrm{H}(A_0)] \subseteq \mathrm{H}(A_d)$ by Hypothesis (c). Thus $\pi_0^d[\mathrm{H}(A_0)] \subseteq h_d^{-1}[K]$. Hence $\pi_0^d(x) \in h_d^{-1}[K]$ implying $h_d(\pi_0^d(x)) \in K$. But $h_d \circ \pi_0^d = h_0$ by Theorem 1.5.4. Conclude $h_0(x) \in K$ as was required to show.

THEOREM 1.5.12 *Let $\mathcal{H}$ be a subgroup-closed and quotient-closed class of groups. Suppose $(\{A_d\}_{d \in D}, \{\pi_d^e \mid d \leqslant e\})$ is a direct system over $(D,\leqslant)$ and L is its direct limit. Suppose each $A_d \in \mathcal{H}$. Then L is locally $\mathcal{H}$.* (See Exercise 10 for a similar but nonidentical result.)
Proof: Let B be a finitely generated subgroup of L, and seek to show $B \in \mathcal{H}$. Thus $B = \langle [(x_1,d_1)], [(x_2,d_2)], \ldots, [(x_n,d_n)] \rangle$. It follows from the Moore–Smith property that there exists $e \in D$ such that $d_k \leqslant e$ for all k, $1 \leqslant k \leqslant n$. By the quotient-closed hypothesis on $\mathcal{H}$, $h_e[A_e] \in \mathcal{H}$. Now note that for $1 \leqslant k \leqslant n$,

$$[(x_k,d_k)] = h_{d_k}(x_k) = h_e \circ \pi_{d_k}^e(x_k) \in h_e[A_e],$$

hence all generators of B belong to $h_e[A_e]$ which implies $B \subseteq h_e[A_e]$. Conclude then that $B \in \mathcal{H}$ because $\mathcal{H}$ is subgroup-closed.

THEOREM 1.5.13 *Suppose $(\{A_d\}_{d \in D}, \{\pi_d^e \mid d \leqslant e\})$ is a direct system over $(D,\leqslant)$ and L is its direct limit. Suppose further that $\pi_d^e[A_d] \lhd A_e$ for all $d \leqslant e$ and that each A_d is solvable (various solvability lengths allowed). Then L is* *hyperabelian.*

Proof: First note that the hypothesis implies $h_d[A_d] \lhd L$ for all d by Theorem 1.5.7. Seek to show that an arbitrary nontrivial quotient group L/H has n nontrivial normal abelian subgroup B/H.

There exists $[(x,d)] \in L \setminus H$ and so $h_d[A_d] \not\subseteq H$. Since both $h_d[A_d]$ and H are normal in L so is $h_d[A_d]H$. Also $h_d[A_d]H/H \cong h_d[A_d]/(h_d[A_d] \cap H)$. But A_d is solvable so $h_d[A_d]$ is also and hence so too is $h_d[A_d]/(h_d[A_d] \cap H)$ by Theorem A.6.5. Thus $h_d[A_d]H/H$ is solvable of length, say, n. Now let $B/H = (h_d[A_d]H/H)^{(n-1)}$. Thus B/H is nontrivial (lest $h_d[A_d]H/H$ have solvability length smaller than n) and is abelian because $(B/H)' = \{1\}$. Finally B/H is a fully invariant, hence characteristic, subgroup of $h_d[A_d]H/H$ by Theorem A.6.2. But $h_d[A_d]H/H \lhd L/H$ so $B/H \lhd L/H$ by Theorem A.10.6.

EXERCISES

1. Let $(\{A_d\}_{d \in D}, \{\pi_d^e \mid d \leqslant e\})$ be a direct system over $(D, \leqslant)$ and let L be its direct limit. Suppose G is a group and for each $d \in D$, $g_d: A_d \to G$ is a homomorphism such that $g_e \circ \pi_d^e = g_d$ for all $d \leqslant e$. Show there is a unique homomorphism $\phi: L \to G$ such that $\phi \circ h_d = g_d$ for all $d \in D$.

In Exercises 2–10 show the given classes are preserved by direct limits.

2. abelian
3. perfect
4. torsion-free
5. R-groups
6. $\{G \mid$ elements of relatively prime order in G commute$\}$
7. divisible
8. semidivisible
9. $\{G \mid G$ has no nontrivial finite quotient groups$\}$
10. locally $\mathcal{H}$ where $\mathcal{H}$ is any quotient-closed class
11. Let $\mathcal{H}$ be any class of groups. Show the class locally $\mathcal{H}$ is preserved by direct limits under the additional assumption that all bonding maps in the direct systems are injective.
12. Show that if all the bonding maps of a direct system are surjective, then all associated maps are surjective.
13. Let $(\{A_d\}_{d \in D}, \{\pi_d^e \mid d \leqslant e\})$ and $(\{B_d\}_{d \in D}, \{\bar{\pi}_d^e \mid d \leqslant e\})$ be direct systems over the same directed set $(D, \leqslant)$, let L and $\bar{L}$ be their respective direct limits, and let $\{h_d\}_{d \in D}$ and $\{\bar{h}_d\}_{d \in D}$ be the respective associated maps. Suppose for all $d \in D$ that $g_d: A_d \to B_d$ is a homomorphism such that $\bar{\pi}_d^e \circ g_d = g_e \circ \pi_d^e$ for all $d \leqslant e$.
    (a) Show there exists a unique homomorphism $g: L \to \bar{L}$ such that $g \circ h_d = \bar{h}_d \circ g_d$ for all $d \in D$.

(b) Show g is injective if each g_d is injective.
(c) Show g is surjective if each g_d is surjective.

14. Let $G = \Sigma\{A_i\}_{i\in I}$. Let $D = \{J \mid J \subseteq I\}$ (the power set of I). First show $(D,\subseteq)$ is a directed set. For all $d \in D$, let $H_d = \mathrm{D}_G(d)$. For $d, e \in D$ such that $d \subseteq e$, let $\pi_d^e{:}\mathrm{D}_G(d) \to \mathrm{D}_G(e)$ be the inclusion map. Show $(\{H_d\}_{d\in D}, \{\pi_d^e \mid d \subseteq e\})$ is a direct system over $(D,\subseteq)$. Show its direct limit is isomorphic to G.
15. Let $(\{A_d\}_{d\in D}, \{\pi_d^e \mid d \leqslant e\})$ be a direct system over $(D,\leqslant)$ and let L be its direct limit. Suppose $\pi_d^e[\mathrm{Z}_m(A_d)] \subseteq \mathrm{Z}_m(A_e)$ for all natural numbers m and for all $d \leqslant e$, and that each A_d is nilpotent (various classes of nilpotency allowed). Show L is hypercentral.
16. Show that the hypotheses of Theorem 1.5.11 imply A_d is subdirectly irreducible for all $d \in D$.

1.6 Countable Central Pushouts

In Section 1.1 we constructed the central pushout of the central pushout system (A,H,K,h,k). In other words we constructed the central pushout for two groups H and K. Here we will attempt to define central pushouts for an infinite number of groups. The plan will be to take the groups two at a time and construct central pushouts as per Section 1.1, then form a direct system out of these central pushouts, and finally to take the direct limit of the direct system.

For simplicity, we will consider only countably many groups. These groups then may be indexed by the set of natural numbers N, leading to the following definition.

DEFINITION $(A, \{H_i\}_{i\in N}, \{f_i\}_{i\in N})$ is a **countable central pushout system** if and only if A is an abelian group, $\{H_i\}_{i\in N}$ is an indexed set of groups, and each $f_i{:}A \to \mathrm{Z}(H_i)$ is an injective homomorphism.

Suppose $(A, \{H_i\}_{i\in N}, \{f_i\}_{i\in N})$ is a countable central pushout system. Let $G_0 = H_0$ and let $k_0 = f_0$. Note that (A,H_0,H_1,k_0,f_1) is a central pushout system, and let G_1 be its central pushout. Thus (Theorem 1.1.3) the following diagram

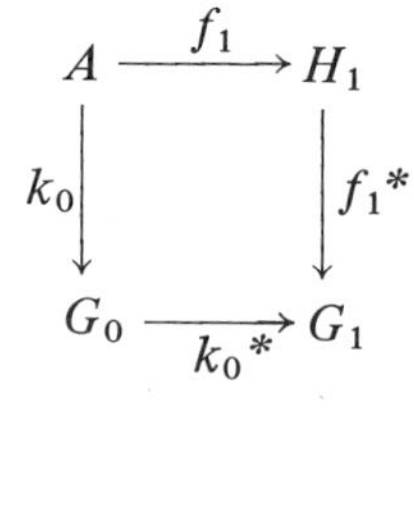

commutes. Note that $k_0{}^*\circ k_0 : A \to G_1$. Moreover $k_0{}^*\circ k_0 : A \to Z(G_1)$ by Theorems 1.1.7 and 1.1.8. Furthermore $k_0{}^*$ is injective (Theorem 1.1.2) hence so is $k_0{}^*\circ k_0$. Denote $k_0{}^*\circ k_0$ by k_1. We thus have a new central pushout system, (A, G_1, H_2, k_1, f_2). Let G_2 be its central pushout. This means commutativity of the following diagram.

$$\begin{array}{ccc} A & \xrightarrow{f_2} & H_2 \\ {\scriptstyle k_1}\downarrow & & \downarrow{\scriptstyle f_2{}^*} \\ G_1 & \xrightarrow[k_1{}^*]{} & G_2 \end{array}$$

Again by Theorems 1.1.2, 1.1.7, and 1.1.8, $k_1{}^*$ is injective and $k_1{}^*\circ k_1 : A \to Z(G_2)$. Denote $k_1{}^*\circ k_1$ by k_2. Thus k_2 is injective.

Continue in this fashion, defining $k_{n+1} = k_n{}^*\circ k_n$ and letting G_n be the central pushout of $(A, G_{n-1}, H_n, k_{n-1}, f_n)$. We then have each $k_n{}^* : G_n \to G_{n+1}$ is injective and for each n the following diagram commutes.

$$\begin{array}{ccc} A & \xrightarrow{f_n} & H_n \\ {\scriptstyle k_{n-1}}\downarrow & & \downarrow{\scriptstyle f_n{}^*} \\ G_{n-1} & \xrightarrow[k_{n-1}{}^*]{} & G_n \end{array}$$

For $n < m$, let $\pi_n^m = k_{m-1}{}^*\circ k_{m-2}{}^*\circ \ldots \circ k_{n+1}{}^*\circ k_n{}^*$ and define π_n^n to be the identity homomorphism on G_n. Clearly then $s \leqslant n$ and $n \leqslant m$ imply $\pi_s^m = \pi_n^m \circ \pi_s^n$. Thus $(\{G_i\}_{i\in N}, \{\pi_n^m \mid n \leqslant m\})$ is a direct system over $(N, \leqslant)$. Let L be its direct limit. This group L will be called the **central pushout** (or sometimes the **countable central pushout**) of $(A, \{H_i\}_{i\in N}, \{f_i\}_{i\in N})$.

As in Section 1.5, denote the set of associated maps for the direct limit L by $\{h_i\}_{i\in N}$. For all $i \in N$ then, $f_i{}^* : H_i \to G_i$ and $h_i : G_i \to L$. Thus $h_i \circ f_i{}^* : H_i \to L$.

THEOREM 1.6.1 *Let $(A, \{H_i\}_{i\in N}, \{f_i\}_{i\in N})$ be a countable central pushout system and let L be its central pushout. Then each π_n^m is injective, each h_n is injective, and each $h_n \circ f_n{}^*$ is injective.*

Proof: Each $f_n{}^*$ is injective by Theorem 1.1.2, as is each $k_n{}^*$. Thus each π_n^m is injective by definition of π_n^m. By Theorem 1.5.6, injectivity of each π_n^m implies injectivity of each h_n. Finally $h_n \circ f_n{}^*$ is the composition of injectives, so is injective itself.

THEOREM 1.6.2 *Let $(A, \{H_i\}_{i\in N}, \{f_i\}_{i\in N})$ be a countable central pushout system and let L be its central pushout. Suppose $B \lhd H_n$ for some n. Then $h_n \circ f_n{}^*[B] \lhd L$.*

Proof: First note that $f_n{}^*[B] \lhd G_n$ by Theorem 1.1.4. It is claimed that

$$\pi_n^m [f_n{}^*[B]] \lhd G_m \quad \text{for all } n \leqslant m. \tag{1}$$

If $n = m$ this is clear since π_n^m is the identity map. If $n < m$, then $\pi_n^m = k_{m-1}{}^* \circ k_{m-2}{}^* \circ \ldots \circ k_n{}^*$ so Expression (1) holds by repeated applications of Theorem 1.1.4. This claim and Theorem 1.5.7 imply the result.

THEOREM 1.6.3 *Let $\mathcal{H}$ be a class of groups which is preserved by both*
(a) *central pushouts (the Section 1.1 construction), and*
(b) *direct limits,*
then $\mathcal{H}$ is preserved by countable central pushouts.
Proof: This is clear from the construction of countable central pushouts.

NOTE: Condition (b) above can clearly be replaced by the weaker condition requiring that the class $\mathcal{H}$ be preserved by direct limits when all bonding maps in the direct systems are injective.

THEOREM 1.6.4 *The following classes of groups are preserved by countable central pushouts.*
(a) *torsion*
(b) *p-groups*
(c) *groups of exponent n*
Proof: By Theorems 1.1.6, 1.5.10, and 1.6.3.

THEOREM 1.6.5 *Let $(A, \{H_i\}_{i \in N}, \{f_i\}_{i \in N})$ be a countable central pushout system and let L be its central pushout. Suppose each $f_i{:}A \to Z(H_i)$ is an isomorphism. Then each $k_i{:}A \to Z(G_i)$ is also an isomorphism.*
Proof: Since each k_i is injective it suffices to show it is onto $Z(G_i)$. This will be done by induction on i. When $i = 0$, $k_i = f_0$ so k_i is surjective by hypothesis. Assume now the induction hypothesis $k_{i-1}{:}A \to Z(G_{i-1})$ is surjective. Recall G_i is the central pushout of the central pushout system $(A, G_{i-1}, H_i, k_{i-1}, f_i)$. By the induction hypothesis and the hypothesis of the theorem, k_{i-1} and f_i are both isomorphism onto $Z(G_{i-1})$ and $Z(H_i)$ respectively. Apply Theorem 1.1.9 to get $Z(G_i) = R(G_i)$. But by Theorem 1.1.7, $R(G_i) = k_{i-1}{}^*[k_{i-1}[A]] = k_i[A]$. Hence k_i is onto $Z(G_i)$ as required.

THEOREM 1.6.6 *Let $(A, \{H_i\}_{i \in N}, \{f_i\}_{i \in N})$ be a countable central pushout system and let L be its central pushout. Suppose each $f_i{:}A \to Z(H_i)$ is an isomorphism. Then*
(a) *$k_r{}^*[Z_n(G_r)] \subseteq Z_n(G_{r+1})$ for all natural numbers r and n.*
(b) *$\pi_r^s[Z_n(G_r)] \subseteq Z_n(G_s)$ for all $n \in N$ and $r \leqslant s$.*
Proof: By Theorem 1.6.5 each $k_r{:}A \to Z(G_r)$ is an isomorphism. Therefore Part (a) follows from Theorem 1.1.11. Part (b) follows from Part (a) because π_r^s is a composition of k functions (with suitable subscripts on k) when $r < s$ and is the identity map when $r = s$.

NOTATION Let $(A, \{H_i\}_{i\in N}, \{f_i\}_{i\in N})$ be a countable central pushout system and let L be its central pushout. Then $\mathrm{R}(L)$ will denote $h_0 \circ f_0[A]$.

THEOREM 1.6.7 *Let $(A, \{H_i\}_{i\in N}, \{f_i\}_{i\in N})$ be a countable central pushout system and let L be its central pushout. Then* $\mathrm{R}(L) \subseteq \mathrm{Z}(L)$.
Proof: Let $x \in \mathrm{R}(L)$. Thus $x = h_0 \circ f_0(a)$ for some $a \in A$. Let $y \in L$ and seek to show $xy = yx$. Since $y \in L$, $y = [(p,n)] = h_n(p)$ for some n and some $p \in G_n$. Then $0 \leqslant n$ implies $h_0 = h_n \circ \pi_0^n$. Hence $x = h_0 \circ f_0(a) = h_n \circ k_{n-1}{}^* \circ k_{n-2}{}^* \circ \ldots \circ k_0{}^* \circ f_0(a)$. But $k_0{}^* \circ f_0 = k_0{}^* \circ k_0 = k_1$ and $k_1{}^* \circ k_1 = k_2$ etcetera. Thus $x = h_n \circ k_n(a)$. Thus $xy = h_n(k_n(a)p) = h_n(pk_n(a))$ (because $k_n(a) \in \mathrm{Z}(G_n)$) $= yx$.

THEOREM 1.6.8 *Let $(A, \{H_i\}_{i\in N}, \{f_i\}_{i\in N})$ be a countable central pushout system and let L be its central pushout. Suppose each $f_i{:}A \to \mathrm{Z}(H_i)$ is an isomorphism. Then*
(a) $\mathrm{R}(L) = \mathrm{Z}(L)$.
(b) $A \cong \mathrm{Z}(L)$.
Proof: By Theorem 1.6.7, $\mathrm{R}(L) \subseteq \mathrm{Z}(L)$. To show the reverse inclusion, let $[(p,n)] = h_n(p)$ belong to the center of L. First it is claimed that $p \in \mathrm{Z}(G_n)$. Given $g \in G_n$, $h_n(p)h_n(g) = h_n(g)h_n(p)$ so $h_n(pg) = h_n(gp)$ which implies $pg = gp$ by injectivity of h_n (Theorem 1.6.1), establishing the claim. By application of Theorem 1.6.5, there exists $a \in A$ such that $p = k_n(a)$. Hence $p = k_{n-1}{}^* \circ k_{n-2}{}^* \circ \ldots \circ k_0{}^* \circ k_0(a) = \pi_0^n(k_0(a))$. Thus $[(p,n)] = h_n(p) = h_n \circ \pi_0^n(k_0(a)) = h_0 \circ k_0(a) = h_0 \circ f_0(a) \in \mathrm{R}(L)$. This proves Part (a). Part (b) follows from Part (a) and the fact that $h_0 \circ f_0{:}A \to \mathrm{R}(L)$ is an isomorphism.

THEOREM 1.6.9 *Let $(A, \{H_i\}_{i\in N}, \{f_i\}_{i\in N})$ be a countable central pushout system such that each $f_i{:}A \to \mathrm{Z}(H_i)$ is an isomorphism and let L be its central pushout. Suppose each H_i is class-n nilpotent. Then L is class-n nilpotent.*
Proof: First it is claimed that G_i is class-n nilpotent for all $i \in N$. This will be shown by induction. When $i = 0$, $G_i = H_0$ which is class-n nilpotent by hypothesis, so induction begins. Assume now the induction hypothesis, G_{i-1} is class-n nilpotent. Then G_i is the central pushout of $(A, G_{i-1}, H_i, k_{i-1}, f_i)$. But both $k_{i-1}{:}A \to \mathrm{Z}(G_{i-1})$ and $f_i{:}A \to \mathrm{Z}(H_i)$ are isomorphisms (the former by Theorem 1.6.5), so Theorem 1.1.6, the induction hypothesis, and the fact that H_i is class-n nilpotent apply to show G_i is class-m nilpotent for some $m \leqslant n$. But as G_{i-1} is class-n nilpotent and can be embedded in G_i, $n \leqslant m$ by Theorem A.6.5. Thus $n = m$ proving the claim.

Now L is the direct limit of $(\{G_i\}_{i\in N}, \{\pi_r^s \mid r \leqslant s\})$. Furthermore by Theorem 1.6.6, $\pi_r^s[\mathrm{Z}_m(G_r)] \subseteq \mathrm{Z}_m(G_s)$ for all $m \in N$ and all $r \leqslant s$. Thus Theorem 1.5.9 may be applied to show L is nilpotent of class less than or equal to n. But H_0 can be embedded in L ($h_0{:}H_0 \to L$ is injective by Theorem 1.6.1) and H_0 is class-n nilpotent so L must be class-n nilpotent as well.

LEMMA *Let G be a group. Suppose that for all $x \in G$, there exists some nat-*

ural number n such that $x \in Z_n(G)$. Then G is hypercentral.

Proof: Seek to show that a nontrivial quotient group G/H has a nontrivial center. Since $H \neq G$, there exists $x \in G \setminus H$. By hypothesis, $x \in Z_n(G)$ for some n. Let $e:G \to G/H$ be the natural map. Then $e[Z_n(G)] \subseteq Z_n(G/H)$ by the Lemma to Theorem A.5.2 so $Hx \in Z_n(G/H)$. This and $x \notin H$ means $Z_n(G/H)$ is nontrivial. It is easily seen then that $Z(G/H)$ is nontrivial.

THEOREM 1.6.10 *Let $(A, \{H_i\}_{i \in N}, \{f_i\}_{i \in N})$ be a countable central pushout system such that each $f_i:A \to Z(H_i)$ is an isomorphism, and let L be its central pushout. Suppose each H_i is nilpotent (various classes of nilpotency allowed). Then L is hypercentral.*

Proof: First it is claimed that for all i, $t \in Z_n(G_i)$ implies $h_i(t) \in Z_n(L)$. This will be shown by induction on n. When $n = 0$, $Z_n(G_i)$ is trivial so $t \in Z_n(G_i)$ implies $t = 1$, hence $h_i(t) = 1 \in Z_n(L)$. Thus induction begins. Assume now the induction hypothesis, for all i, $t \in Z_{n-1}(G_i)$ implies $h_i(t) \in Z_{n-1}(L)$, and suppose $t \in Z_n(G_i)$. It must be shown that $h_i(t)^{-1}y^{-1}h_i(t)y \in Z_{n-1}(L)$ for all $y \in L$. But for $y \in L$, $y = [(q,k)] = h_k(q)$ for some k and some $q \in G_k$. It may be assumed that $i \leqslant k$ (if $k \leqslant i$, then $y = h_i(\pi_k^i(q))$ so replace q by $\pi_k^i(q)$ in the following). Applying Theorem 1.6.6 yields $\pi_i^k[Z_n(G_i)] \subseteq Z_n(G_k)$. Therefore $t \in Z_n(G_i)$ implies $\pi_i^k(t) \in Z_n(G_k)$. This and $q \in G_k$ imply $\pi_i^k(t)^{-1}q^{-1}\pi_i^k(t)q \in Z_{n-1}(G_k)$. The induction hypothesis now yields $h_k(\pi_i^k(t)^{-1}q^{-1}\pi_i^k(t)q) \in Z_{n-1}(L)$. But $h_k \circ \pi_i^k = h_i$ (Theorem 1.5.4) so $h_i(t)^{-1}h_k(q)^{-1}h_i(t)h_k(q) \in Z_{n-1}(L)$. That is, $h_i(t)^{-1}y^{-1}h_i(t)y \in Z_{n-1}(L)$ as required. This establishes the claim.

Since each H_i is nilpotent by hypothesis, it follows from Theorem 1.1.6 that each G_i is nilpotent.

Now let $x \in L$. Then $x = [(t,i)] = h_i(t)$ for some i and some $t \in G_i$. By the preceding paragraph, $G_i = Z_n(G_i)$ for some n, so $t \in Z_n(G_i)$. Now apply the claim established in the first paragraph to conclude $x \in Z_n(L)$. Thus for all $x \in L$ there exists some n such that $x \in Z_n(L)$. Application of the lemma completes the proof.

EXERCISES

In Exercises 1–3 show the given classes are preserved by countable central pushouts.

1. divisible
2. semidivisible
3. perfect
4. Let $(A, \{H_i\}_{i \in N}, \{f_i\}_{i \in N})$ be a countable central pushout system and let L be its central pushout. If each H_i is solvable (various solvability lengths allowed), show L is hyperabelian.

1.7 Restricted Wreath Products

The wreath product is a somewhat complicated but enormously useful construction which has a legion of applications to both finite and infinite group theory. Actually there are two wreath product constructions. The one to be considered in this section is based on restricted direct products. The other, which is based on direct products, will be described in the exercises at the end of the chapter.

NOTATION If A and B are groups let $\mathrm{F}[A,B] = \Sigma\{A_b\}_{b\in B}$ where $A_b = A$ for all $b \in B$. Note that $\mathrm{F}[A,B]$ is a group.

NOTATION If A and B are groups, $f \in \mathrm{F}[A,B]$, and $b \in B$, let f^b denote the function defined by $f^b(\beta) = f(\beta b)$ for all $\beta \in B$.

NOTATION If A and B are groups and $b \in B$ let γ_b denote the function defined by $\gamma_b(f) = f^b$ for all $f \in \mathrm{F}[A,B]$.

LEMMA *Let A and B be groups. Then*
(a) $f^{xy} = (f^y)^x$ *for all* $f \in \mathrm{F}[A,B]$ *and all* $x,y \in B$.
(b) $\gamma_{xy} = \gamma_x \circ \gamma_y$ *for all* $x,y \in B$.
Proof: For all $\beta \in B$, $(f^y)^x(\beta) = f^y(\beta x) = f(\beta xy) = f^{xy}(\beta)$. This proves Part (a). Part (b) follows immediately.

THEOREM 1.7.1 *Let A and B be groups. Then* $\gamma_b \in \mathrm{Aut}(\mathrm{F}[A,B])$ *for all* $b \in B$.
Proof: First it must be shown that $\gamma_b(f)$ belongs to $\mathrm{F}[A,B]$ whenever f does. So suppose $\mathrm{S}(f)$ is finite, say $\mathrm{S}(f) = \{\beta_1,\beta_2,\ldots,\beta_n\}$. Then the set $H = \{\beta_1 b^{-1}, \beta_2 b^{-1},\ldots,\beta_n b^{-1}\}$ is finite. Furthermore $y \notin H$ implies $yb \notin \mathrm{S}(f)$ so $1 = f(yb) = f^b(y)$ and hence $y \notin \mathrm{S}(f^b)$. Thus $\mathrm{S}(f^b) \subseteq H$ and so $\mathrm{S}(f^b)$ is finite so $\gamma_b(f) = f^b \in \mathrm{F}[A,B]$ as required.

To show γ_b is a homomorphism let $f,g \in \mathrm{F}[A,B]$. It must be shown that $(fg)^b = f^b g^b$. But for all $\beta \in B$, $(fg)^b(\beta) = (fg)(\beta b) = f(\beta b)g(\beta b) = f^b(\beta)g^b(\beta) = (f^b g^b)(\beta)$. Thus γ_b is a homomorphism.

By the lemma (Part (b)), $\gamma_b \circ \gamma_{b^{-1}} = \gamma_1$ which is clearly the identity map on $\mathrm{F}[A,B]$. It follows, therefore, that γ_b is surjective.

Again by the lemma, $\gamma_{b^{-1}} \circ \gamma_b$ is also the identity. Hence γ_b is injective as well.

Because of Theorem 1.7.1, γ is a function from B into $\mathrm{Aut}(\mathrm{F}[A,B])$ and because of the Lemma to Theorem 1.7.1, Part (b), it is a homomorphism as well. This proves the following theorem.

THEOREM 1.7.2 *Let A and B be groups, then* $(B,\gamma,\mathrm{F}[A,B])$ *is a semidirect system.*

DEFINITION AND NOTATION Let A and B be groups. Then A wr B is the **restricted wreath product** of A by B if and only if A wr B is the semidirect product of $(B,\gamma,\mathrm{F}[A,B])$.

Thus the underlying set of A wr B is $\mathrm{F}[A,B] \times B$, multiplication is defined by

$$(f,b)(g,c) = (fg^b, bc),$$

$(1,1)$ is the identity element, and inversion is described by

$$(f,b)^{-1} = ((f^{-1})^{b^{-1}}, b^{-1}).$$

The next theorem describes the powers of an element in a wreath product.

THEOREM 1.7.3 *If A and B are groups and $(f,b) \in A$ wr B, then*

$$(f,b)^n = (ff^bf^{b^2} \ldots f^{b^{n-1}}, b^n)$$

for all positive integers n.

Proof: This will be shown by induction, noting that induction begins when $n = 1$. Assume the induction hypothesis,

$$(f,b)^{n-1} = (ff^bf^{b^2} \ldots f^{b^{n-2}}, b^{n-1}).$$

Then $(f,b)^n = (f,b)^{n-1}(f,b) =$

$$(ff^bf^{b^2} \ldots f^{b^{n-2}}, b^{n-1})(f,b) = (ff^bf^{b^2} \ldots f^{b^{n-1}}, b^n).$$

DEFINITION AND NOTATION Let A and B be groups. Then $\mathrm{Diag}(\mathrm{F}[A,B])$ is the **diagonal** of $\mathrm{F}[A,B]$ if and only if $\mathrm{Diag}(\mathrm{F}[A,B]) = \{f \mid f \in \mathrm{F}[A,B]$ and $f(p) = f(q)$ for all $p,q \in B\}$.

Note that if B is infinite, then $\mathrm{Diag}(\mathrm{F}[A,B])$ is trivial because for any $f \in \mathrm{Diag}(\mathrm{F}[A,B])$ there exists some $q \notin \mathrm{S}(f)$ so for all $p \in B$, $f(p) = f(q) = 1$, and hence $f = 1$. The following theorem describes $\mathrm{Diag}(\mathrm{F}[A,B])$ when B is finite.

THEOREM 1.7.4 *Let A and B be groups with B finite. Then* $\mathrm{Diag}(\mathrm{F}[A,B]) \cong A$. *In particular there is an isomorphism $\phi: A \to \mathrm{Diag}(\mathrm{F}[A,B])$ such that for all $a \in A$, $\phi(a)$ is the constant function determined by a.*

Proof: For all $a \in A$ let $k_a: B \to A$ be the constant function determined by a. That is, $k_a(b) = a$ for all $b \in B$. Since B is finite, there is no problem about k_a having finite support so $k_a \in \mathrm{F}[A,B]$ and clearly $k_a \in \mathrm{Diag}(\mathrm{F}[A,B])$.

Define $\phi: A \to \mathrm{Diag}(\mathrm{F}[A,B])$ by $\phi(a) = k_a$. To show ϕ is a homomorphism it must be shown that $k_{xy} = k_xk_y$ for all $x,y \in A$. But for all $b \in B$, $(k_xk_y)(b) = k_x(b)k_y(b) = xy = k_{xy}(b)$, so $k_{xy} = k_xk_y$ as required.

To see that ϕ is injective, suppose $\phi(x) = \phi(y)$. Then $k_x = k_y$ so $x = k_x(1) = k_y(1) = y$.

Finally to see that ϕ is surjective let $d \in \mathrm{Diag}(\mathrm{F}[A,B])$ and let $d(1) = w$. Then $d(b) = w$ for all $b \in B$ by definition of diagonal. Therefore $k_w(b) = w = d(b)$ for all $b \in B$, so $d = k_w = \phi(w)$.

The following result is probably the most famous application of restricted wreath products. Used in conjunction with Theorem 1.7.8, it supplies a wealth of infinite p-groups with trivial centers. This is noteworthy in view of the fact that finite p-groups must have nontrivial centers by Theorem A.1.1.

THEOREM 1.7.5 *Let A and B be groups such that A is nontrivial and B is infinite. Then A wr B has a trivial center.*

Proof: Suppose (f,b) belongs to the center. First it will be shown that $f = 1$. To do this it suffices to show $\mathrm{S}(f)$ is empty. Assume then that $s \in \mathrm{S}(f)$ and seek a contradiction. Since $\mathrm{S}(f)$ is finite and B is infinite there exists $\beta \in B \setminus \mathrm{S}(f)$. Let $w = \beta^{-1}s$. Then since (f,b) belongs to the center, $(f,b)(1,w) = (1,w)(f,b)$ so $(f1^b, bw) = (1f^w, wb)$ implying $f = f^w$. Thus $f(\beta) = f^w(\beta) = f(\beta w) = f(\beta\beta^{-1}s) = f(s)$. But $1 = f(\beta)$ since $\beta \notin \mathrm{S}(f)$. Therefore $1 = f(s)$ contrary to the fact that $s \in \mathrm{S}(f)$. This contradiction shows $f = 1$.

To complete the proof it must be shown that $b = 1$. By hypothesis there exists $a \in A$ such that $a \neq 1$. Define $g: B \to A$ by

$$g(\beta) = \begin{cases} a & \text{if } \beta = 1, \\ 1 & \text{if } \beta \neq 1. \end{cases}$$

Note g has finite support, so $g \in \mathrm{F}[A,B]$. Thus $(g,1) \in A$ wr B. Now by the above $f = 1$ so $(1,b)$ belongs to the center. Hence $(1,b)(g,1) = (g,1)(1,b)$ implying $(1g^b, b) = (g1, b)$, hence $g = g^b$. Consequently $g(b^{-1}) = g^b(b^{-1})$, that is, $g(b^{-1}) = g(1) = a$. If $b^{-1} \neq 1$, then $g(b^{-1}) = 1$ by definition of g. Hence $a = 1$, which is false. Conclude $b^{-1} = 1$ so $b = 1$.

THEOREM 1.7.6 *If $\mathcal{H}$ is a class of groups such that*
(a) *$\mathcal{H}$ is extension closed, and*
(b) *$\mathcal{H}$ is preserved by restricted direct products,*
then $\mathcal{H}$ is preserved by restricted wreath products.

Proof: Let $A, B \in \mathcal{H}$. Then $\mathrm{F}[A,B] \in \mathcal{H}$ and hence $\mathrm{F}[A,B]^* \in \mathcal{H}$. This and $B \in \mathcal{H}$ and $(A \text{ wr } B)/\mathrm{F}[A,B]^* \cong B$ imply A wr $B \in \mathcal{H}$.

LEMMA *The class of semidivisible groups is extension closed.*

Proof: Suppose G/H and H are semidivisible. To show G is semidivisible, let n be a positive integer and let $g \in G$. Then $Hg \in G/H$ so $Hg = (Hy_1)^n(Hy_2)^n \ldots (Hy_k)^n$ for some $y_1, y_2, \ldots, y_k \in G$. Thus $Hg = Hy_1{}^n y_2{}^n \ldots y_k{}^n$ implying $g(y_1{}^n y_2{}^n \ldots y_k{}^n)^{-1} \in H$. Consequently there exist $h_1, h_2, \ldots, h_m$ such that $g(y_1{}^n y_2{}^n \ldots y_k{}^n)^{-1} = h_1{}^n h_2{}^n \ldots h_m{}^n$ by semidivisibility of H. Conclude that $g = h_1{}^n h_2{}^n \ldots h_m{}^n y_1{}^n y_2{}^n \ldots y_k{}^n$ proving G is semidivisible.

THEOREM 1.7.7 *The class of semidivisible groups is preserved by restricted wreath products.*
Proof: By Theorem 1.4.5, Theorem 1.7.6, and the lemma.

LEMMA
(a) *The class of torsion groups is extension closed.*
(b) *The class of p-groups is extension closed.*
Proof: Suppose G/H and H are both torsion, and seek to show G is torsion. Let $x \in G$. Then Hx belongs to G/H so has finite order, say n. Thus $(Hx)^n = 1$ which implies $x^n \in H$. This and the fact that H is torsion imply $o(x^n) = m$ for some m so $1 = (x^n)^m = x^{nm}$. Thus x has finite order completing the proof of Part (a). Part (b) is similar.

THEOREM 1.7.8 *The following classes are preserved by restricted wreath products.*
(a) *torsion groups*
(b) *p-groups*
(c) *torsion-free groups*
Proof: Parts (a) and (b) by Theorem 1.4.6, Theorem 1.7.6, and the lemma. Part (c) by Theorem 1.4.7, Theorem 1.7.6, and the Lemma to Theorem 1.2.19.

THEOREM 1.7.9 *The class of solvable groups is preserved by restricted wreath products.*
Proof: By Theorem 1.7.6, the Corollary to Theorem 1.4.8, and Theorem A.6.6.

The last two theorems in this section are of a special nature. They will be used to explicate the properties of an infinite simple group in Chapter 6 (Example 6.15).

THEOREM 1.7.10 *Let A be a group, let $a \in A$, and let n be a positive integer. Then there exists an injection $e:A \to A$ wr $Z(n)$ such that $e(a) = x^n$ for some $x \in A$ wr $Z(n)$.*
Proof: Let $k:Z(n) \to A$ denote the constant function determined by a. That is, $k(b) = a$ for all $b \in Z(n)$. By Theorem 1.7.4, there is an isomorphism $\phi:A \to \mathrm{Diag}(\mathrm{F}[A,Z(n)])$ such that $\phi(a) = k$. Let $s:\mathrm{F}[A,Z(n)] \to \mathrm{F}[A,Z(n)]^*$ be the usual injection, $s(f) = (f,0)$ and let $e = s \circ \phi$. Thus $e:A \to A$ wr $Z(n)$ is an injection, and $e(a) = (k,0)$. To complete the proof then it suffices to show $(k,0)$ can be expressed as an nth power in A wr $Z(n)$.

Define $f \in \mathrm{F}[A,Z(n)]$ by

$$f(t) = \begin{cases} a & \text{if } t = 0, \\ 1 & \text{otherwise.} \end{cases}$$

Let $h = f^1$. By Theorem 1.7.3, $(h,1)^n = (hh^1h^2 \ldots h^{n-1},0)$ so it suffices to show $k = hh^1h^2 \ldots h^{n-1}$. For all $b \in Z(n)$, $(hh^1h^2 \ldots h^{n-1})(b) = h(b)h^1(b)h^2(b) \ldots h^{n-1}(b) = h(b)h(b+1)h(b+2) \ldots h(b+n-1) = f(b+1)f(b+2) \ldots f(b+n)$.

By the definition of f, exactly one of these factors is a, the remainder being 1. Hence the product is a so $(hh_1h_2...h_{n-1})(b) = a = k(b)$ for all $b \in Z(n)$ showing $hh_1h_2...h_{n-1} = k$ as required.

THEOREM 1.7.11 *If n and m are natural numbers such that $1 \leqslant n$ and $2 \leqslant m$, then there exists G, a, and b such that G is a finite group, a and b belong to G, $o(a) = o(b) = m$ and $o(ab) = n$.*

Proof: Let $G = Z(n)$ wr $Z(m)$. Note: When b is a number, the notation f^b is ambiguous. It can either mean f raised to the bth power or it can denote the wreath-product notation $f^b(\beta) = f(\beta + b)$. In this proof all instances of f^b refer to the wreath-product notation. Define $f \in \mathrm{F}[Z(n),Z(m)]$ by

$$f(x) = \begin{cases} 1 & \text{if } x = 1, \\ n-1 & \text{if } x = 2, \\ 0 & \text{if } x \neq 1,2. \end{cases}$$

Then for all $x \in Z(m)$, $(f^m + f^{m-1} + ... + f^1)(x) = f^m(x) + f^{m-1}(x) + ... + f^1(x) = f(x+m) + f(x+m-1) + ... + f(x+1)$. By definition of f, exactly one of these summands is 1, and exactly one is $n-1$, the remainder being 0. Therefore summing yields $f^m + f^{m-1} + ... + f^1 = 0$. Noting that $f^m = f^0 = f$, this becomes $f^{m-1} + f^{m-2} + ... + f^1 + f = 0$.

By Theorem 1.7.3, $(f,1)^i = (f + f^1 + f^2 + ... + f^{i-1}, i)$ for all positive integers i. For $1 \leqslant i < m$, $(f,1)^i$ has nontrivial second coordinate so is itself nontrivial. But $(f,1)^m = (f + f^1 + ... + f^{m-1}, 0) = (0,0)$ by the above. Hence $(f,1)$ has order m. Let $a = (f,1)$.

Let $b = (0,m-1)$. Clearly $o(b) = m$.

Finally $ab = (f,1)(0,m-1) = (f,m) = (f,0)$. Thus $o(ab)$ is the same as $o(f)$ in the group $\mathrm{F}[Z(n),Z(m)]$. Then as $kf(1) = k$ for all natural numbers k, $o(f) = n$, completing the proof.

EXERCISES

1. Show the class of perfect groups is preserved by restricted wreath products.
2. Show $\{G \mid G$ has no nontrivial finite quotient groups$\}$ is preserved by restricted wreath products.
3. Suppose $\mathcal{H}$ is an extension-closed class of groups. Show $A \in \mathcal{H}$, $B \in \mathcal{H}$, and B is finite imply A wr $B \in \mathcal{H}$.
4. Suppose $\alpha: A \to C$ is a homomorphism. Define $\alpha^*: \mathrm{F}[A,B] \to \mathrm{F}[C,B]$ by $\alpha^*(f)(b) = \alpha(f(b))$ for all $b \in B$. Define $\alpha^{**}: A$ wr $B \to C$ wr B by $\alpha^{**}(f,b) = (\alpha^*(f),b)$. Show
   (a) α^{**} is a homomorphism.
   (b) α is injective if and only if α^{**} is injective.
   (c) α is surjective if and only if α^{**} is surjective.
5. Suppose $A,B,C,$ and D are groups and $\alpha: A \to C$ and $\beta: C \to D$ are homomorphisms. Show $\beta^{**} \circ \alpha^{**} = (\beta \circ \alpha)^{**}$.
6. Let $\{A_i\}_{i \in I}$ be an indexed set of groups and let B be a group. Show

$\mathsf{X}\{A_i\}_{i\in I}$ wr B can be embedded in $\mathsf{X}\{A_i \text{ wr } B\}_{i\in I}$.

7. (a) Let $\mathcal{H}$ be an extension-closed class of groups that is preserved by restricted direct products. Show A is residually $\mathcal{H}$ and $B \in \mathcal{H}$ imply A wr B is residually $\mathcal{H}$.
   (b) Show A is residually solvable and B is solvable imply A wr B is residually solvable.
8. Show that the class of groups of exponent n is not preserved by restricted wreath products.
9. (a) Show that Z(Diag(F$[A,B)$] $\cong$ Z(A wr B) for any groups A and B such that A is nontrivial.
   (b) Show that Z(A) $\cong$ Z(A wr B) for any groups A and B such that A is nontrivial and B is finite.
10. Let A be a nontrivial group and let B be an abelian group. Let $x \in B$. Define $\alpha: A \text{ wr } B \to A \text{ wr } B$ by $\alpha(f,b) = (f^x, b)$. Show
    (a) α is an automorphism.
    (b) B can be embedded in Aut(A wr B).

1.8 Homomorphism Groups

The construction to be described below is a relatively simple one which uses two abelian groups as materials and results in a third abelian group. Its applications, unfortunately, are rather limited, and in fact it will be used in this book to construct only one example (or rather one class of examples) namely the p-adic integers (Example 5.5).

NOTATION If A and B are abelian groups let $\text{Hom}(A,B) = \{f \mid f: A \to B \text{ is a homomorphism}\}$.

Let $B_a = B$ for all $a \in A$. Then Hom(A,B) is a subset of $\mathsf{X}\{B_a\}_{a\in A}$. This gives an addition for Hom(A,B), namely that of $\mathsf{X}\{B_a\}_{a\in A}$. In other words for $f,g \in$ Hom(A,B), $(f+g)(a) = f(a) + g(a)$ for all $a \in A$.

THEOREM 1.8.1 *If A and B are abelian groups, then* Hom(A,B) *is an abelian group, and in particular, is a subgroup of* $\mathsf{X}\{B_a\}_{a\in A}$.

Proof: Let $f,g \in$ Hom(A,B), let $h = f - g$, and seek to show $h \in$ Hom(A,B). Clearly h is a function on A into B. To show it is a homomrophism, let $x,y \in A$. Then $h(x+y) = f(x+y) - g(x+y) = f(x) + f(y) - g(x) - g(y) = f(x) - g(x) + f(y) - g(y)$ (since B is abelian) $= h(x) + h(y)$ so $h \in$ Hom(A,B).

Because of Theorem 1.8.1, the identity element of Hom(A,B) is the same as that of $\mathsf{X}\{B_a\}_{a\in A}$ (namely the function 0 defined by $0(a) = 0$ for all $a \in A$) and inverses are also the same as in $\times\{B_a\}_{a\in A}$ (namely the inverse of f, $-f$, is defined by $(-f)(a) = -f(a)$ for all $a \in A$).

The next two theorems give some of the structure of Hom(A,B).

THEOREM 1.8.2 *If A and B are abelian groups such that A is divisible, then* Hom(A,B) *is torsion-free.*

Proof: Suppose $f \in$ Hom(A,B) such that $o(f) = n$. Thus $nf = 0$. Seek to show $f(a) = 0$ for all $a \in A$. For such a, however, there exists $a^* \in A$ satisfying $a = na^*$ by divisibility of A. Then $(nf)(a^*) = 0(a^*) = 0$. Thus $0 = nf(a^*) = f(na^*) = f(a)$.

THEOREM 1.8.3 *Let A and B be abelian groups, and let p be a prime. If B is a p-group, then* Hom(A,B) *is q-divisible for all primes $q \neq p$.*

Proof: Let $f \in$ Hom(A,B) and seek an element $h \in$ Hom(A,B) such that $f = qh$. For all $a \in A$, $f(a) \in B$ so by Theorem A.8.6 there is a unique y satisfying $qy = f(a)$. Define h by $h(a) = y$. Then $qh = f$ immediately so all that remains to be shown is that h is a homomorphism. Let $a, a^* \in A$. Thus $f(a) = qy$, $f(a^*) = qy^*$, and $f(a + a^*) = qy^{**}$ for unique $y, y^*, y^{**} \in A$. But then $q(y + y^*) = qy + qy^* = f(a) + f(a^*) = f(a + a^*) = qy^{**}$ implying $y + y^* = y^{**}$ by uniqueness. Thus $h(a) + h(a^*) = y + y^* = y^{**} = h(a + a^*)$.

EXERCISES

1. Let A, B, and C be abelian groups and let $\phi: B \to C$ be a homomorphism. Define ϕ_*: Hom$(A,B) \to$ Hom(A,C) by $\phi_*(f)(a) = \phi(f(a))$. Show
   (a) ϕ_* is a homomorphism.
   (b) ϕ is injective implies ϕ_* is injective. (An analogous statement need not hold for surjectivity.)
2. Let A, B, C, and D be abelian groups and let $\phi: B \to C$ and $\beta: C \to D$ be homomorphisms. Show $(\beta \circ \phi)_* = \beta_* \circ \phi_*$.
3. Let A, B, and C be abelian groups and let $\phi: B \to C$ be a homomorphism. Define ϕ^*: Hom$(C,A) \to$ Hom(B,A) by $\phi^*(f)(b) = f(\phi(b))$. Show
   (a) ϕ^* is a homomorphism.
   (b) ϕ is surjective implies ϕ^* is injective. (When ϕ is injective, ϕ^* need not be surjective.)
4. Let A, B, C, and D be abelian groups and let $\phi: B \to C$ and $\beta: C \to D$ be homomorphisms. Show $(\beta \circ \phi)^* = \phi^* \circ \beta^*$.
5. Let $\{A_i\}_{i\in I}$ be an indexed set of abelian groups and let B be an abelian group. Show Hom$(\Sigma\{A_i\}_{i\in I}, B) \cong \times\{\text{Hom}(A_i, B)\}_{i\in I}$.
6. Let $\{B_i\}_{i\in I}$ be an indexed set of abelian groups and let A be an abelian group. Show Hom$(A, \times\{B_i\}_{i\in I}) \cong \times\{\text{Hom}(A, B_i)\}_{i\in I}$.

MISCELLANEOUS EXERCISES

I Pullbacks

DEFINITION (A,H,K,h,k) is a **pullback system** if and only if $A, H,$ and K are groups and $h:H \to A$ and $k:K \to A$ are surjective homomorphisms.

DEFINITION If (A,H,K,h,k) is a pullback system, then G is the **pullback** of (A,H,K,h,k) if and only if $G = \{(x,y) \mid (x,y) \in H \times K \text{ and } h(x) = k(y)\}$.

I.1. Show pullbacks are groups.

DEFINITION AND NOTATION Let (A,H,K,h,k) be a pullback system and let G be its pullback. Then the **associated maps** $h^*:G \to H$ and $k^*:G \to K$ are defined as follows: $h^*(x,y) = x$ and $k^*(x,y) = y$ for all $(x,y) \in G$.

I.2. Show the associated maps are surjective homomorphisms.

I.3. State and prove an analog of Theorem 1.1.3 for pullbacks.

I.4. Let (A,H,K,h,k) be a pullback system and let G be its pullback. Show $B \lhd G$ implies both $h^*[B] \lhd H$ and $k^*[B] \lhd K$.

I.5. If $\mathcal{H}$ is a subgroup-closed class of groups such that $B,C \in \mathcal{H}$ implies $B \times C \in \mathcal{H}$, it is clear that $\mathcal{H}$ is preserved by pullbacks. Name five classes of groups that meet these conditions.

I.6. Show that the class of divisible R-groups is preserved by pullbacks.

I.7. Let (A,H,K,h,k) be a pullback system and let G be its pullback.
- (a) Show $Z(G) \subseteq Z(H) \times Z(K)$.
- (b) If $Z(H) \subseteq \ker h$ and $Z(K) \subseteq \ker k$, show $Z(G) = Z(H) \times Z(K)$.

I.8. Let (A,H,K,h,k) be a pullback system such that $Z(H) = \ker h$ and $Z(K) = \ker k$ and let G be its pullback. Show $G/Z(G) \cong A$.

I.9. Let (A,H,K,h,k) be a pullback system such that $Z(H) = \ker h$ and $Z(K) = \ker k$. If A, H, and K are finite, show that card(G) is given by the formula card(G) = card$(Z(H))$card$(Z(K))$card(A).

I.10.
- (a) Suppose B is a group satisfying the ascending chain condition on normal subgroups. Show $Z(B)$ does likewise.
- (b) Let (A,H,K,h,k) be a pullback system such that $Z(H) = \ker h$ and $Z(K) = \ker k$ and let G be its pullback. If H and K satisfy the ascending chain condition on normal subgroups, show G does likewise.
- (c) Repeat Parts (a) and (b) for the descending chain condition on normal subgroups.

II Inverse Limits

DEFINITION If $(D,\leqslant)$ is a directed set, then $(\{A_d\}_{d\in D}, \{\pi_d^e \mid d \leqslant e\})$ is an **inverse system** over $(D,\leqslant)$ if and only if

(a) A_d is a group for all $d \in D$.
(b) $\pi_d^e : A_e \to A_d$ is a homomorphism for all $d \leqslant e$. (These homomorphisms are called **bonding maps.**)
(c) $\pi_d^d : A_d \to A_d$ is the identity map for all $d \in D$.
(d) If $d \leqslant e$ and $e \leqslant i$, then $\pi_d^e \circ \pi_e^i = \pi_d^i$.

DEFINITION If $(\{A_d\}_{d \in D}, \{\pi_d^e \mid d \leqslant e\})$ is an inverse system over $(D, \leqslant)$, then G is its **inverse limit** if and only if $G = \{f \mid f \in \times\{A_d\}_{d \in D}$ and $\pi_d^e(f(e)) = f(e)$ for all $d \leqslant e\}$. Thus G is a subset of $\times\{A_d\}_{d \in D}$ and hence inherits a multiplication.

II.1. Show inverse limits are groups.

DEFINITION AND NOTATION Let $(\{A_d\}_{d \in D}, \{\pi_d^e \mid d \leqslant e\})$ be an inverse system over $(D, \leqslant)$ and let G be its inverse limit. Let $\pi_e : \times\{A_d\}_{d \in D} \to A_e$ be the eth coordinate projection. Then ϕ_e is the eth **associated map** of the inverse limit if and only if ϕ_e is the restriction of π_e to G. Thus $\phi_e : G \to A_e$ and ϕ_e is a homomorphism.

II.2. Let $(\{A_d\}_{d \in D}, \{\pi_d^e \mid d \leqslant e\})$ be an inverse system over $(D, \leqslant)$ and let G be its inverse limit. Show $\pi_d^e \circ \phi_e = \phi_d$ for all $d \leqslant e$.
II.3. Let $(\{A_d\}_{d \in D}, \{\pi_d^e \mid d \leqslant e\})$ be an inverse system over $(D, \leqslant)$ and let G be its inverse limit. Suppose B is a group and for each $d \in D$, $f_d : B \to A_d$ is a homomorphism such that $\pi_d^e \circ f_e = f_d$ for all $d \leqslant e$. Show there is a unique homomorphism $\beta : B \to G$ such that $\phi_d \circ \beta = f_d$ for all $d \in D$.
II.4. Show that if all the bonding maps of an inverse system are injective, then all associated maps are injective.
II.5. If $\mathcal{H}$ is a subgroup-closed class of groups that is preserved by direct products, it is clear that $\mathcal{H}$ is preserved by inverse limits. Name five classes of groups that meet these conditions.
II.6. Show that the class of divisible R-groups is preserved by inverse limits.

III (Unrestricted) Wreath Products

NOTATION If A and B are groups let $\bar{\mathrm{F}}[A, B] = \times\{A_b\}_{b \in B}$ where $A_b = A$ for all $b \in B$.

NOTATION If A and B are groups, $f \in \bar{\mathrm{F}}[A, B]$, and $b \in B$, let f^b denote the function defined by $f^b(\beta) = f(\beta b)$ for all $\beta \in B$.

NOTATION If A and B are groups and $b \in B$, let γ_b denote the function defined by $\gamma_b(f) = f^b$ for all $f \in \bar{\mathrm{F}}[A, B]$.

III.1. Let A and B be groups. Show $(B, \gamma, \bar{\mathrm{F}}[A, B])$ is a semidirect system.

DEFINITION AND NOTATION Let A and B be groups. Then A WR B is the **wreath product** of A by B if and only if A WR B is the semidirect product of $(B, \gamma, \bar{\mathrm{F}}[A,B])$.

III.2. (a) If $\mathcal{H}$ is an extension-closed class of groups that is also preserved by direct products, show $\mathcal{H}$ is preserved by wreath products. In particular, show the class of torsion-free groups is preserved by wreath products.

(b) Show the class of solvable groups is preserved by wreath products.

In Exercises III.3–III.6 state and prove analogs of the given exercises for (unrestricted) wreath products.

III.3. Section 1.7, Exercise 4.
III.4. Section 1.7, Exercise 5.
III.5. Section 1.7, Exercise 6.
III.6. Section 1.7, Exercise 7.

SUGGESTIONS FOR FURTHER READING

Free products and *free products with amalgamated subgroups* are covered in Rotman [31], pages 262–276. *Ext,* and important construction for abelian groups, is given in MacLane [24], Chapter III, Section 2. See Gorenstein [12], pages 27 and 28, for a generalization of the semidirect product construction called the *partial semidirect product.*

2

Free Groups and Presentations

A presentation is a description of a group in terms of elements that generate the group and relations satisfied by these generators. Despite the fact that the picture a presentation gives of a group is often an obscure one, the presentation concept has proven itself to be most useful in various branches of group theory.

Presentations are covered in Section 2.2. They are preceded by a necessary preliminary concept in Section 2.1, and are followed at the end of Section 2.2 by a discussion of their use throughout the remainder of the book.

2.1 Free Groups

Let X be any set. We begin by choosing a set X' such that X and X' are equivalent (there is a bijection from one to the other) and disjoint. (Such a set can always be found. For example, if we let ∞ denote anything that is not an element of $\cup X$, then $X' = \{\{A,\infty\} \mid A \in X\}$ will satisfy the requirements. This is seen as follows: Define $g{:}X \to X'$ by $g(A) = \{A,\infty\}$. It is clear that g is surjective; g is one-to-one because if $g(A) = g(B)$, then $\{A,\infty\} = \{B,\infty\}$ so $B \in \{A,\infty\}$ and $A \in \{B,\infty\}$. Thus $B = A$ or $B = \infty$, and $A = B$ or $A = \infty$. Since $B = \infty$ and $A = \infty$ combine to $A = B$, the result is that $A = B$ in any case. Thus g is one-to-one. To see that X and X' are disjoint, suppose $T \in X \cap X'$. Since $T \in X'$, $T = \{A,\infty\}$ for some $A \in X$. Consequently $\infty \in T \in X$ implying $\infty \in \cup X$ which has been assumed false.)

Thus X and X' are equivalent disjoint sets. Continue to let $g{:}X \to X'$ denote a bijection. Let U denote the union $U = X \cup X'$. For all $A \in U$, let A' be

$$A' = \begin{cases} g(A) & \text{if } A \in X, \\ g^{-1}(A) & \text{if } A \in X'. \end{cases}$$

Thus we have $A'' = A$ for all $A \in U$.

Let $\emptyset$ denote the empty set and let U^n denote $U \times U \times \ldots \times U$ to n factors. Let $W(X) = [\cup\{U^n \mid n \in N, n \neq 0\}] \cup \{\emptyset\}$. That is, $W(X)$ is the collection of all n-tuples of elements of U for all positive integers n, plus the "zero-tuple" $\emptyset$. The elements of $W(X)$ will be called **words.** If $T \in W(X)$, we define len(T) to be the **length** of T if and only if

$$\text{len}(T) = \begin{cases} 0 & \text{if } T = \emptyset, \\ n & \text{if } T \text{ is an } n\text{-tuple.} \end{cases}$$

A **reduced** word is defined to be either $\emptyset$ or an n-tuple $(A_1, A_2, \ldots, A_n)$ such that no coordinate is adjacent to its prime. That is, there is no i such that $A_i = A_{i+1}'$ or $A_i' = A_{i+1}$.

We may reduce an n-tuple by simply eliminating all adjacent occurences of A and A'. It then becomes either $\emptyset$ or a reduced m-tuple where $m \leqslant n$. If T is a word, denote the reduced word formed from T by T_r. Lest there be any question as to whether or not T_r is uniquely determined by T, we will agree to reduce T in the following manner: If T contains one or more pairs of adjacent letters A and A', first eliminate the leftmost pair, and consider the resulting word. If it contains one or more pairs of adjacent letters B and B', eliminate the leftmost pair. Continue until no more pairs can be eliminated and call the result T_r.

NOTATION $W_r(X)$ will denote the set of reduced words.

We define a group operation on $W_r(X)$ by the following rules:

(a) $\emptyset T = T$ for all $T \in W_r(X)$.
(b) $T\emptyset = T$ for all $T \in W_r(X)$.
(c) $(A_1, A_2, \ldots, A_n)(B_1, B_2, \ldots, B_m) = (A_1, A_2, \ldots, A_n, B_1, B_2, \ldots, B_m)_r$.

THEOREM 2.1.1 $W_r(X)$, *under the group operation defined above, is a group with identity element* $\emptyset$.

Proof: The bulk of the proof will be to show associativity. It is first claimed that

$$(RS)T = R(ST) \tag{1}$$

for all 1-tuples T. Let $R = (A_1, A_2, \ldots, A_k)$, let $S = (B_1, B_2, \ldots, B_m)$, and let $T = (C)$. The claim will be shown by induction on len(S) noting that induction begins when len(S) = 0 (when $S = \emptyset$) because of the above Rules (a) and (b) for multiplication.

Assume now the induction hypothesis $(PQ)T = P(QT)$ for all P and Q such that len(Q) $\leqslant m$. If $V \in W_r(X)$ is an n-tuple, say $V = (E_1, E_2, \ldots, E_n)$, denote $(E_2, E_3, \ldots, E_n)$ by ${}_*V$ and $(E_1, E_2, \ldots, E_{n-1})$ by V_*.

CASE ONE, $A_k = B_1'$ and $B_m \neq C'$

Thus $RS = (A_1, A_2, \ldots, A_{k-1}, B_2, B_3, \ldots, B_m)_r$. Clearly the product $(R_*)({}_*S)$ is the same thing so $RS = (R_*)({}_*S)$. Observing that len$({}_*S) = m - 1 < m$ yields

$((R_*)(_*S))T = R_*((_*S)T)$ by the induction hypothesis. But $B_m \neq C'$ implies $(_*S)T = (B_2, B_3, \ldots, B_m, C)$. Thus $(RS)T = R_*((_*S)T) = R_*(B_2, B_3, \ldots, B_m, C)$ which equals $(A_1, A_2, \ldots, A_{k-1}, B_2, B_3, \ldots, B_m, C)_r$. But on the other hand, $R(ST) = R(B_1, B_2, \ldots, B_m, C) = (A_1, A_2, \ldots, A_{k-1}, B_2, B_3, \ldots, B_m, C)_r$ also. Thus Equation (1) holds.

CASE TWO, $A_k = B_1'$ and $B_m = C'$

If len$(S) = 1$, then $S = (B_1)$ and $B_m = B_1$. Then $A_k = B_1'$ implies $RS = (A_1, A_2, \ldots, A_{k-1})_r$ and $B_m = C'$ implies $ST = \emptyset$. Thus $(RS)T = (A_1, A_2, \ldots, A_{k-1})(C)$ which equals $(A_1, A_2, \ldots, A_{k-1}, C)_r$ while $R(ST) = R\emptyset = R = R_r = (A_1, A_2, \ldots, A_k)_r$. Since $B_1 = C'$, $B_1' = C'' = C$. Thus $A_k = C$, hence $R(ST) = (A_1, A_2, \ldots, A_{k-1}, C)_r = (RS)T$ as required. Assume now $2 \leqslant$ len(S). As in Case one $(RS)T = R_*((_*S)T)$. Thus $(RS)T = R_*(B_2, B_3, \ldots, B_{m-1}) = (A_1, A_2, \ldots, A_{k-1}, B_2, B_3, \ldots, B_{m-1})_r$. On the other hand, $R(ST) = R(B_1, B_2, \ldots, B_{m-1}) = (A_1, A_2, \ldots, A_k, B_1, B_2, \ldots, B_{m-1})_r = (A_1, A_2, \ldots, A_{k-1}, B_2, B_3, \ldots, B_{m-1})_r$. So again Equation (1) holds.

CASE THREE, $A_k \neq B_1'$ and $B_m = C'$

Then $(RS)T = (A_1, A_2, \ldots, A_k, B_1, B_2, \ldots, B_m)(C) = (A_1, A_2, \ldots, A_k, B_1, B_2, \ldots, B_{m-1})_r$ while $R(ST) = (A_1, A_2, \ldots, A_k)(B_1, B_2, \ldots, B_{m-1}) = (A_1, A_2, \ldots, A_k, B_1, B_2, \ldots, B_{m-1})_r = (RS)T$.

CASE FOUR, $A_k \neq B_1'$ and $B_m \neq C'$

Then $(RS)T = (A_1, A_2, \ldots, A_k, B_1, B_2, \ldots, B_m, C)_r = R(ST)$.

This completes the proof or the claim in all cases. With the aid of this claim it will now be shown that Equation (1) holds for all n-tuples T. This will be done by induction on len(T), noting that induction begins when len$(T) = 0$ by the multiplication rules. Assume now the induction hypothesis $(RS)M = R(SM)$ for all M such that len$(M) <$ len(T). Let $T = (C_1, C_2, \ldots, C_n)$. Then $T = M(C_n)$ where $M = T_*$. Let $Q = RS$ and $P = SM$. Then $Q(M(C_n)) = (QM)(C_n)$ by the claim. But $QM = (RS)M = R(SM)$ by induction. Thus $QM = RP$, consequently $Q(M(C_n)) = (RP)(C_n) = R(P(C_n))$ by the claim. However, $P(C_n) = (SM)(C_n) = S(M(C_n))$ (by the claim) $= ST$. Collecting yields $R(ST) = Q(M(C_n)) = QT = (RS)T$.

Conclude that multiplication in $W_r(X)$ is associative. The remainder of the proof is quite easy. By Rules (a) and (b) in the definition of multiplication conclude that $\emptyset$ is the identity element of $W_r(X)$. Finally the existence of inverses must be shown. Let $R \in W_r(X)$. If $R = \emptyset$, then R is its own inverse. Assume then that R is an n-tuple, $R = (A_1, A_2, \ldots, A_n)$. Let $S = (A_n', A_{n-1}', \ldots, A_1')$. Then clearly $RS = SR = \emptyset$. So $S = R^{-1}$.

The group $W_r(X)$ is known as the **free group** on X.

THEOREM 2.1.2 *For all* $A \in X \cup X'$, (A') *is the inverse of* (A) *in* $W_r(X)$.

Proof: Recall that the method of reducing an n-tuple is by eliminating adjacent occurences of any B and B'. Thus $(A, A') = \emptyset$, and hence $(A)(A') = \emptyset$. Similarly

$(A')(A) = \emptyset$. Conclude the result.

Define $\gamma: X \to W_r(X)$ to be the function that sends A into the 1-tuple (A).

THEOREM 2.1.3 *If G is any group and $g: X \to G$ is a function, then there is a unique homomorphism $g^*: W_r(X) \to G$ such that*

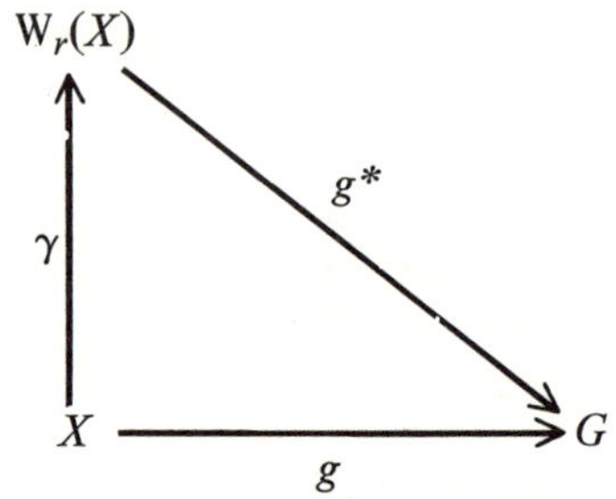

commutes.

Proof: Let g^* be the function sending $\emptyset$ into 1 (the identity element of G) and sending any n-tuple $(A_1, A_2, \ldots, A_n)$ into $g(B_1)^{e_1} g(B_2)^{e_2} \ldots g(B_n)^{e_n}$ where

$$B_i = \begin{cases} A_i & \text{if } A_i \in X \\ A_i' & \text{if } A_i \in X' \end{cases} \quad \text{and} \quad e_i = \begin{cases} 1 & \text{if } A_i \in X \\ -1 & \text{if } A_i \in X'. \end{cases}$$

Thus for all $A \in X$, $g^* \circ \gamma(A) = g^*((A)) = g(A)^1 = g(A)$. So the diagram commutes. To show g^* is a homomorphism let $R = (A_1, A_2, \ldots, A_n)$ and $S = (A_{n+1}, A_{n+2}, \ldots, A_m)$ belong to $W_r(X)$. First assume $A_n \neq A_{n+1}'$. Then

$$\begin{aligned} g^*(RS) &= g^*((A_1, A_2, \ldots, A_n, A_{n+1}, A_{n+2}, \ldots, A_m)) \\ &= g(B_1)^{e_1} g(B_2)^{e_2} \ldots g(B_n)^{e_n} g(B_{n+1})^{e_{n+1}} g(B_{n+2})^{e_{n+2}} \ldots g(B_m)^{e_m} \\ &= [g(B_1)^{e_1} g(B_2)^{e_2} \ldots g(B_n)^{e_n}]\,[g(B_{n+1})^{e_{n+1}} g(B_{n+2})^{e_{n+2}} \ldots g(B_m)^{e_m}] \\ &= g^*((A_1, A_2, \ldots, A_n)) g^*((A_{n+1}, A_{n+2}, \ldots, A_m)). \end{aligned}$$

Next suppose $A_n = A_{n+1}'$ and also that $A_{n-1} = A_{n+2}'$, $A_{n-2} = A_{n+3}'$, ..., $A_{n-k} = A_{n+k+1}'$ (where k, of course, might be zero). Then

$$\begin{aligned} g^*(RS) &= g^*((A_1, A_2, \ldots, A_{n-k-1}, A_{n+k+2}, A_{n+k+3}, \ldots, A_m)) \\ &= g(B_1)^{e_1} g(B_2)^{e_2} \ldots g(B_{n-k-1})^{e_{n-k-1}} \\ &\quad g(B_{n+k+2})^{e_{n+k+2}} g(B_{n+k+3})^{e_{n+k+3}} \ldots g(B_m)^{e_m}. \end{aligned}$$

On the other hand,

$$g^*(R) g^*(S) = g(B_1)^{e_1} g(B_2)^{e_2} \ldots g(B_n)^{e_n} g(B_{n+1})^{e_{n+1}} g(B_{n+2})^{e_{n+2}} \ldots g(B_m)^{e_m}.$$

Now suppose $0 \leqslant i \leqslant k$. Either A_{n-i} belongs to X or belongs to X'. In the first case $B_{n-i} = A_{n-i}$ and $e_{n-i} = 1$. Also $A_{n-i} = A_{n+i+1}'$ and the case hypothesis imply $A_{n+i+1}' \in X$ so $B_{n+i+1} = A_{n+i+1}' = A_{n-i}$ and $e_{n+i+1} = -1$. Therefore $g(B_{n-i})^{e_{n-i}} g(B_{n+i+1})^{e_{n+i+1}} = g(A_{n-i}) g(A_{n-i})^{-1} = 1$. The second case, $A_{n-i} \in X'$, yields $g(B_{n-i})^{e_{n-i}} g(B_{n+i+1})^{e_{n+i+1}} = 1$. This holds for all i such that $0 \leqslant i \leqslant k$. Conclude that $1 = g(B_{n-k})^{e_{n-k}} g(B_{n-k+1})^{e_{n-k+1}} \ldots g(B_n)^{e_n} g(B_{n+1})^{e_{n+1}} \ldots$

$g(B_{n+k+1})^{e_{n+k+1}}$. When this is substituted into the expression for $g^*(R)g^*(S)$ it yields $g^*(R)g^*(S) = g(B_1)^{e_1}g(B_2)^{e_2}\ldots g(B_{n-k-1})^{e_{n-k-1}}g(B_{n+k+2})^{e_{n+k+2}}$ $g(B_{n+k+3})^{e_{n+k+3}}\ldots g(B_m)^{e_m} = g^*(RS)$ proving g^* is a homomorphism.

Finally it will be shown that g^* is unique. Let $k:\mathrm{W}_r(X) \to G$ be any homomorphism such that $k\circ\gamma = g$, and seek to show $k = g^*$. For all $A \in X$, $k((A)) = k\circ\gamma(A) = g(A) = g^*\circ\gamma(A) = g^*((A))$. Thus k and g^* agree on the set $\{(A) \mid A \in X\}$. But clearly this set generates $\mathrm{W}_r(X)$ so k and g^* agree on a set of generators and hence are identical.

EXERCISES

1. Suppose the sets X and Y are equivalent. Show $\mathrm{W}_r(X) \cong \mathrm{W}_r(Y)$.
2. Suppose F is some group and $k:X \to F$ is a function such that if G is any group and $g:X \to G$ is any function, then there is a unique homomorphism $\bar{g}:F \to G$ making the diagram

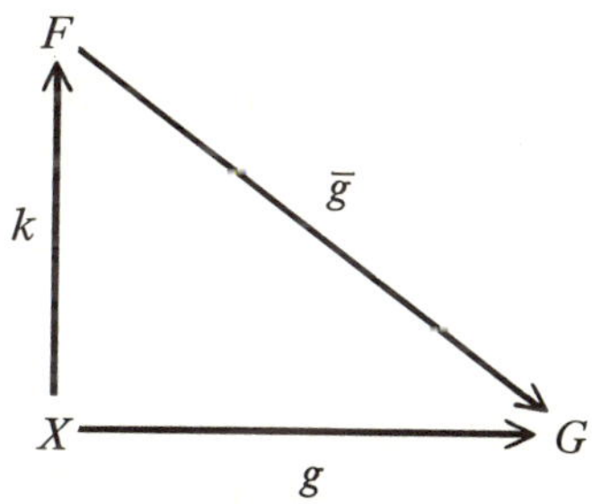

commute. Prove $F \cong \mathrm{W}_r(X)$.
3. Show $\mathrm{card}(X) = 1$ implies $\mathrm{W}_r(X) \cong Z$.
4. Show $2 \leqslant \mathrm{card}(X)$ implies $\mathrm{W}_r(X)$ is nonabelian.
5. Let W denote $\mathrm{W}_r(X)$. Suppose A is an abelian group and $f:X \to A$ is a function. Show there exists a unique homomorphism $f^*:W/W' \to A$ such that the following diagram commutes.

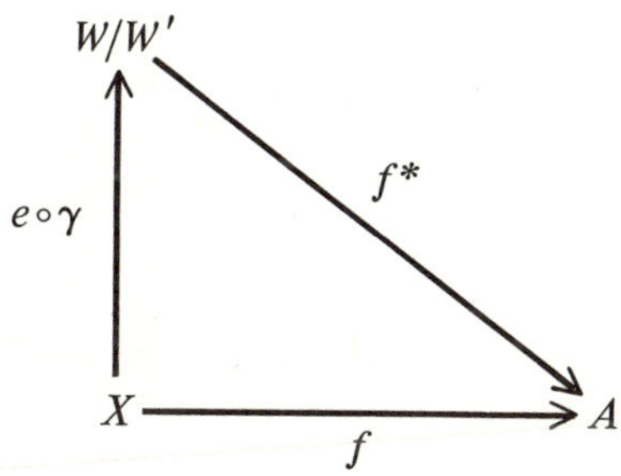

where $e:W \to W/W'$ is the natural map. The group W/W' is called the **free abelian group** on X.

2.2 *Presentations*

DEFINITION $(X \mid \mathcal{R})$ is a **presentation** if and only if $\mathcal{R} \subseteq W_r(X)$. ($X$ here can be any set.)

DEFINITION AND NOTATION If $(X \mid \mathcal{R})$ is a presentation, then $G(X \mid \mathcal{R})$ is the group of $(X \mid \mathcal{R})$ if and only if $G(X \mid \mathcal{R}) = W_r(X)/N$ where $N = \cap \{H \mid H \lhd W_r(X)$ and $\mathcal{R} \subseteq H\}$. It is clear that $N \lhd W_r(X)$. Elements of X are called the **generators** and elements of $\mathcal{R}$ are called the **relations** of $G(X \mid \mathcal{R})$.

We adopt four conventions at this point to avoid cumbersome notation and make presentations more meaningful. An example of how these conventions are used follows Theorem 2.2.1.

Convention 1 If X is a finite set, we will list the elements of X to the left of the vertical bar in a presentation. Similarly if $\mathcal{R}$ is finite, we will list its elements to the right of the vertical bar.
Convention 2 For $A \in X \cup X'$ we will use the exponents 1 and -1 as follows: $A^1 = A$ and $A^{-1} = A'$.
Convention 3 We will use the notation $A_1^{e_1}A_2^{e_2} \ldots A_m^{e_m} = 1$ instead of $(A_1^{e_1}, A_2^{e_2}, \ldots, A_m^{e_m}) \in \mathcal{R}$ whenever $e_i = \pm 1$ for $1 \leqslant i \leqslant m$. Justification for this convention is given in Theorem 2.2.3.
Convention 4 We will collect like terms in relations and use exponents. Formally this process is described by the rules: Use A^{ek} instead of $A^eA^e \ldots A^e$ (k times) whenever k is a positive integer and $e = \pm 1$. Use $(A_1^{e_1}A_2^{e_2})^k$ instead of $A_1^{e_1}A_2^{e_2}A_1^{e_1}A_2^{e_2} \ldots A_1^{e_1}A_2^{e_2}$ ($A_1^{e_1}A_2^{e_2}$ repeated k times) whenever k is a positive integer and $e_1, e_2 = \pm 1$, etcetera.

THEOREM 2.2.1 *For any nonzero integer k, let*

$$e = \begin{cases} 1 & \text{if } k \text{ is positive,} \\ -1 & \text{if } k \text{ is negative.} \end{cases}$$

Then $A^k = A^eA^e \ldots A^e$ (*$|k|$ times*).
Proof: Suppose k is positive. Then $|k| = k$ and $e = 1$ so $A^eA^e \ldots A^e$ ($|k|$ times) $= A^eA^e \ldots A^e$ (k times) $= A^{ek}$ (by Convention 4) $= A^k$. Next suppose k is negative. Then $|k| = -k$ and $e = -1$ so $A^eA^e \ldots A^e$ ($|k|$ times) $= A^eA^e \ldots A^e$ ($-k$ times) $= A^{(-1)(-k)} = A^k$.

As an illustration of how the four conventions are used we consider the presentation of a certain group of cardinality 16. Let

$$G = G(\{A,B\} \mid \{(A,A,A,A), (B,B,B,B), (A,B,A,B), (A',B,A',B)\}).$$

By Convention 1 we can list both the generators and relations of G to get

$$G = G(A,B \mid (A,A,A,A), (B,B,B,B), (A,B,A,B), (A',B,A',B)).$$

Next we can change the primes in the last relation to the exponent − 1 by Convention 2. Thus

$$G = \mathrm{G}(A,B \mid (A,A,A,A), (B,B,B,B), (A,B,A,B), (A^{-1},B,A^{-1},B)).$$

Applying Convention 3 yields

$$G = \mathrm{G}(A,B \mid AAAA = 1, BBBB = 1, ABAB = 1, A^{-1}BA^{-1}B = 1).$$

Finally we collect terms by Convention 4 and arrive at the preferred form for G,

$$G = \mathrm{G}(A,B \mid A^4 = 1, B^4 = 1, (AB)^2 = 1, (A^{-1}B)^2 = 1).$$

We need one final item of notation. For $A \in X \cup X'$ we have $(A) \in \mathrm{W}_r(X)$ and so $N(A)$ (where $N = \cap\{H \mid H \lhd \mathrm{W}_r(X)$ and $\mathcal{R} \subseteq H\}$ as in the definition of $\mathrm{G}(X \mid \mathcal{R})$) belongs to $\mathrm{G}(X \mid \mathcal{R})$. It is desirable to have some special notation to show that the group element $N(A)$ corresponds to the set element A. Many authors refuse to distinguish (in notation) A and its corresponding element of $\mathrm{G}(X \mid \mathcal{R})$, writing the same symbol for each. This approach makes up for what it lacks in technical exactness by being the most convenient notation possible. It is hoped that the approach we will take below is only slightly less convenient.

We continue to use capital letters* for the set X and its elements, and will use lowercase letters to denote the elements of $\mathrm{G}(X \mid \mathcal{R})$ corresponding to the capital-letter elements of X. Thus

$$a = N(A).$$

THEOREM 2.2.2 *If $A \in X \cup X'$ and k is a positive integer, then*

$$a^{ek} = N(A^e, A^e, \ldots, A^e) \qquad (k \text{ coordinates})$$

where $e = \pm 1$.

Proof: First suppose $e = 1$. Then $a = N(A)$ implies $a^k = [N(A)]^k = N(A)^k = N(A,A,\ldots,A)$ (k coordinates). Thus the result holds. Next suppose $e = -1$. Then $a = N(A)$ implies $a^{-1} = [N(A)]^{-1} = N(A)^{-1} = N(A')$ (Theorem 2.1.2) $= N(A^e)$ by Convention 2. Hence $a^{-k} = (a^{-1})^k = [N(A^e)]^k = N(A^e)^k = N(A^e, A^e, \ldots, A^e)$ (k coordinates) as required.

The following theorem says that group elements $a_1, a_2, \ldots, a_n \in \mathrm{G}(X \mid \mathcal{R})$ "satisfy the same relations" as their corresponding set elements $A_1, A_2, \ldots, A_n \in X \cup X'$. This, of course, is the motivation for using the conventions to recast relations into their preferred form in the first place. The reader should note

*In particular, small-sized capital letters $A, B, C, \ldots$ have been and will continue to be used in this chapter for the elements of X, while regular-sized capitals will be used in the later chapters.

that this theorem will be used many times when presentations are dealt with, and will usually be used without explicit reference.

THEOREM 2.2.3 *If* $A_1^{k_1}A_2^{k_2}\ldots A_m^{k_m} = 1$ *where each* k_i *is a nonzero integer, then* $a_1^{k_1}a_2^{k_2}\ldots a_m^{k_m} = 1$ *in the group* $G(X|\mathcal{R})$.

Proof: By Theorem 2.2.1, $A_1^{k_1}A_2^{k_2}\ldots A_m^{k_m} = 1$ means that $A_1^{e_1}A_1^{e_1}\ldots A_1^{e_1}A_2^{e_2}A_2^{e_2}\ldots A_2^{e_2}\ldots A_m^{e_m}A_m^{e_m}\ldots A_m^{e_m} = 1$ where each term A_i is repeated $|k_i|$ times and where $e_i = 1$ for positive k_i and -1 for negative k_i. By Convention 3 this means $T \in \mathcal{R}$ where

$$T = (A_1^{e_1}, A_1^{e_1}, \ldots, A_1^{e_1}, A_2^{e_2}, A_2^{e_2}, \ldots, A_2^{e_2}, \ldots, A_m^{e_m}, A_m^{e_m}, \ldots, A_m^{e_m})$$

(here again each term A_i is repeated $|k_i|$ times). Now $a_i^{k_i} = a_i^{e_i|k_i|} = N(A_i^{e_i}, A_i^{e_i}, \ldots, A_i^{e_i})$ ($|k_i|$ coordinates) by Theorem 2.2.2. Therefore $a_1^{k_1}a_2^{k_2}\ldots a_m^{k_m} = N(A_1^{e_1}, A_1^{e_1}, \ldots, A_1^{e_1})N(A_2^{e_2}, A_2^{e_2}, \ldots, A_2^{e_2})\ldots N(A_m^{e_m}, A_m^{e_m}, \ldots, A_m^{e_m})$ which is seen to be the coset (with respect to N) determined by the word T. But $T \in \mathcal{R} \subseteq N$ so this coset is 1. That is, $a_1^{k_1}a_2^{k_2}\ldots a_m^{k_m} = 1$.

As an illustration of how Theorem 2.2.3 is used, consider the example dealt with previously, namely $G = G(A, B\,|\,A^4 = 1, B^4 = 1, (AB)^2 = 1, (A^{-1}B)^2 = 1)$. Theorem 2.2.3 says that $a^4 = 1, b^4 = 1, (ab)^2 = 1$, and $(a^{-1}b)^2 = 1$ in G.

The following theorem provides justification for why the elements of X are called *generators* of $G(X|\mathcal{R})$.

THEOREM 2.2.4 $\{a|\,A \in X\}$ *generates* $G(X|\mathcal{R})$.

Proof: Clearly the set $\{(A)|\,A \in X\}$ generates $W_r(X)$. Thus $\{N(A)|\,A \in X\}$ generates $G(X|\mathcal{R})$ by Theorem A.10.4.

We now prove an important theorem which says that if the group H "satisfies the same relations" as $G(X|\mathcal{R})$, it must be a homomorphic image of $G(X|\mathcal{R})$.

Theorem 2.2.5 (Von Dyck's theorem) *Let* $(X|\mathcal{R})$ *be a presentation and let* $G = G(X|\mathcal{R})$. *Suppose* H *is some group, generated by a set* S, *where* $f{:}X \to S$ *is a surjective function. Suppose further that any relation* $A_1^{k_1}A_2^{k_2}\ldots A_m^{k_m} = 1$ *where each* k_i *is a nonzero integer implies* $f(A_1)^{k_1}f(A_2)^{k_2}\ldots f(A_m)^{k_m} = 1$ *in the group* H. *Then there exists a surjective homomorphism* $d{:}G \to H$ *such that* $d(a) = f(A)$ *for all* $A \in X$.

Proof: Since $f{:}X \to H$ is a function, application of Theorem 2.1.3 gives a homomorphism f^* such that $f^* \circ \gamma = f$. First it is claimed that f^* is onto H. Let $y \in H$. Then since S is a generating set, $y = s_1^{c_1}s_2^{c_2}\ldots s_n^{c_n}$ where each $s_i \in S$ and c_i is an integer. Thus there exists $A_i \in X$ such that $f(A_i) = s_i$ for all i. Hence $\gamma(A_1)^{c_1}\gamma(A_2)^{c_2}\ldots\gamma(A_n)^{c_n} \in W_r(X)$ and $f^*(\gamma(A_1)^{c_1}\gamma(A_2)^{c_2}\ldots\gamma(A_n)^{c_n}) = f(A_1)^{c_1}f(A_2)^{c_2}\ldots f(A_n)^{c_n} = y$. Thus f^* is onto H as claimed.

Recall from the definition of G that $G = W_r(X)/N$ where $N = \cap\{H|H$

is normal in $W_r(X)$ and $\mathcal{R} \subseteq H\}$. It is now claimed that $N \subseteq \ker f^*$. To show this it suffices, by the definition of N, to show $\mathcal{R} \subseteq \ker f^*$. So let $T \in \mathcal{R}$. If $T = \emptyset$, then $T \in \ker f^*$ because $\emptyset$ is the identity element of $W_r(X)$. Assume then $T \neq \emptyset$. Hence $T = (A_1{}^{e_1}, A_1{}^{e_1}, \ldots, A_1{}^{e_1}, A_2{}^{e_2}, A_2{}^{e_2}, \ldots, A_2{}^{e_2}, \ldots, A_m{}^{e_m}, A_m{}^{e_m}, \ldots, A_m{}^{e_m})$ where each $e_i = \pm 1$ and where each term A_i is repeated, say, t_i times. Thus $T = (A_1)^{e_1 t_1}(A_2)^{e_2 t_2} \ldots (A_m)^{e_m t_m} = \gamma(A_1)^{e_1 t_1}\gamma(A_2)^{e_2 t_2} \ldots \gamma(A_m)^{e_m t_m}$ and so $f^*(T) = f(A_1)^{e_1 t_1} f(A_2)^{e_2 t_2} \ldots f(A_m)^{e_m t_m}$. But by Conventions 3 and 4, $T \in \mathcal{R}$ implies $A_1{}^{e_1 t_1} A_2{}^{e_2 t_2} \ldots A_m{}^{e_m t_m} = 1$. Applying the hypothesis therefore yields $f(A_1)^{e_1 t_1} f(A_2)^{e_2 t_2} \ldots f(A_m)^{e_m t_m} = 1$. Thus $f^*(T) = 1$ completing the proof that $N \subseteq \ker f^*$.

Hence $f^*[N] = \{1\}$. Applying the Corollary to Theorem A.10.1 shows that the function $f'': W_r(X)/N \to H$ that sends the coset determined by a word T into $f^*(T)$ is a surjective homomorphism. Finally note that for all $A \in X$, $f''(a) = f''(N(A)) = f^*((A)) = f^* \circ \gamma(A) = f(A)$. Conclude that $d = f''$ is the required homomorphism.

For finite groups Von Dyck's theorem may be sharpened as follows:

THEOREM 2.2.6 *Let $(X \mid \mathcal{R})$ be a presentation and let $G = G(X \mid \mathcal{R})$. Suppose H is a group such that $\operatorname{card}(G) \leqslant n \leqslant \operatorname{card}(H)$ for some natural number n and $H = \langle S \rangle$ where $f: X \to S$ is a surjective function. Suppose further that any relation $A_1{}^{k_1} A_2{}^{k_2} \ldots A_m{}^{k_m} = 1$, each k_i a nonzero integer, implies $f(A_1)^{k_1} f(A_2)^{k_2} \ldots f(A_m)^{k_m} = 1$ in the group H. Then*

(a) *There exists an isomorphism $d: G \to H$ such that $d(a) = f(A)$ for all $A \in X$.*

(b) $\operatorname{card}(G) = \operatorname{card}(H) = n$.

(c) $G = \{d^{-1}(h) \mid h \in H\}$.

Proof: Since the hypotheses of Theorem 2.2.5 are satisfied, there exists a surjective homomorphism $d: G \to H$ such that $d(a) = f(A)$ for all $A \in X$. Then $G/\ker d \cong H$. But $\operatorname{card}(G/\ker d) \leqslant \operatorname{card}(G)$ so $\operatorname{card}(H) \leqslant \operatorname{card}(G)$. This and the hypothesis $\operatorname{card}(G) \leqslant n \leqslant \operatorname{card}(H)$ imply $\operatorname{card}(G) = \operatorname{card}(H) = n$ proving Part (*b*). But $\operatorname{card}(G) = \operatorname{card}(H)$ forces $\operatorname{card}(\ker d) = 1$ since G is finite. Thus $\ker d = \{1\}$ so d is an isomorphism proving Part (a). Part (c) follows immediately.

The reader encountering presentations here for the first time may be somewhat awed by their seemingly magical quality. It seems as if all one needs to do is write down some generators, then write any relations among these that happen to come to mind, and ZAP! almost out of thin air comes the existence of a group having these generators and relations. While this is indeed true, there are several reasons why it cannot be regarded as the object of unrestrained enthusiasm it appears to be at first glance.

First of all there is the problem that when searching for examples of groups with exotic properties, these properties may not be able to be expressed in terms of generators and relations. What, for instance, presentation can one consider for an example of a simple group with no maximal subgroups?

Secondly, as mentioned at the beginning of this chapter, the picture a presentation gives of a group is often an obscure one. Given a group $G = \mathrm{G}(X \mid \mathcal{R})$, with no further information about G than its presentation, the properties of G may be difficult or impossible to deduce. It may not be at all obvious, for instance, whether or not G is finite or even whether or not G is trivial. Consider $G = \mathrm{G}(A \mid A^2 = 1, A^3 = 1)$. The two relations force $a = 1$ so G is trivial. This is fairly obvious, but with more generators and more complicated relations among them it becomes less and less obvious.

Proving that a group $G = \mathrm{G}(X \mid \mathcal{R})$ is nontrivial usually involves an application of Von Dyck's theorem. If a nontrivial group H can be shown to be a homomorphic image of G using Von Dyck's theorem, then, of course, G itself is nontrivial. Faced with an unfriendly-looking presentation, there is no algorithm for deciding what group H to try to work with. One simply calls on all his knowledge of examples of groups and hopes for the best. The task of showing that a group $G = \mathrm{G}(X \mid \mathcal{R})$ is infinite is handled similarly – find an infinite group H and show H is a homomorphic image of G by Von Dyck's theorem. (An alternate procedure that frequently works is to find a sequence of finite groups $\{H_n\}_{n \in N}$, each of which is a homomorphic image of G by Von Dyck's theorem, such that the cardinalities of the groups in the sequence increase without bound.)

PREVIEW

The following is a preview of how presentations will be used in this book.

The first eleven groups of Chapter 4 will be defined by presentations. After $\mathrm{G}(X \mid \mathcal{R})$ is given in each case, a second group, F, will be constructed, then Theorem 2.2.6 will be used to show that $\mathrm{G}(X \mid \mathcal{R}) \cong F$. This will yield a highly desirable "concrete representation" for $\mathrm{G}(X \mid \mathcal{R})$. In order to apply Theorem 2.2.6 it will have to be shown somehow or another that $\mathrm{card}(\mathrm{G}(X \mid \mathcal{R})) \leqslant n$ for some natural number n. All but one of the first eleven groups in Chapter 4 are solvable and getting such a bound on the cardinality of $\mathrm{G}(X \mid \mathcal{R})$ will turn out to be not too difficult. It will involve defining a subgroup K of $\mathrm{G}(X \mid \mathcal{R})$ by means of certain generators or their combinations, then showing K is normal. Thus finding a bound for $\mathrm{card}(\mathrm{G}(X \mid \mathcal{R}))$ reduces to finding bounds for the cardinalities of the less complicated groups K and $\mathrm{G}(X \mid \mathcal{R})/K$. A procedure such as this one works well for solvable groups which have an abundant supply of normal subgroups, but will be useless when working with a simple group.

Thus in Example 4.11, which is a simple group, some other procedure will have to be used. Here two subgroups, H and B will be defined and it will be shown that $\mathrm{G}(X \mid \mathcal{R}) = HB$. Once this is done it is a relatively easy matter to get a bound for $\mathrm{card}(\mathrm{G}(X \mid \mathcal{R}))$ in terms of $\mathrm{card}(H)$ and $\mathrm{card}(B)$. Showing that $\mathrm{G}(X \mid \mathcal{R}) = HB$, however, will involve rather extensive calculations. It should be pointed out that there does exist a "cookbook" procedure, called the method of coset enumeration, for handling problems such as Example 4.11 (see Coxeter and Moser [6]). The coset enumeration method also greatly cuts down on the amount of calculation. It has not been included in this book because explaining it and then proving that the method works would have taken more than the space presently devoted to its one potential application, Example 4.11.

The last three finite groups in Chapter 4 are not defined by presentations. Presentations exist for these groups (indeed every group has a presentation) and will be given without proof in the notes to the groups.

The Miscellaneous Exercises to Chapter 4, in addition to providing further examples of solvable groups defined by presentations, will also give an example of a finite simple group, and ask the reader to show that the group is perfect. Problems of this kind can be solved with utter disregard to the cardinality of the group in question. They are solved by working with the relations until it has been shown that every generator of the group is a commutator or a product of commutators.

The use of presentations in the last two chapters will be much less extensive than their use in Chapter 4. None of the (numbered) examples in these two chapters will be defined by presentations, but presentations for several of them will be found in the Exercises. The Miscellaneous Exercises to Chapter 6 will also have some groups defined by presentations.

EXERCISES

1. For all positive integers n, show $G(A \mid A^n = 1) \cong Z(n)$.
2. Show $G(A, B \mid A^2 = 1, B^2 = 1, (AB)^2 = 1) \cong Z(2) \times Z(2)$.

SUGGESTIONS FOR FURTHER READING

The book Magnus, Karass, and Solitar [25] is almost entirely devoted to free groups, presentations, and related matters.

3

Matrices

Matrices find many and varied applications to different branches of mathematics and the mathematically-based sciences. Group theory is no exception. Indeed matrices provide a way for constructing examples of groups that would be difficult if not impossible to exhibit otherwise.

Several families of matrix groups will be described in Section 3.2. Preceding this is a brief discussion of matrices themselves.

3.1 Matrices

If R is a ring (we will be mostly concerned with commutative rings with identity – fields in particular – but at the moment require nothing further of R than that it be a ring) and n is a nonzero natural number, we formally define an n by n matrix as a function from $\{1,2,3,\ldots,n\} \times \{1,2,3,\ldots,n\}$ into R. Thus if M is an n by n matrix, M assigns a ring element to every pair (i,j) such that $1 \leqslant i,j \leqslant n$. In matrix theory it is traditional to denote the ring element assigned to (i,j) by $M_{i,j}$ instead of the usual notation for functions, $M(i,j)$. Moreover, when confusion is not likely to occur by deletion of the comma, it is traditional to use M_{ij} instead of $M_{i,j}$. The ring elements $\{M_{ij} | 1 \leqslant i,j \leqslant n\}$ are called the **entries** of M.

There is a very pleasant representation for matrices which is unavailable to arbitrary functions. A matrix can be represented by writing its entries in rows and columns so that the entry M_{ij} goes into the ith row and jth column. Thus, for example, if R is the ring of integers and if M is the 2 by 2 matrix determined by $M_{11} = 6$, $M_{12} = 8$, $M_{21} = -4$, and $M_{22} = 0$, then M can conveniently be represented as

$$\begin{pmatrix} 6 & 8 \\ -4 & 0 \end{pmatrix}$$

In practice this method of representing matrices has proved so convenient that in most cases, especially when n is small, the formal definition of a matrix may be all but forgotten about and a matrix may be considered as a representation of ring elements in the square-array form shown above.

Much of the matrix theory developed in this book will be concerned only with 2 by 2 matrices. The theorems presented do have generalizations to the n by n case but their proofs are considerably more complicated. Fortunately while 2 by 2 matrices are the simplist kind that can be considered (the 1 by 1 case being trivial) they will be powerful enough to yield many and various examples of groups.

We will, however, present some theorems for n by n matrices where n is arbitrary. In particular, special n by n matrices called unitriangular matrices will be defined shortly and their properties developed.

We proceed with some notation, definitions, and theorems for matrices. In the remainder of this chapter if explicit mention of the ring R is not made, it will be assumed that R is any commatative ring with identity.

First we define a multiplication for matrices. Given two n by n matrices A and B, we define the product matrix (denoted by AB) by specifying all of its entries as follows: $(AB)_{ij} = \Sigma_{t=1}^{n} A_{it}B_{tj}$. This says that the entry in the ith row and jth column of the product matrix is the sum $A_{i1}B_{1j} + A_{i2}B_{2j} + \ldots + A_{in}B_{nj}$.

For the 2 by 2 case this specializes to

$$\begin{pmatrix} A_{11} & A_{12} \\ A_{21} & A_{22} \end{pmatrix}\begin{pmatrix} B_{11} & B_{12} \\ B_{21} & B_{22} \end{pmatrix} = \begin{pmatrix} A_{11}B_{11}+A_{12}B_{21} & A_{11}B_{12}+A_{12}B_{22} \\ A_{21}B_{11}+A_{22}B_{21} & A_{21}B_{12}+A_{22}B_{22} \end{pmatrix}.$$

Although it can be easily shown by example that matrix multiplication is not commutative, the following theorem shows that it is associative.

THEOREM 3.1.1 *If A,B, and C are n by n matrices, then $A(BC) = (AB)C$.*
Proof: It must be shown that $(A(BC))_{ij} = ((AB)C)_{ij}$ for all $i,j \leqslant n$. By definition of multiplication, $((AB)C)_{ij} = \Sigma_{k=1}^{n}(AB)_{ik}C_{kj}$. Furthermore, for $1 \leqslant k \leqslant n$, $(AB)_{ik}C_{kj} = (\Sigma_{t=1}^{n}A_{it}B_{tk})C_{kj} = \Sigma_{t=1}^{n}A_{it}B_{tk}C_{kj}$ because multiplication is distributive over addition in the ring R. Thus $((AB)C)_{ij} = \Sigma_{k=1}^{n}(\Sigma_{t=1}^{n}A_{it}B_{tk}C_{kj}) = \Sigma_{t=1}^{n}(\Sigma_{k=1}^{n}A_{it}B_{tk}C_{kj}) = \Sigma_{t=1}^{n}A_{it}(\Sigma_{k=1}^{n}B_{tk}C_{kj}) = \Sigma_{t=1}^{n}A_{it}(BC)_{tj} = (A(BC))_{ij}$.

For all n there is a particular n by n matrix deserving of special attention. It is denoted by I and defined by

$$I_{ij} = \begin{cases} 1 & \text{if } i = j, \\ 0 & \text{if } i \neq j. \end{cases}$$

This matrix is called the (n by n) identity matrix. When we consider groups of matrices in Section 3.2, we will see that I becomes the identity element of these groups. It should be kept in mind that there is (unfortunately) no provision in our notation to distinguish between the n by n identity matrix and the m by m identity matrix when $n \neq m$. This will cause no great difficulities however. When I is used it will always be clear from the context which I is meant.

THEOREM 3.1.2 *If A is an n by n matrix, then $IA = A$ and $AI = A$.*
Proof: For $1 \leqslant i,j \leqslant n$, $(IA)_{ij} = \Sigma_{k=1}^{n} I_{ik}A_{kj}$ which equals $(\Sigma_{k=1}^{i-1} I_{ik}A_{kj}) + I_{ii}A_{ij} + (\Sigma_{k=i+1}^{n} I_{ik}A_{kj})$. Using the definition of I, this becomes $(IA)_{ij} = 0 + A_{ij} + 0 = A_{ij}$. Conclude $IA = A$. The proof that $AI = A$ is similar.

NOTATION If $1 \leqslant i,j \leqslant n$ where $i \neq j$ and $x \in R$, then $\Theta_n(i,j,x)$ will denote the matrix defined by

$$\Theta_n(i,j,x)_{hk} = \begin{cases} 1 & \text{if } h = k, \\ x & \text{if } (h,k) = (i,j), \\ 0 & \text{otherwise.} \end{cases}$$

Note that $\Theta_n(i,j,0) = I$ for all i and j such that $1 \leqslant i,j \leqslant n$ and $i \neq j$.

DEFINITION For $1 \leqslant i,j \leqslant n$ where $i \neq j$ and $x \in R$ such that $x \neq 0$, $\Theta_n(i,j,x)$ is called the n by n **transvection** determined by i, j, and x.

Thus a transvection is a matrix M which differs from the identity matrix in exactly one entry, M_{ij}, where $i \neq j$.

THEOREM 3.1.3 *If $1 \leqslant i,j \leqslant n$ where $i \neq j$ and if $x,y \in R$, then*

$$\Theta_n(i,j,x)\Theta_n(i,j,y) = \Theta_n(i,j,x+y).$$

Proof: Let A denote $\Theta_n(i,j,x)$ and let B denote $\Theta_n(i,j,y)$. It must be shown that (1) $(AB)_{hh} = 1$ for all $h \leqslant n$, (2) $(AB)_{ij} = x + y$, and (3) $(AB)_{hk} = 0$ if $h \neq k$ and $(h,k) \neq (i,j)$.

Proof of (1): $(AB)_{hh} = \Sigma_{t=1}^{n} A_{ht}B_{th}$. First suppose $h = i$. Then $(AB)_{hh} = A_{ii}B_{ii} + A_{ij}B_{ji}$. But $i \neq j$ implies $B_{ji} = 0$ so $(AB)_{hh} = A_{ii}B_{ii} = 1$. Secondly suppose $h \neq i$. Then $(h,t) \neq (i,j)$ so $(AB)_{hh} = A_{hh}B_{hh} = 1$.

Proof of (2): $(AB)_{ij} = \Sigma_{t=1}^{n} A_{it}B_{tj} = A_{ii}B_{ij} + A_{ij}B_{jj} = B_{ij} + A_{ij} = y + x$.

Proof of (3): Assume now that $h \neq k$ and $(h,k) \neq (i,j)$. Then $(AB)_{hk} = \Sigma_{t=1}^{n} A_{ht}B_{tk}$. First suppose $h = i$. Then $(AB)_{hk} = A_{ii}B_{ik} + A_{ij}B_{jk}$. But $k \neq j$ (lest $(h,k) = (i,j)$) and also $k \neq i$ (lest $h = k$) so $B_{ik} = 0$. Furthermore $k \neq j$ and $j \neq i$ (by hypothesis) imply $B_{jk} = 0$. Conclude $(AB)_{hk} = 0$. Secondly suppose $h \neq i$. Then $(AB)_{hk} = A_{hh}B_{hk} = B_{hk} = 0$.

COROLLARY *If $1 \leqslant i,j \leqslant n$ and $i \neq j$ and $x \in R$, then $\Theta_n(i,j,x)^m = \Theta_n(i,j,mx)$ for all positive integers m.*

Next we consider triangular and unitriangular matrices.

DEFINITION The n by n matrix A is **triangular** if and only if $A_{ij} = 0$ for all i and j such that $1 \leqslant i < j \leqslant n$. (These matrices are also commonly called lower triangular.)

DEFINITION The n by n matrix A is **unitriangular** if and only if
(a) A is triangular.
(b) $A_{ii} = 1$ for all i such that $1 \leqslant i \leqslant n$.

THEOREM 3.1.4 *If $j < i$, then $\Theta_n(i,j,x)$ is unitriangular.*
Proof: Let $M = \Theta_n(i,j,x)$. Then for $1 \leqslant h \leqslant n$, $M_{hh} = 1$ and for $1 \leqslant h < k \leqslant n$, $M_{hk} = 0$ (M_{hk} cannot equal x since $h < k$ implies $(h,k) \neq (i,j)$ because $j < i$). Thus M is unitriangular.

If A is an n by n matrix where $2 \leqslant n$, A can be transformed into an $n - 1$ by $n - 1$ matrix by deleting any row and any column. Let $A(k \mid k')$ denote the matrix formed from A by deletion of row k and column k'. We will mostly be concerned with $A(n \mid n)$ whose definition can be restated as follows: $A(n \mid n)_{ij} = A_{ij}$ for $1 \leqslant i,j \leqslant n - 1$.

THEOREM 3.1.5
(a) *If A is an n by n triangular matrix, then $A(n \mid n)$ is also triangular.*
(b) *If A is an n by n unitriangular matrix, then $A(n \mid n)$ is also unitriangular.*
Proof: Suppose A is triangular. Then for $i < j$, $A(n \mid n)_{ij} = A_{ij} = 0$. Thus $A(n \mid n)$ is triangular, proving Part (a).

Suppose now A is unitriangular. Then $A(n \mid n)_{ij} = 0$ for $i < j$ as above and in addition $A(n \mid n)_{ii} = A_{ii} = 1$ for all pertinent i. Therefore A is unitriangular.

THEOREM 3.1.6
(a) *The product of n by n triangular matrices is triangular.*
(b) *The product of n by n unitriangular matrices is unitriangular.*
Proof: Let A and B be n by n triangular matrices. It must be shown that $i < j$ implies $(AB)_{ij} = 0$. For $i < j$, $(AB)_{ij} = \Sigma_{k=1}^{n} A_{ik}B_{kj}$. But for $k \leqslant n$, either $k < j$ or $j \leqslant k$. In the first case $B_{kj} = 0$ because B is triangular. In the second case $i < k$ because $i < j$. Thus $A_{ik} = 0$ because A is triangular. In either case, therefore, $A_{ik}B_{kj} = 0$. Conclude $(AB)_{ij} = 0$ as required proving Part (a).

To show Part (b) suppose A and B are n by n unitriangular matrices. Then AB is triangular by Part (a) so all that must be shown is $(AB)_{ii} = 1$ for all $i \leqslant n$. For such i, $(AB)_{ii} = \Sigma_{k=1}^{n} A_{ik}B_{ki}$. Since A and B are triangular, $A_{ik} = 0$ for all $k > i$ and $B_{ki} = 0$ for all $k < i$. Thus $(AB)_{ii} = A_{ii}B_{ii} = 1$.

The next two theorems give more detailed information on the product of two triangular matrices.

THEOREM 3.1.7 *If A and B are n by n triangular matrices, then $j \leqslant i$ implies $(AB)_{ij} = \Sigma_{t=j}^{i} A_{it}B_{tj}$.*

Proof: For such i and j the product $(AB)_{ij} = \Sigma_{t=1}^{n} A_{it}B_{tj} = S_1 + S_2 + S_3$ where $S_1 = \Sigma_{t=1}^{j-1} A_{it}B_{tj}$, $S_2 = \Sigma_{t=j}^{i} A_{it}B_{tj}$, and $S_3 = \Sigma_{t=i+1}^{n} A_{it}B_{tj}$. But $t \leqslant j - 1$ implies $t < j$ and hence $B_{tj} = 0$ since B is triangular. Thus $S_1 = 0$. Also $i + 1 \leqslant t \leqslant n$ implies $i < t$ and hence $A_{it} = 0$ since A is triangular. Thus $S_3 = 0$. Conclude that $(AB)_{ij} = S_2$.

THEOREM 3.1.8 *Let A and B be n by n triangular matrices, let $C = A(n\,|\,n)$ and $D = B(n\,|\,n)$. Then*

$$(AB)_{ij} = \begin{cases} (CD)_{ij} & 1 \leqslant i,j \leqslant n-1, \\ (\Sigma_{t=1}^{n-1} A_{nt}D_{tj}) + A_{nn}B_{nj} & i = n \text{ and } 1 \leqslant j \leqslant n-1, \\ 0 & 1 \leqslant i \leqslant n-1 \text{ and } j = n, \\ A_{nn}B_{nn} & i = j = n. \end{cases}$$

Proof: By definition of multiplication, $(AB)_{ij} = (\Sigma_{t=1}^{n-1} A_{it}B_{tj}) + A_{in}B_{nj}$.

If $1 \leqslant i,j \leqslant n - 1$, then $i < n$ so $A_{in} = 0$ and hence $A_{in}B_{nj} = 0$. Thus $(AB)_{ij} = \Sigma_{t=1}^{n-1} A_{it}B_{tj} = \Sigma_{t=1}^{n-1} C_{it}D_{tj} = (CD)_{ij}$.

If $i = n$ and $1 \leqslant j \leqslant n - 1$, then $(AB)_{ij} = (\Sigma_{t=1}^{n-1} A_{nt}B_{tj}) + A_{nn}B_{nj}$ which is equal to $(\Sigma_{t=1}^{n-1} A_{nt}D_{tj}) + A_{nn}B_{nj}$.

If $1 \leqslant i \leqslant n - 1$ and $j = n$, then $t < j$ for all $t \leqslant n - 1$, hence $B_{tj} = 0$ for all such t. Therefore $(AB)_{ij} = A_{in}B_{nj} = A_{in}B_{nn} = 0$ because $A_{in} = 0$ since $i < n$.

Finally if $i = j = n$, then again $B_{tj} = 0$ for all $t \leqslant n - 1$. Therefore $(AB)_{ij} = A_{in}B_{nj} = A_{nn}B_{nn}$.

Theorem 3.1.8 becomes more meaningful when the n by n matrices involved are represented in so-called "block" form. Theorem 3.1.7 then states

$$\left(\begin{array}{c|c} C & 0 \\ \hline u & A_{nn} \end{array}\right)\left(\begin{array}{c|c} D & 0 \\ \hline w & B_{nn} \end{array}\right) = \left(\begin{array}{c|c} CD & 0 \\ \hline uD + A_{nn}w & A_{nn}B_{nn} \end{array}\right)$$

where u and w are the vectors $u = (A_{n1}, A_{n2}, \ldots, A_{n,n-1})$ and $w = (B_{n1}, B_{n2}, \ldots, B_{n,n-1})$, and where $uD + A_{nn}w$ is the vector whose jth coordinate is the sum $(\Sigma_{t=1}^{n-1} A_{nt}D_{tj}) + A_{nn}B_{nj}$.

COROLLARY 1 *Let A and B be n by n triangular matrices. Then $(AB)(n\,|\,n) = A(n\,|\,n)B(n\,|\,n)$.*

Proof: Let $C = A(n\,|\,n)$ and $D = B(n\,|\,n)$. Then for $1 \leqslant i,j \leqslant n - 1$, $(AB)(n\,|\,n)_{ij} = (AB)_{ij} = (CD)_{ij}$ by Theorem 3.1.8. Thus $(AB)(n\,|\,n) = CD$.

COROLLARY 2 *Let A and B be n by n triangular matrices such that $A_{nj} = B_{nj} = 0$ for all $j < n$. Then $(AB)_{nj} = 0$ for all $j < n$.*

Proof: Let $D = B(n|n)$. Then by the theorem, $(AB)_{nj} = (\Sigma_{t=1}^{n-1} A_{nt}D_{tj}) + A_{nn}B_{nj}$. But $A_{nt} = 0$ for all $t \leqslant n - 1$ and $B_{nj} = 0$ also. Thus $(AB)_{nj} = 0$.

LEMMA *For* $1 \leqslant k \leqslant n - 1$, *let* $F(k) = \Theta_n(n,1,x_1)\Theta_n(n,2,x_2)\ldots\Theta_n(n,k,x_k)$. *Then*

(a) $j \leqslant k$ *implies* $F(k)_{nj} = x_j$.

(b) $k < j \leqslant n - 1$ *implies* $F(k)_{nj} = 0$.

(c) $j < i < n$ *implies* $F(k)_{ij} = 0$.

Proof: This will be shown by induction on k. When $k = 1$, $F(k) = \Theta_n(n,1,x_1)$. Therefore $F(k)_{n1} = x_1$; $F(k)_{nj} = 0$ for $k < j \leqslant n - 1$; and $F(k)_{ij} = 0$ for $j < i < n$. Thus induction begins.

Assume now the induction hypothesis, namely that Properties (a), (b), and (c) hold for $F(k - 1)$. Note that $F(k) = F(k - 1)\Theta_n(n,k,x_k)$. Then for $j \leqslant k$, $F(k)_{nj} = \Sigma_{t=1}^{n} F(k - 1)_{nt}\Theta_n(n,k,x_k)_{tj}$. Now either $j = k$ or $j \leqslant k - 1$. If $j = k$, then $\Theta_n(n,k,x_k)_{tj}$ is nonzero only when $t = n$ or j. Thus $F(k)_{nj} = F(k - 1)_{nj} + \Theta_n(n,k,x_k)_{nj}$. But $j = k$ implies $k - 1 < j$, and hence $F(k - 1)_{nj} = 0$ by the induction hypothesis (Property (b)). Hence $F(k)_{nj} = \Theta_n(n,k,x_k)_{nj} = x_k$. On the other hand if $j \leqslant k - 1$, then $F(k - 1)_{nj} = x_j$ by the induction hypothesis (Property (a)). Also $\Theta_n(n,k,x_k)_{tj}$ is nonzero only when $t = j$. Thus $F(k)_{nj} = F(k - 1)_{nj}$ which is equal to x_j. This establishes Property (a) for $F(k)$.

Next suppose $k < j \leqslant n - 1$. Then $\Theta_n(n,k,x_k)_{tj}$ is nonzero only when $t = j$. Again $F(k)_{nj} = \Sigma_{t=1}^{n} F(k - 1)_{nt}\Theta_n(n,k,x_k)_{tj}$. Therefore $F(k)_{nj} = F(k - 1)_{nj}$. But $k < j \leqslant n - 1$ implies $k - 1 < j \leqslant n - 1$, so $F(k - 1)_{nj} = 0$ by the induction hypothesis (Property (b)). Thus $F(k)_{nj} = 0$, establishing Property (b) for $F(k)$.

Finally suppose $j < i < n$. Then $F(k)_{ij} = \Sigma_{t=j}^{i} F(k - 1)_{it}\Theta_n(n,k,x_k)_{tj}$ by Theorem 3.1.7. But $t \leqslant i < n$ implies $t < n$ so $\Theta_n(n,k,x_k)_{tj}$ is nonzero only when $t = j$. Thus $F(k)_{ij} = F(k - 1)_{ij}$ which is zero by the induction hypothesis (Property (c)).

THEOREM 3.1.9 *Let n be a natural number such that* $2 \leqslant n$. *If A is an n by n unitriangular matrix, then A can be expressed as the product*

$$\begin{aligned} A = {} & \Theta_n(2,1,A_{21}) \\ & \Theta_n(3,1,A_{31})\Theta_n(3,2,A_{32}) \\ & \cdot \\ & \cdot \\ & \cdot \\ & \Theta_n(n,1,A_{n1})\Theta_n(n,2,A_{n2})\ldots\Theta_n(n,n-1,A_{n,n-1}). \end{aligned}$$

Proof: This will be shown by induction on n. When $n = 2$ it is clear that $A = \Theta_n(2,1,A_{21})$ so induction begins.

Suppose now that $2 \leqslant n$. Define matrices R and S by

$$R_{ij} = \begin{cases} 1 & i = j, \\ 0 & j < i < n, \\ A_{ij} & j < i = n, \end{cases} \qquad S_{ij} = \begin{cases} 1 & i = j, \\ A_{ij} & j < i < n, \\ 0 & j < i = n. \end{cases}$$

Note that $S(n|n) = A(n|n)$ and that $R(n|n) = I$. It is claimed that $SR = A$. By Theorem 3.1.8,

$$(SR)_{ij} = \begin{cases} (A(n|n)I)_{ij} = A_{ij} & 1 \leqslant i,j \leqslant n-1 \\ (\Sigma_{t=1}^{n-1} S_{nt}I_{tj}) + R_{ij} = S_{nj} + R_{ij} = R_{ij} = A_{ij} & i = n \text{ and } 1 \leqslant j \leqslant n-1 \\ 0 & 1 \leqslant i \leqslant n-1 \text{ and } j = n \\ S_{nn}R_{nn} = 1 & i = j = n \end{cases}$$

thus proving the claim.

Now consider $S(n|n)$. By the induction hypothesis (and since $S(n|n)_{ij} = A_{ij}$ for all $i < n$),

$$\begin{aligned} S(n|n) = \Theta(2,1,A_{21}) & \\ & \Theta(3,1,A_{31})\Theta(3,2,A_{32}) \\ & \cdot \\ & \cdot \\ & \cdot \\ & \Theta(n-1,1,A_{n-1,1})\Theta(n-1,2,A_{n-1,2})\ldots\Theta(n-1,n-2,A_{n-1,n-2}) \end{aligned}$$

where each Θ means Θ_{n-1}. Let C be the same product of transvections, but where each Θ means Θ_n instead. For $h \leqslant n-1$, $\Theta_n(h,j,x)(n|n) = \Theta_{n-1}(h,j,x)$. Hence (Corollary 1 to Theorem 3.1.8), $C(n|n) = S(n|n)$. Now consider the last rows of these two matrices. For $j < n$, $S_{nj} = 0$ by definition of S. Also, for $j < n$, each $\Theta_n(h,j,x)_{nj} = 0$ so $C_{nj} = 0$ by Corollary 2 to Theorem 3.1.8. From the facts that $C(n|n) = S(n|n)$ and $S_{nj} = C_{nj} = 0$ for $j < n$, conclude that $S = C$.

Let M be the matrix $M = \Theta_n(n,1,A_{n1})\Theta_n(n,2,A_{n2})\ldots\Theta_n(n,n-1,A_{n,n-1})$. Then by the lemma, $M_{ij} = 0$ for $j < i < n$ and $M_{ij} = A_{nj}$ for $j < i = n$. That is, $M = R$. This, $SR = A$, and $S = C$ combine to show $A = CM$ completing the proof.

Our final topic in this section is the determinant of a 2 by 2 matrix.

DEFINITION AND NOTATION If A is a 2 by 2 matrix, then $\det(A)$ is the **determinant** of A if and only if $\det(A) = A_{11}A_{22} - A_{21}A_{12}$.

THEOREM 3.1.10 $\det(AB) = \det(A)\det(B)$ *for all* 2 *by* 2 *matrices A and B.*

Proof: For convenience, subcripts will be avoided by letting $A = \begin{pmatrix} a & b \\ c & d \end{pmatrix}$ and $B = \begin{pmatrix} u & v \\ y & z \end{pmatrix}$. Then $AB = \begin{pmatrix} a & b \\ c & d \end{pmatrix}\begin{pmatrix} u & v \\ y & z \end{pmatrix} = \begin{pmatrix} au+by & av+bz \\ cu+dy & cv+dz \end{pmatrix}$ so $\det(AB) = (au + by)(cv + dz) - (cu + dy)(av + bz) = aucv + audz + bycv + bydz - cuav - cubz - dyav - dybz$. Canceling the first and fifth terms (which can be done since the ring R is commutative) and also the fourth and eighth terms, and transposing the third and seventh terms yields $\det(AB) = audz - dyav - cubz + bycv$. Factoring ad from the first two terms (again recall R is commutative) and factoring cb from the last two yields $\det(AB) = ad(uz - yv) - cb(uz - yv) = (ad - cb)(uz - yv)$ which is equal to $\det(A)\det(B)$.

3.2 Matrix Groups

Under suitable conditions, collections of matrices form groups. These groups, once constructed, can be used to form further groups by the methods given in Chapter 1, and by the simple process of taking quotient groups. This section will be concerned with defining and investigating the properties of certain groups of matrices and some of the groups they give rise to.

Roughly the order of topics will be as follows: First we will consider n by n unitriangular matrices over a ring with identity, then n by n triangular matrices over a field. Finally some groups of special 2 by 2 matrices will be examined, first over an arbitrary ring with identity, then over a field.

NOTATION If R is a commutative ring with identity and n is a natural number such that $1 \leqslant n$, then $UT(n,R)$ will denote the set of n by n unitriangular matrices with entries in R.

THEOREM 3.2.1 *Let R be a commutative ring with identity and let n be a natural number such that $1 \leqslant n$. Then $UT(n,R)$ is a group.*

Proof: By Theorems 3.1.6, Part (b); 3.1.1; and 3.1.2 respectively, $UT(n,R)$ is closed under products, has an associative multiplication, and has an identity element (I). Thus it suffices to show $UT(n,R)$ has inverses. This will be done by induction on n. (For an alternate proof see Exercise I.1.) When $n = 1$, $UT(n,R)$ is clearly trivial and hence a group so induction begins. Assume now $2 \leqslant n$ and that $UT(n-1,R)$ is a group by induction. Let $A \in UT(n,R)$.

Let $C = A(n|n)$. Then C is unitriangular by Theorem 3.1.5 so by induction it has an inverse in $UT(n-1,R)$, call it D. Define an n by n matrix F as follows

$$F_{ij} = \begin{cases} D_{ij} & 1 \leqslant i,j < n, \\ -\sum_{t=1}^{n-1} A_{nt}D_{tj} & i = n \text{ and } 1 \leqslant j < n, \\ 0 & 1 \leqslant i < n \text{ and } j = n, \\ 1 & i = j = n. \end{cases}$$

First it will be shown that F is unitriangular. Suppose $i < j$. If $j < n$, then $F_{ij} = D_{ij} = 0$ (because D is unitriangular), while if $j = n$, then $i < n$ so $F_{ij} = 0$ also. Next consider the main diagonal entries. If $i < n$, then $F_{ii} = D_{ii} = 1$ while if $i = n$, then $F_{ii} = 1$ also. Thus F is unitriangular as claimed.

Note that $F(n|n) = D$. Therefore by Theorem 3.1.8,

$$(AF)_{ij} = \begin{cases} (CD)_{ij} & 1 \leqslant i,j < n, \\ (\sum_{t=1}^{n-1} A_{nt}D_{tj}) + A_{nn}F_{nj} & i = n \text{ and } 1 \leqslant j < n, \\ 0 & 1 \leqslant i < n \text{ and } j = n, \\ A_{nn}F_{nn} & i = j = n. \end{cases}$$

But $CD = I$ so $(CD)_{ij} = 1$ or 0 depending on whether or not $i = j$; $A_{nn} = 1$ and $F_{nj} = -\sum_{t=1}^{n-1} A_{nt}D_{tj}$, so $(\sum_{t=1}^{n-1} A_{nt}D_{tj}) + A_{nn}F_{nj} = 0$; and $A_{nn} = F_{nn} = 1$. Thus $(AF)_{ij} = 1$ for $i = j$ while $(AF)_{ij} = 0$ otherwise. Conclude $AF = I$.

It has now been shown that each $A \in UT(n,R)$ has a right inverse. Appeal to the Lemma to Theorem 1.2.1 completes the proof.

We would like to show that the unitriangular groups are nilpotent. This involves a description of their ascending central series (an interesting result in its own right) and to facilitate such a description the following notation is used.

NOTATION If R is a commutative ring with identity and n,m are natural numbers such that $1 \leqslant n$, then $[Z]_m(UT(n,R))$ will denote $\{A \mid A \in UT(n,R)$ and $0 < i - j < n - m$ implies $A_{ij} = 0\}$.

Visually, the requirement on A for membership in $[Z]_m(UT(n,R))$ is that A be of the form

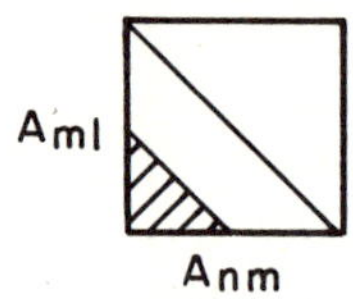

Here elements in the shaded area may be arbitrary but all other elements (excepting those on the main diagonal) are zero.

LEMMA 1 *Let R be a commutative ring with identity and let n,m be natural numbers such that $1 \leqslant n$. Suppose $A \in [Z]_m(UT(n,R))$ and $B \in UT(n,R)$. Then*
(a) $0 < i - j < n - m$ *implies* $(AB)_{ij} = B_{ij}$.
(b) $0 < i - j < n - m$ *implies* $(BA)_{ij} = B_{ij}$.
Proof: By Theorem 3.1.7, $(AB)_{ij} = \Sigma_{t=j}^{i} A_{it}B_{tj} = (\Sigma_{t=j}^{i-1} A_{it}B_{tj}) + A_{ii}B_{ij}$. But for all t such that $j \leqslant t \leqslant i - 1$, $0 < i - t$ since $t < i$ and $i - t \leqslant i - j$ because $j \leqslant t$. Thus $0 < i - t \leqslant i - j$ and so $0 < i - t < n - m$. This and $A \in [Z]_m(UT(n,R))$ imply $A_{it} = 0$. Hence $(AB)_{ij} = A_{ii}B_{ij} = B_{ij}$, proving Part (a). Part (b) is similar.

LEMMA 2 *Let R be a commutative ring with identity and let n,m be natural numbers such that $1 \leqslant n$. Then $[Z]_m(UT(n,R)) \lhd UT(n,R)$.*
Proof: For convenience let $[Z]_m$ denote $[Z]_m(UT(n,R))$. First it will be shown that $A,B \in [Z]_m$ implies $AB \in [Z]_m$. Let $0 < i - j < n - m$ and seek to show $(AB)_{ij} = 0$. Since $A \in [Z]_m$, $(AB)_{ij} = B_{ij}$ (Lemma 1) $= 0$ since $B \in [Z]_m$. Thus $[Z]_m$ is closed under products.

Next closure under inverses will be shown. Suppose $0 < i - j < n - m$. Then by Lemma 1, $(AA^{-1})_{ij} = A^{-1}{}_{ij}$. But $AA^{-1} = I$ so $A^{-1}{}_{ij} = I_{ij} = 0$. Thus $[Z]_m$ is closed under inverses and hence is a subgroup of $UT(n,R)$.

Finally it must be shown that $A \in [Z]_m$ implies $B^{-1}AB \in [Z]_m$ for all $B \in UT(n,R)$. Suppose $0 < i - j < n - m$. Then by Theorem 3.1.7, $(B^{-1}AB)_{ij} = \Sigma_{t=j}^{i} B^{-1}{}_{it}(AB)_{tj} = B^{-1}{}_{ij}(AB)_{jj} + (\Sigma_{t=j+1}^{i} B^{-1}{}_{it}(AB)_{tj})$. But for all t such that $j + 1 \leqslant t \leqslant i$, $0 < t - j$ and $t - j \leqslant i - j$ which implies $0 < t - j < n - m$. Hence $(AB)_{tj} = B_{tj}$ by Lemma 1. Also note that $(AB)_{jj} = 1 = B_{jj}$. Thus $(B^{-1}AB)_{ij} =$

$B^{-1}{}_{ij}B_{jj} + (\Sigma_{t=j+1}^{i} B^{-1}{}_{it}B_{tj}) = \Sigma_{t=j}^{i} B^{-1}{}_{it}B_{tj} = (B^{-1}B)_{ij}$ (by a second application of Theorem 3.1.7) $= I_{ij} = 0$ as required.

THEOREM 3.2.2 *Let R be a commutative ring with identity and let n,m be natural numbers such that $1 \leqslant n$. Then $[\mathrm{Z}]_m(UT(n,R)) = \mathrm{Z}_m(UT(n,R))$.*

Proof: For convenience let Z_m denote $\mathrm{Z}_m(UT(n,R))$ and let $[\mathrm{Z}]_m$ denote $[\mathrm{Z}]_m(UT(n,R))$. The proof will be by induction on m. When $m = 0$, Z_m is trivial. To show that $[\mathrm{Z}]_m$ is likewise, let $A \in [\mathrm{Z}]_m$. Because A is unitriangular, $A_{ii} = 1$ for all i, and $A_{ij} = 0$ for all $i < j$. Next consider A_{ij} when $j < i$. Then $0 < j < i \leqslant n$ implies $0 < i - j \leqslant n - j < n$. Consequently $A_{ij} = 0$ since $A \in [\mathrm{Z}]_m$. Thus $A = I$ so induction begins.

Assume now the induction hypothesis $\mathrm{Z}_{m-1} = [\mathrm{Z}]_{m-1}$. It will be shown first that $\mathrm{Z}_m \subseteq [\mathrm{Z}]_m$. Let $A \in \mathrm{Z}_m$, suppose $0 < i - j < n - m$, and seek to show that $A_{ij} = 0$.

CASE ONE, $i < n$. Then $i + 1 \leqslant n$. Let $B = \Theta_n(i + 1, i, 1)$. Then $B \in UT(n,R)$ by Theorem 3.1.4. Let $L = A^{-1}B^{-1}AB$. Note that regardless of what t is, $(t,j) \neq (i + 1,i)$ because $i \neq j$. Hence $B_{tj} = 0$ unless $t = j$. Thus $(AB)_{i+1,j} = \Sigma_{t=1}^{n} A_{i+1,t}B_{tj}$ which equals $A_{i+1,j}B_{jj} = A_{i+1,j}$. Next note that $(BAL)_{i+1,j} = \Sigma_{t=1}^{n} B_{i+1,t}(AL)_{tj} = B_{i+1,i}(AL)_{ij} + B_{i+1,i+1}(AL)_{i+1,j} = (AL)_{ij} + (AL)_{i+1,j}$. By Theorem 3.1.7, $(AL)_{ij} = \Sigma_{t=j}^{i} A_{it}L_{tj} = A_{ij}L_{jj} + (\Sigma_{t=j+1}^{i} A_{it}L_{tj})$. But $j + 1 \leqslant t \leqslant i$ implies $0 < t - j \leqslant i - j < n - m < n - (m - 1)$. Since $A \in \mathrm{Z}_m$, $L \in \mathrm{Z}_{m-1}$ so $L \in [\mathrm{Z}]_{m-1}$ by induction. Therefore the last battery of inequalities implies $L_{tj} = 0$. Hence $(AL)_{ij} = A_{ij}L_{jj} = A_{ij}$. Similarly $(AL)_{i+1,j} = A_{i+1,j}$. Thus $(BAL)_{i+1,j} = A_{ij} + A_{i+1,j}$. From the definition of L it follows that $BAL = AB$. Thus $(AB)_{i+1,j} = A_{ij} + A_{i+1,j}$. Consequently $A_{i+1,j} = A_{ij} + A_{i+1,j}$ implying $A_{ij} = 0$.

CASE TWO, $i = n$. Then $n - j < n - m$ so $m + 1 \leqslant j$. Assuming $1 \leqslant m$ (the case $m = 0$ was treated in the first paragraph) yields $2 \leqslant m + 1 \leqslant j$ so $1 \leqslant j - 1$. Let $B = \Theta_n(j,j-1,1)$. Then $B \in UT(n,R)$ by Theorem 3.1.4. Let $L = A^{-1}B^{-1}AB$. Note that regardless of what t is, $(i,t) \neq (j,j-1)$ because $i \neq j$. Hence $B_{it} = 0$ unless $t = i$. Thus $(AB)_{i,j-1} = \Sigma_{t=1}^{n} A_{it}B_{t,j-1} = A_{ij}B_{j,j-1} + A_{i,j-1}B_{j-1,j-1} = A_{ij} + A_{i,j-1}$. Also for all h such that $1 \leqslant h \leqslant n$, $(BA)_{ih} = \Sigma_{t=1}^{n} B_{it}A_{th} = B_{ii}A_{ih} = A_{ih}$. By Theorem 3.1.7, $(BAL)_{i,j-1} = \Sigma_{t=j-1}^{i} (BA)_{it}L_{t,j-1} = (BA)_{i,j-1}L_{j-1,j-1} + (\Sigma_{t=j}^{i} (BA)_{it}L_{t,j-1})$. But $j \leqslant t \leqslant i$ implies $0 < t - (j - 1) \leqslant i - j + 1 < n - m + 1 = n - (m - 1)$. Since $A \in \mathrm{Z}_m$, $L \in \mathrm{Z}_{m-1}$ so $L \in [\mathrm{Z}]_{m-1}$ by induction. Therefore the last battery of inequalities implies $L_{t,j-1} = 0$. Consequently $(BAL)_{i,j-1} = (BA)_{i,j-1}L_{j-1,j-1} = (BA)_{i,j-1} = A_{i,j-1}$. But $BAL = AB$ so $(AB)_{i,j-1} = A_{i,j-1}$. Thus $A_{ij} + A_{i,j-1} = A_{i,j-1}$ implying $A_{ij} = 0$. This completes the demonstration that $\mathrm{Z}_m \subseteq [\mathrm{Z}]_m$.

Conversely let $A \in [\mathrm{Z}]_m$ and seek to show $L = A^{-1}B^{-1}AB \in \mathrm{Z}_{m-1}$ for all $B \in UT(n,R)$. By induction it suffices to show $L \in [\mathrm{Z}]_{m-1}$. So suppose that $0 < i - j < n - (m - 1)$ and seek to show $L_{ij} = 0$. Since $A \in [\mathrm{Z}]_m$, Lemma 2 implies $A^{-1} \in [\mathrm{Z}]_m$ and $B^{-1}AB \in [\mathrm{Z}]_m$. Thus, again by Lemma 2, $L \in [\mathrm{Z}]_m$. The last battery of inequalities implies either $0 < i - j < n - m$ or $0 < i - j = n - m$. If the first case holds, then since $L \in [\mathrm{Z}]_m$, $L_{ij} = 0$ as required. Next suppose $i - j = n - m$. Therefore by Theorem 3.1.7, $(BAL)_{ij} = \Sigma_{t=j}^{i} (BA)_{it}L_{tj} = (BA)_{ij}L_{jj} +$

$(\Sigma_{t=j+1}^{i-1}(BA)_{it}L_{tj}) + (BA)_{ii}L_{ij}$. But $j + 1 \leqslant t \leqslant i - 1$ implies $1 \leqslant t - j \leqslant i - j - 1$, so $0 < t - j < n - m$. This and $L \in [Z]_m$ imply $L_{tj} = 0$. Hence $(BAL)_{ij} = (BA)_{ij} + L_{ij}$. However $BAL = AB$ so $(AB)_{ij} = (BA)_{ij} + L_{ij}$. By Lemma 1, $(AB)_{ij} = B_{ij} = (BA)_{ij}$. Thus $B_{ij} = B_{ij} + L_{ij}$ implying $L_{ij} = 0$ as required.

THEOREM 3.2.3 *Let R be a commutative ring with identity and let n be a natural number such that* $1 \leqslant n$. *Then* $UT(n,R)$ *is nilpotent of class* $n - 1$.

Proof: Since $UT(1,R)$ is trivial and hence class-0 nilpotent, it may be assumed that $2 \leqslant n$. For convenience let U denote $UT(n,R)$. By Theorem 3.2.2, $Z_{n-1}(U)$ is equal to $[Z]_{n-1}(U)$. But $[Z]_{n-1}(U) = U$ because if $A \in U$, then $0 < i - j < n - (n - 1)$ implies $A_{ij} = 0$ vacuously. Thus $U = Z_{n-1}(U)$. To complete the proof it must be shown that $Z_{n-2}(U) \neq U$. Let $A = \Theta_n(2,1,1)$. Then $0 < 2 - 1 < n - (n - 2)$, but $A_{21} \neq 0$. Thus $A \notin [Z]_{n-2}(U)$ and hence $A \notin Z_{n-2}(U)$.

Next we would like to identify the commutator subgroup of $UT(n,R)$. Following a preliminary theorem and its lemma, this is done in Theorem 3.2.5.

LEMMA *Let R be a commutative ring with identity, let n be a natural number such that* $2 \leqslant n$, *and let* $1 \leqslant j \leqslant n - 2$. *Then for all* $x \in R$, $\Theta_n(n,j,x)$ *is a commutator in* $UT(n,R)$.

Proof: For convenience denote the transvection $\Theta_n(n,j,x)$ by T. It will be shown that $T = A^{-1}B^{-1}AB$ where $A = \Theta_n(n,j + 1,-x)$ and $B = \Theta_n(j + 1,j,-1)$. To do this it suffices to show $BAT = AB$. First note that $T(n|n) = I$ and that $A(n|n) = I$. Consequently $(AT)(n|n) = I$ as well. Let $1 \leqslant h,k \leqslant n$.

CASE ONE $h,k \leqslant n - 1$. Then, by Theorem 3.1.8, $(AB)_{hk} = (IB(n|n))_{hk} = B(n|n)_{hk}$, while $(BAT)_{hk} = (B(n|n)(AT)(n|n))_{hk} = (B(n|n)I)_{hk} = B(n|n)_{hk}$ also.

CASE TWO $h = n$ and $1 \leqslant k \leqslant n - 1$. Then, by Theorem 3.1.8, $(AB)_{hk} = (\Sigma_{t=1}^{n-1}A_{nt}B(n|n)_{tk}) + B_{nk}$. But for $t \leqslant n - 1$, A_{nt} is nonzero only when $t = j + 1$. Also $B_{nk} = 0$ since $j + 1 \neq n$ by hypothesis on j. Consequently $(AB)_{hk} = A_{n,j+1}B(n|n)_{j+1,k}$

$$= \begin{cases} (-x)(-1) = x & \text{if } k = j \\ (-x)(1) = -x & \text{if } k = j + 1 \\ (-x)(0) = 0 & \text{if } k \neq j, j + 1 \end{cases}$$

On the other hand, $(B(AT))_{hk} = (\Sigma_{t=1}^{n-1}B_{nt}(AT)(n|n)_{tk}) + B_{nn}(AT)_{nk}$. Since $(AT)(n|n) = I$, $(AT)(n|n)_{tk}$ is nonzero only when $t = k$. Also $B_{nn} = 1$. Thus $(BAT)_{hk} = B_{nk} + (AT)_{nk}$. But $B_{nk} = 0$ since $j + 1 \neq n$ and $k \neq n$, so $(BAT)_{hk} = (AT)_{nk}$. Another application of Theorem 3.1.8 yields $(AT)_{hk} = (\Sigma_{t=1}^{n-1}A_{nt}I_{tk}) + T_{nk} = A_{nk} + T_{nk}$

$$= \begin{cases} -x & \text{if } k = j + 1, \\ 0 & \text{if } k \neq j + 1, \end{cases} + \begin{cases} x & \text{if } k = j, \\ 0 & \text{if } k \neq j, \end{cases} = \begin{cases} x & \text{if } k = j, \\ -x & \text{if } k = j + 1, \\ 0 & \text{if } k \neq j, j + 1. \end{cases}$$

CASE THREE $1 \leqslant h \leqslant n - 1$ and $k = n$. Then, by Theorem 3.1.8, $(AB)_{hk} = 0$ while $(BAT)_{hk} = 0$.
CASE FOUR $h = k = n$. Then, by Theorem 3.1.8, $(AB)_{hk} = A_{nn}B_{nn} = 1$ while $(BAT)_{hk} = B_{nn}(AT)_{nn} = 1$.

THEOREM 3.2.4 *Let R be a commutative ring with identity, let n be a natural number such that* $2 \leqslant n$, *let h and j be such that* $h \leqslant n$ *and* $j \leqslant h - 2$. *Then for all* $x \in R$, $\Theta_n(h,j,x)$ *is a commutator in* $UT(n,R)$.
Proof: This will be shown by induction on n. Note that when $n = 2$ the conclusion holds vacuously since $j \leqslant h - 2$ would imply $j = 0$. Thus induction begins.

Assume now the theorem holds for $n - 1$ and seek to show it also holds for n. Suppose $h \leqslant n$ and $j \leqslant h - 2$. Then either $h = n$, or $h \leqslant n - 1$. In the first case, $\Theta_n(h,j,x) = \Theta_n(n,j,x)$ which is a commutator by the lemma. Assume then that $h \leqslant n - 1$. Therefore $\Theta_{n-1}(h,j,x)$ is a commutator in $UT(n - 1,R)$ by the induction hypothesis. Thus $\Theta_{n-1}(h,j,x) = A^{-1}B^{-1}AB$ for some $A,B \in UT(n - 1,R)$. Consequently $BA\Theta_{n-1}(h,j,x) = AB$. Define R and S by

$$R_{ij} = \begin{cases} 1 & i = j, \\ A_{ij} & j < i < n, \\ 0 & j < i = n, \end{cases} \quad \text{and} \quad S_{ij} = \begin{cases} 1 & i = j, \\ B_{ij} & j < i < n, \\ 0 & j < i = n. \end{cases}$$

Note that $R(n|n) = A$ and $S(n|n) = B$. Also note that $\Theta_n(h,j,x)(n|n) = \Theta_{n-1}(h,j,x)$ Therefore by Corollary 1 to Theorem 3.1.8, $(SR\Theta_n(h,j,x))(n|n) = BA\Theta_{n-1}(h,j,x)$ while $(RS)(n|n) = AB$. Thus

$$(SR\Theta_n(h,j,x))(n|n) = (RS)(n|n). \tag{1}$$

Furthermore, for all $u < n$, $S_{nu} = 0, R_{nu} = 0$, and $\Theta_n(h,j,x)_{nu} = 0$. Thus by Corollary 2 to Theorem 3.1.8, $(SR\Theta_n(h,j,x))_{nu} = 0$ and $(RS)_{nu} = 0$ for all $u < n$. This and Line (1) clearly imply $SR\Theta_n(h,j,x) = RS$ and so $\Theta_n(h,j,x) = R^{-1}S^{-1}RS$.

THEOREM 3.2.5 *Let R be a commutative ring with identity. Let n be a natural number such that* $2 \leqslant n$. *Then* $UT(n,R)' = Z_{n-2}(UT(n,R))$.
Proof: Let $A \in Z_{n-2}(UT(n,R))$. By Theorem 3.1.9,

$$\begin{aligned} A = {} & \Theta_n(2,1,A_{21}) \\ & \Theta_n(3,1,A_{31})\Theta_n(3,2,A_{32}) \\ & \cdot \\ & \cdot \\ & \cdot \\ & \Theta_n(n,1,A_{n1})\Theta_n(n,2,A_{n2})\ldots\Theta_n(n,n-1,A_{n,n-1}). \end{aligned}$$

But for $2 \leqslant i \leqslant n - 1, 0 < i - (i - 1) < n - (n - 2)$. This implies $A_{i,i-1} = 0$ for all such i because $A \in Z_{n-2}(UT(n,R)) = [Z]_{n-2}(UT(n,R))$ (Theorem 3.2.2). Thus $\Theta_n(2,1,A_{21}) = \Theta_n(3,2,A_{32}) = \ldots = \Theta_n(n,n-1,A_{n,n-1}) = I$. Hence

$$
\begin{aligned}
A = {} & \Theta_n(3,1,A_{31}) \\
& \Theta_n(4,1,A_{41})\Theta_n(4,2,A_{42}) \\
& \cdot \\
& \cdot \\
& \cdot \\
& \Theta_n(n,1,A_{n1})\Theta_n(n,2,A_{n2}) \ldots \Theta_n(n,n-2,A_{n,n-2}).
\end{aligned}
$$

By Theorem 3.2.4, therefore, $A \in UT(n,R)'$. This shows $Z_{n-2}(UT(n,R)) \subseteq UT(n,R)'$. Conversely $UT(n,R) = Z_{n-1}(UT(n,R))$ by Theorem 3.2.3. This implies $UT(n,R)' \subseteq Z_{n-2}(UT(n,R))$ completing the proof.

THEOREM 3.2.6 *Let R be a commutative ring with identity and let n be a natural number such that $2 \leqslant n$. Define $e_n{:}R \to UT(n,R)$ by $e_n(x) = \Theta_n(n,1,x)$. Then e_n is an isomorphism of the additive group of R onto $Z(UT(n,R))$.*

Proof: For convenience denote e_n by e and denote $UT(n,R)$ by U. First it will be shown that Im $e \subseteq Z(U)$. For $x \in R$ it suffices by Theorem 3.2.2 to show $0 < i - j < n - 1$ implies $e(x)_{ij} = 0$. If $i = n$ and $j = 1$, then $i - j < n - 1$ would imply $n - 1 < n - 1$ which is false. Thus either $i \neq n$ or $j \neq 1$. Also $i \neq j$ because $0 < i - j$. Therefore $e(x)_{ij} = \Theta_n(n,1,x)_{ij} = 0$.

Next it will be shown that e is a homomorphism. Let x,y belong to R. Then $(e(x)e(y))_{n1} = \Sigma_{t=1}^{n} e(x)_{nt}e(y)_{t1} = e(x)_{n1}e(y)_{11} + e(x)_{nn}e(y)_{n1} = x + y = e(x+y)_{n1}$. Clearly $(e(x)e(y))_{hh} = 1$ for all h and so $(e(x)e(y))_{hh} = e(x+y)_{hh}$. Finally for $h \neq k$ and $(h,k) \neq (n,1)$, $(e(x)e(y))_{hk} = \Sigma_{t=1}^{n} e(x)_{ht}e(y)_{tk}$. But for $t \neq h,k$, the product $e(x)_{ht}e(y)_{tk}$ is nonzero only when $(h,t) = (n,1)$ and $(t,k) = (n,1)$. As this only happens when $(h,k) = (n,1)$, each term $e(x)_{ht}e(y)_{tk} = 0$ when $t \neq h,k$. Furthermore, since $(h,k) \neq (n,1)$ and $h \neq k$, the terms $e(x)_{hh}e(y)_{hk}$ and $e(x)_{hk}e(y)_{kk}$ are both zero. Thus $(e(x)e(y))_{hk} = 0 = e(x+y)_{hk}$. Hence e is a homomorphism.

To show injectivity suppose $e(x) = I$. Therefore $\Theta_n(n,1,x)_{n1} = I_{n1} = 0$. Hence $x = 0$.

Finally it will be shown that e is onto $Z(U)$. Let $A \in Z(U)$. It is claimed that $A = \Theta_n(n,1,A_{n1})$. Let $1 \leqslant h,k \leqslant n$. First suppose $k < h$ and $(h,k) \neq (n,1)$. Then either $h \neq n$ or $k \neq 1$. If $h \neq n$, $h < n$ so $h - k < n - k$. But $1 \leqslant k$ implies $n - k \leqslant n - 1$, hence $h - k < n - 1$. If $k \neq 1$, then $2 \leqslant k$ so $h - k \leqslant h - 2$ and again $h - k < n - 1$. Thus $k < h$ and $(h,k) \neq (n,1)$ imply $0 < h - k < n - 1$. This and $A \in [Z]_1(U)$ (by Theorem 3.2.2) imply $A_{hk} = 0 = \Theta_n(n,1,A_{n1})_{hk}$. It is clear that $A_{n1} = \Theta_n(n,1,A_{n1})_{n1}$ and also that $h = k$ implies $A_{hk} = 1 = \Theta_n(n,1,A_{n1})_{hk}$ so $A = \Theta_n(n,1,A_{n1})$ as claimed. Thus $A = e(A_{n1})$.

THEOREM 3.2.7 *Let R be a commutative ring with identity. Let n be a natural number such that $2 \leqslant n$. Then $UT(n,R)/Z_{n-2}(UT(n,R)) \cong R \times R \times \ldots \times R$ ($n - 1$ times).*

Proof: For convenience denote $UT(n,R)$ by U and denote $R \times R \times \ldots \times R$

($n - 1$ times) by R^{n-1}. Define $g: U \to R^{n-1}$ by $g(A) = (A_{21}, A_{32}, \ldots, A_{n,n-1})$. If $A, B \in U$, then for $1 \leqslant i \leqslant n - 1$ the ith coordinate of $g(AB)$ is $(AB)_{i+1,i}$. By Theorem 3.1.7, $(AB)_{i+1,i} = A_{i+1,i}B_{ii} + A_{i+1,i+1}B_{i+1,i} = A_{i+1,i} + B_{i+1,i}$ which is the ith coordinate of $g(A) + g(B)$. This proves g is a homomorphism. To show surjectivity, let $y = (x_1, x_2, \ldots, x_{n-1}) \in R^{n-1}$. Define M as

$$M_{ij} = \begin{cases} x_j & \text{if } i = j + 1, \\ 1 & \text{if } i = j, \\ 0 & \text{otherwise.} \end{cases}$$

Then $M \in U$ and $g(M) = y$. Thus $U/\ker g \cong R^{n-1}$ so to complete the proof it suffices to show $\ker g = Z_{n-2}(U)$.

Let $A \in Z_{n-2}(U)$, so $A \in [Z]_{n-2}(U)$ by Theorem 3.2.2. For $1 \leqslant i \leqslant n - 1$ the ith coordinate of $g(A)$ is $A_{i+1,i}$. But $0 < (i + 1) - i < n - (n - 2)$ so $A_{i+1,i} = 0$ since $A \in [Z]_{n-2}(U)$. Hence $g(A)$ is the identity element of R^{n-1} and so $A \in \ker g$. Conversely let $A \in \ker g$. Using Theorem 3.2.2 again it suffices to show $0 < i - j < n - (n - 2)$ implies $A_{ij} = 0$. From the condition on i and j it follows that $i = j + 1$. But $A_{j+1,j}$ is the jth coordinate of $g(A)$, hence is zero as required.

THEOREM 3.2.8 *Let R be a commutative ring with identity. Let n be a natural number such that $2 \leqslant n$. Then* $\operatorname{card}(UT(n,R)) = \operatorname{card}(R)^{n(n-1)/2}$.

Proof: Below the main diagonal, there are $1 + 2 + 3 + \ldots + n - 1 = n(n - 1)/2$ entries, each of which may be filled in $\operatorname{card}(R)$ ways. Thus $\operatorname{card}(UT(n,R)) = \operatorname{card}(R)^{n(n-1)/2}$.

For all positive integers n define $d_n: UT(n,R) \to UT(2n - 1, R)$ as follows:

$$d_n(A)_{ij} = \begin{cases} 1 & \text{if } i = j, \\ 0 & \text{if } i < j, \\ 0 & \text{if } j < i \text{ and } i \text{ or } j \text{ is even}, \\ A_{(i+1)/2,(j+1)/2} & \text{if } j < i \text{ and } i \text{ and } j \text{ are odd.} \end{cases}$$

THEOREM 3.2.9 *Let R be a commutative ring with identity. Let n be a positive integer. Then $d_n: UT(n,R) \to UT(2n - 1, R)$ is an injective homomorphism and furthermore* $\operatorname{Im} d_n \subseteq UT(2n - 1, R)'$.

Proof: Let $A, B \in UT(n,R)$. It must be shown that $(d_n(A)d_n(B))_{ij}$ is equal to $d_n(AB)_{ij}$ for all $i, j \leqslant 2n - 1$.

CASE ONE, $i = j$. Then $d_n(AB)$ is unitriangular so $d_n(AB)_{ij} = 1$. Also the product $d_n(A)d_n(B)$ is unitriangular since its two factors are, so $(d_n(A)d_n(B))_{ij} = 1$.

CASE TWO, $i < j$. Similarly to Case one, $d_n(AB)_{ij} = 0 = (d_n(A)d_n(B))_{ij}$.

CASE THREE, $j < i$ and i or j is even. Then $d_n(AB)_{ij} = 0$ while $(d_n(A)d_n(B))_{ij}$ is equal to $\Sigma_{t=1}^{2n-1} d_n(A)_{it} d_n(B)_{tj}$. But by case hypothesis, either $d_n(A)_{it}$ or $d_n(B)_{tj}$ equals zero since either i or j is even; so $(d_n(A)d_n(B))_{ij} = 0 = d_n(AB)_{ij}$.

CASE FOUR, $j < i$ and i and j are both odd. In this case $(d_n(A)d_n(B))_{ij} =$

$\Sigma_{t=1}^{2n-1} d_n(A)_{it} d_n(B)_{tj} = \Sigma_{t=1}^{2n-1} A_{(i+1)/2,(t+1)/2} B_{(t+1)/2,(j+1)/2}$ where both summations are taken over the odd integers only. But $1 \leqslant t \leqslant 2n - 1$ and t odd implies $2 \leqslant t + 1 \leqslant 2n$ and $t + 1$ even. Hence $1 \leqslant (t + 1)/2 \leqslant n$ and $(t + 1)/2$ is an integer. Set $s = (t + 1)/2$. Then $(d_n(A)d_n(B))_{ij} = \Sigma_{s=1}^{n} A_{(i+1)/2,s} B_{s,(j+1)/2} = (AB)_{(i+1)/2,(j+1)/2} = d_n(AB)_{ij}$.

Conclude d_n is a homomorphism.

To show injectivity, suppose $d_n(A) = I$. It suffices to show $j < i$ implies $A_{ij} = 0$. When $j < i \leqslant n$, $2j - 1 < 2i - 1 \leqslant 2n - 1$. Therefore $I_{2i-1,2j-1} = 0$ and thus $d_n(A)_{2i-1,2j-1} = 0$. Hence $A_{ij} = 0$ as required.

To show $\operatorname{Im} d_n \subseteq UT(2n - 1,R)'$ let $A \in UT(n,R)$. By Theorems 3.2.5 and 3.2.2, $UT(2n - 1,R)' = [Z]_{2n-3}(UT(2n - 1,R))$, so it must be shown that $d_n(A)$ belongs to $[Z]_{2n-3}(UT(2n - 1,R))$. But if $0 < i - j < (2n - 1) - (2n - 3)$, then $i - j = 1$ and hence i or j must be even, hence $d_n(A)_{ij} = 0$. Therefore $d_n(A) \in [Z]_{2n-3}(UT(2n - 1,R))$ as required.

THEOREM 3.2.10 *Let R be a commutative ring with identity. Let n be a positive integer. Then* $d_n[Z(UT(n,R))] \subseteq Z(UT(2n - 1,R))$.

Proof: Let $A \in Z(UT(n,R))$. Using Theorem 3.2.2, it suffices to show that $0 < i - j < (2n - 1) - 1$ implies $d_n(A)_{ij} = 0$. From the first inequality it follows that $i \neq j$. If $i < j$ or if either i or j is even then $d_n(A)_{ij} = 0$ immediately. Assume then that $j < i$ and both i and j are odd. Then $(i + 1)/2$ and $(j + 1)/2$ are integers and $0 < [(i + 1)/2] - [(j + 1)/2] = (i - j)/2 < (2n - 2)/2 = n - 1$. Consequently $A_{(i+1)/2,(j+1)/2} = 0$ since $A \in [Z]_1(UT(n,R))$. Therefore $d_n(A)_{ij} = 0$.

Since the unitriangular groups are nilpotent (Theorem 3.2.3), they are also solvable by Theorem A.6.7. The following theorem shows that the solvable length of $UT(n,R)$ increases as n does.

THEOREM 3.2.11 *Let R be a commutative ring with identity and let d be a natural number. Denote the solvable length of* $UT(2^d + 1,R)$ *by* m_d. *Then* $d + 1 \leqslant m_d$.

Proof: This will be shown by induction. When $d = 0$, $UT(2^d + 1,R) = UT(2,R)$ which is an abelian group by Theorem 3.2.3 hence has solvable length 1. Thus induction begins since $0 + 1 \leqslant 1$. Assume now the induction hypothesis $(d - 1) + 1 \leqslant m_{d-1}$. By Theorem 3.2.9, $UT(2^{d-1} + 1,R)$ can be embedded in the group $UT(2^d + 1,R)'$. Consequently m_{d-1} is less than or equal to the solvable length of $UT(2^d + 1,R)'$ which is exactly $m_d - 1$. Combining $m_{d-1} \leqslant m_d - 1$ and the induction hypothesis yields $d \leqslant m_d - 1$ implying the result.

Since every $UT(n,R)$ can be embedded in $UT(2n - 1,R)$ (Theorem 3.2.9), it is natural to think about a direct limit. For all $n \in N$, let $k_n = 2^{n+1} - 2^n + 1$. Note that $2k_n - 1 = 2(2^{n+1} - 2^n + 1) - 1 = 2^{n+2} - 2^{n+1} + 1 = k_{n+1}$. Thus

$$d_{k_n}: UT(k_n,R) \to UT(k_{n+1},R).$$

For $n,m \in N$ such that $n < m$ define $\pi_n^m: UT(k_n,R) \to UT(k_m,R)$ to be

$$\pi_n^m = d_{k_{m-1}} \circ d_{k_{m-2}} \circ \cdots \circ d_{k_n}.$$

Also define π_n^n to be the identity homomorphism on $UT(k_n,R)$ for all $n \in N$.

THEOREM 3.2.12 *($\{UT(k_n,R)\}_{n\in N}$, $\{\pi_n^m \mid n \leqslant m\}$) is a direct system over $(N,\leqslant)$.*

Proof: It must be shown that $n \leqslant m \leqslant t$ implies the equation $\pi_m^t \circ \pi_n^m = \pi_n^t$. If $n = m$ (or $m = t$), then π_n^m (or π_m^t) is the identity homomorphism and the equation holds. So assume $n < m < t$. Then

$$\begin{aligned}\pi_m^t \circ \pi_n^m &= (d_{k_{t-1}} \circ d_{k_{t-2}} \circ \cdots \circ d_{k_m}) \circ (d_{k_{m-1}} \circ d_{k_{m-2}} \circ \cdots \circ d_{k_n}) \\ &= d_{k_{t-1}} \circ d_{k_{t-2}} \circ \cdots \circ d_{k_n} \\ &= \pi_n^t.\end{aligned}$$

NOTATION $UT(\infty,R)$ will denote the direct limit of ($\{UT(k_n,R)\}_{n\in N}$, $\{\pi_n^m \mid n \leqslant m\}$).

The groups $UT(\infty,R)$ have two important properties. First they are locally nilpotent (this follows easily from Theorem 3.2.3). The second property is given in the following theorem.

THEOREM 3.2.13 *Let R be a commutative ring with identity. Then $UT(\infty,R)$ is perfect.*

Proof: Let $[(A,n)] \in UT(\infty,R)$. Thus $A \in UT(k_n,R)$, and hence $\pi_n^{n+1}(A) \in UT(k_{n+1},R)'$ by Theorem 3.2.9. Applying Theorem 1.5.4 yields $[(A,n)] = h_n(A) = h_{n+1} \circ \pi_n^{n+1}(A)$ which belongs to $UT(\infty,R)'$ because $\pi_n^{n+1}(A)$ belongs to $UT(k_{n+1},R)'$. Conclude $UT(\infty,R) = UT(\infty,R)'$.

We turn our attention now to triangular matrices. Although the entries on the main diagonal of a triangular matrix need not be the identity element 1 (that is, although the matrix need not be unitriangular) some restriction on these entries must be made to ensure that collections of triangular matrices will form a group. We choose to require the underlying ring to be a field, and to prohibit any main diagonal entry from equaling 0. The field requirement, however, can be eliminated provided the requirement on the main diagonal entries is suitably modified and a generalization to commutative rings with identity will be given as an exercise.

NOTATION If F is a field and n is a natural number such that $1 \leqslant n$, then $T(n,F)$ will denote the set of n by n triangular matrices M, with entries in F, such that $1 \leqslant i \leqslant n$ implies $M_{ii} \neq 0$.

It will be considerably easier to show $T(n,F)$ is a group (Theorem 3.2.14) than it was to show $UT(n,R)$ is a group (Theorem 3.2.1). The reason for this is that Theorem 3.2.1 will be used in the proof.

THEOREM 3.2.14 *If F is a field and n is a natural number such that* $1 \leqslant n$, *then* $T(n,F)$ *is a group.*

Proof: First closure under products is shown. If $A,B \in T(n,F)$, then AB is triangular by Theorem 3.1.6, Part (a). Also for $1 \leqslant i \leqslant n$, $(AB)_{ii} = A_{ii}B_{ii}$ by Theorem 3.1.7. Since $A,B \in T(n,F)$, $A_{ii} \neq 0$ and $B_{ii} \neq 0$. This implies $A_{ii}B_{ii} \neq 0$ since fields have no zero divisors. Thus $(AB)_{ii} \neq 0$ showing $AB \in T(n,F)$.

Multiplication is associative in $T(n,F)$ by Theorem 3.1.1 and I acts as the identity by Theorem 3.1.2. Thus it suffices to show $T(n,F)$ has inverses, and by using the Lemma to Theorem 1.2.1, only the existence of right inverses need be shown. Let $M \in T(n,F)$. Thus for $1 \leqslant i \leqslant n$, $M_{ii} \neq 0$, hence $1/M_{ii} \in F$. Define A by

$$A_{ij} = \begin{cases} 1/M_{ii} & \text{if } j = i, \\ 0 & \text{if } j \neq i. \end{cases}$$

Clearly A is a triangular matrix, hence so is MA. Also for $1 \leqslant i \leqslant n$, by Theorem 3.1.7 again, $(MA)_{ii} = M_{ii}A_{ii} = M_{ii}(1/M_{ii}) = 1$. Consequently MA is a unitriangular matrix. Theorem 3.2.1, therefore, may be applied to produce an inverse for MA, call it B. Thus $(MA)B = I$. By associativity this becomes $M(AB) = I$. Since A and B belong to $T(n,F)$, AB does also by the first paragraph. Thus AB is the required right inverse for M.

THEOREM 3.2.15 *If F is a field and n is a natural number such that* $1 \leqslant n$, *then* $T(n,F)$ *is solvable.*

Proof: First it is claimed that $T(n,F)' \subseteq UT(n,F)$. It suffices to show all commutators are unitriangular, so let $A,B \in T(n,F)$ and let $M = A^{-1}B^{-1}AB$. Then $BAM = AB$. For $1 \leqslant i \leqslant n$, $(AB)_{ii} = A_{ii}B_{ii}$ and $(BAM)_{ii} = B_{ii}A_{ii}M_{ii}$ by Theorem 3.1.7. Thus $B_{ii}A_{ii}M_{ii} = A_{ii}B_{ii}$. Division by B_{ii} then by A_{ii} (note both are nonzero since $A,B \in T(n,F)$) yields $M_{ii} = 1$. Thus $M \in UT(n,F)$ as claimed.

By Theorem 3.2.3, $UT(n,F)$ is nilpotent, hence $T(n,F)'$ is a subgroup of a nilpotent group so it too is nilpotent (Theorem A.5.2). This implies $T(n,F)'$ is solvable by Theorem A.6.7. Also $T(n,F)/T(n,F)'$ is abelian hence solvable. Conclude $T(n,F)$ is solvable by Theorem A.6.6.

The remainder of this section will be devoted to groups of 2 by 2 matrices that are not necessarily triangular. For a given commutative ring with identity, R, the most general matrix group (called in fact the *general linear* group and denoted $GL(2,R)$) is described in the exercises. We consider below a somewhat more restricted group.

NOTATION Let $SL(2,R)$ denote the set of 2 by 2 matrices M, with entries in R, such that $\det(M) = 1$.

NOTE: See the note following the introduction of the notation $PSL(2,F)$ on page 84.

THEOREM 3.2.16 *If R is a commutative ring with identity, then $SL(2,R)$ is a group. Moreover inverses are given by the formula:* $\begin{pmatrix} a & b \\ c & d \end{pmatrix}^{-1} = \begin{pmatrix} d & -b \\ -c & a \end{pmatrix}$.

Proof: If $A,B \in SL(2,R)$, then by Theorem 3.1.10, $\det(AB) = \det(A)\det(B) = 1$ so $SL(2,R)$ is closed under products. Multiplication is associative in $SL(2,R)$ by Theorem 3.1.1, and I acts as the identity by Theorem 3.1.2.

Now let $M = \begin{pmatrix} a & b \\ c & d \end{pmatrix}$ belong to $SL(2,R)$ and let $A = \begin{pmatrix} d & -b \\ -c & a \end{pmatrix}$ be the candidate for M^{-1}. First note that $A \in SL(2,R)$ because $\det(A) = da - cb = \det(M) = 1$. Next note that $MA = \begin{pmatrix} a & b \\ c & d \end{pmatrix}\begin{pmatrix} d & -b \\ -c & a \end{pmatrix} = \begin{pmatrix} ad - bc & -ab + ba \\ cd - dc & -cb + da \end{pmatrix} = I$ because $ad - bc = \det(M) = 1$. Similarly $AM = I$ so $A = M^{-1}$.

NOTE: The above formula for M^{-1} will be used several places in the remainder of the book, usually without explicit reference.

The group $SL(2,R)$ is called the **special linear group** (of 2 by 2 matrices with entries in R). Most of the theory to be developed here for $SL(2,R)$ will be done only in the case when R is a field. Without making this restriction, however, it will still be possible to identify the center of $SL(2,R)$ (Theorem 3.2.20) and to prove the next theorem which is concerned with some particular normal subgroups.

DEFINITION AND NOTATION If R is a commutative ring with identity and J is an ideal of R, then $C(J)$ is the **principal congruence subgroup** of $SL(2,R)$ determined by J if and only if $C(J) = \{A \mid A \in SL(2,R)$ and $A_{11} - 1, A_{12}, A_{21}$, and $A_{22} - 1$ all belong to $J\}$. (The requirement on A for membership in $C(J)$ is often denoted $A \equiv I \bmod J$.)

THEOREM 3.2.17 *Let R be a commutative ring with identity and let J be an ideal. Then*
(a) $C(J) \lhd SL(2,R)$.
(b) *$SL(2,R)/C(J)$ can be embedded in $SL(2,R/J)$.*

Proof: Define $d: SL(2,R) \to SL(2,R/J)$ by $d(A)_{ij} = J + A_{ij}$. Clearly d is a homomorphism and is into $SL(2,R/J)$ (that is, $d(A)$ has determinant 1 when A does). Thus $SL(2,R)/\ker d$ can be embedded in $SL(2,R/J)$. Since kernels of homomorphisms are normal subgroups, showing $\ker d = C(J)$ will prove both parts of the theorem.

Let $A \in C(J)$. Thus $A_{kk} = 1 + j_{kk}$ where $j_{kk} \in J$ for $k = 1$ and 2, and also $A_{12}, A_{21} \in J$. Then $d(A) = \begin{pmatrix} J+1 & J+0 \\ J+0 & J+1 \end{pmatrix} = I$ so $A \in \ker d$. Conversely if $A \in \ker d$, then $I = d(A)$. Hence $J + 1 = d(A)_{11} = J + A_{11}$, $J + 0 = d(A)_{12} = J + A_{12}$, $J + 0 = d(A)_{21} = J + A_{21}$, and $J + 1 = d(A)_{22} = J + A_{22}$ implying $A_{11} - 1, A_{12}, A_{21}$, and $A_{22} - 1$ belong to J so $A \in C(J)$ as required.

THEOREM 3.2.18 *Let F be a field. Then every matrix in $SL(2,F)$ is the product of transvections. In other words $SL(2,F) = \langle M \mid M$ is a transvection$\rangle$.*

Proof: Let $A \in SL(2,F)$. For convenience, subscripts on the entries of A will be avoided by letting $A = \begin{pmatrix} a & b \\ c & d \end{pmatrix}$. First assume $c \neq 0$. Then $d - da$ can be divided by c since F is a field. Let $k = b + (1/c)(d - da)$. Then

$$\begin{pmatrix} 1 & (a-1)/c \\ 0 & 1 \end{pmatrix}\begin{pmatrix} 1 & 0 \\ c & 1 \end{pmatrix}\begin{pmatrix} 1 & k \\ 0 & 1 \end{pmatrix} = \begin{pmatrix} a & (a-1)/c \\ c & 1 \end{pmatrix}\begin{pmatrix} 1 & k \\ 0 & 1 \end{pmatrix} = \begin{pmatrix} a & ak+(1/c)(a-1) \\ c & ck+1 \end{pmatrix}$$

It is claimed that this last matrix is A. First note $ck + 1 = c(b + (1/c)(d - da)) + 1 = cb + d - da + 1 = d + 1 - \det(A) = d$. Secondly $1 = \det(A) = ad - bc$ implies $ad - 1 = bc$ so $(ad - 1)/c = b$. Therefore $ak + (1/c)(a - 1) = a(b + (1/c)(d - da)) + (1/c)(a - 1) = ab + (ad/c)(1 - a) + (1/c)(a - 1) = ab + (1 - a)[(ad/c) - (1/c)] = ab + (1 - a)[(ad - 1)/c] = ab + (1 - a)b = ab + b - ab = b$, establishing the claim. Thus A is the product $\Theta_2(1,2,(a - 1)/c)\Theta_2(2,1,c)\Theta_2(1,2,k)$. By assumption $c \neq 0$ so the middle factor is a transvection. The first and third may or may not be depending on whether or not $(a - 1)/c$ and k are nonzero. But if $(a - 1)/c$ and/or k are zero, then the first and/or third factor is the identity matrix. Thus in any case A is a product of (1,2, or 3) transvections.

Next consider the case $c = 0$. Let $B = \begin{pmatrix} a & c \\ a & d+c \end{pmatrix}$. Then $\det(B) = a(d + c) - ac = ad$. Also note $\det(A) = ad$ because $c = 0$, hence $\det(B) = \det(A) = 1$, showing $B \in SL(2,F)$. By definition of B, $B_{21} = a \neq 0$ ($a = 0$ would imply $1 = \det(A) = ad = 0$). Consequently by the first case considered above, B can be factored as a product of transvections. Finally note that $\Theta_2(2,1,-1)$ is also a transvection and $\Theta_2(2,1,-1)B = \begin{pmatrix} 1 & 0 \\ -1 & 1 \end{pmatrix}\begin{pmatrix} a & c \\ a & d+c \end{pmatrix} = \begin{pmatrix} a & c \\ 0 & d \end{pmatrix} = A$. But since B is a product of transvections and since $\Theta_2(2,1,-1)$ is a transvection, then $A = \Theta_2(2,1,-1)B$ is also a product of transvections, completing the proof.

The following theorem (Theorem 3.2.19) is actually a lemma to the highly important Theorem 3.2.22 which shows that if F meets certain conditions, then $SL(2,F)/Z(SL(2,F))$ is simple.

LEMMA *If F is a field such that* $\mathrm{char}(F) \neq 2$, *then every element of F can be expressed as the difference of two squares.*

Proof: Let $x \in F$. As usual denote the element $1 + 1$ by 2. By hypothesis $2 \neq 0$, so any element of F may be divided by 2. Thus $(x - 1)/2$ and $(x + 1)/2$ both exist in F. Then $((x + 1)/2)^2 - ((x - 1)/2)^2 = ((x^2 + 2x + 1)/4) - ((x^2 - 2x + 1)/4) = 4x/4 = x$.

THEOREM 3.2.19 *Let F be a field such that* $\mathrm{char}(F) \neq 2$. *If $H \lhd SL(2,F)$ such that $\Theta_2(2,1,a) \in H$ for some nonzero a, then $H = SL(2,F)$.*

Proof: It suffices, by Theorem 3.2.18, to show that every transvection belongs to H. So let B be a transvection. Then either $B = \Theta_2(2,1,b)$ or $B = \Theta_2(1,2,b)$ for some nonzero b.

CASE ONE, $B = \Theta_2(2,1,b)$. By hypothesis $a \neq 0$, so $b/a \in F$. By the lemma then, $b/a = y^2 - z^2$ for some y and z. Thus $b = ay^2 - az^2$. It is now claimed that $\Theta_2(2,1,ay^2) \in H$. Surely this is true if $y = 0$ so assume $y \neq 0$. Then $1/y \in F$. Consider the matrix $M = \begin{pmatrix} y & 0 \\ 1 & 1/y \end{pmatrix}$. Note M has determinant 1, so belongs to $SL(2,F)$. Then since $\Theta_2(2,1,a) \in H$ and $H \lhd SL(2,F)$, $M^{-1}\Theta_2(2,1,a)M \in H$. That is,

$$\begin{pmatrix} 1/y & 0 \\ -1 & y \end{pmatrix}\begin{pmatrix} 1 & 0 \\ a & 1 \end{pmatrix}\begin{pmatrix} y & 0 \\ 1 & 1/y \end{pmatrix} = \begin{pmatrix} 1/y & 0 \\ -1+ay & y \end{pmatrix}\begin{pmatrix} y & 0 \\ 1 & 1/y \end{pmatrix} = \begin{pmatrix} 1 & 0 \\ ay^2 & 1 \end{pmatrix}$$

belongs to H, proving the claim. A similar demonstration shows $\Theta_2(2,1,az^2) \in H$. Consequently the product $\Theta_2(2,1,ay^2)\Theta_2(2,1,az^2)^{-1} = B$ belongs to H.

CASE TWO, $B = \Theta_2(1,2,b)$. By Case one, $\Theta_2(1,2,-b) \in H$. Therefore the product $K^{-1}\Theta_2(1,2,-b)K \in H$ for all $K \in SL(2,F)$. In particular $\begin{pmatrix} 0 & 1 \\ -1 & 0 \end{pmatrix} \in SL(2,F)$ hence

$$\begin{pmatrix} 0 & -1 \\ 1 & 0 \end{pmatrix}\begin{pmatrix} 1 & -b \\ 0 & 1 \end{pmatrix}\begin{pmatrix} 0 & 1 \\ -1 & 0 \end{pmatrix} = \begin{pmatrix} 0 & -1 \\ 1 & -b \end{pmatrix}\begin{pmatrix} 0 & 1 \\ -1 & 0 \end{pmatrix} = \begin{pmatrix} 1 & 0 \\ b & 1 \end{pmatrix} = B \in H.$$

The next theorem describes the center of $SL(2,R)$. For convenience we adopt the following notation.

NOTATION Let $-I$ denote

$$\begin{pmatrix} -1 & 0 \\ 0 & -1 \end{pmatrix}.$$

THEOREM 3.2.20 *Let R be a commutative ring with identity. Then $\{I,-I\} = Z(SL(2,R))$.*

Proof: Let $M \in SL(2,R)$. Then

$$(-I)M = \begin{pmatrix} -1 & 0 \\ 0 & -1 \end{pmatrix}\begin{pmatrix} M_{11} & M_{12} \\ M_{21} & M_{22} \end{pmatrix} = \begin{pmatrix} -M_{11} & -M_{12} \\ -M_{21} & -M_{22} \end{pmatrix}$$

while

$$M(-I) = \begin{pmatrix} M_{11} & M_{12} \\ M_{22} & M_{22} \end{pmatrix}\begin{pmatrix} -1 & 0 \\ 0 & -1 \end{pmatrix} = \begin{pmatrix} -M_{11} & -M_{12} \\ -M_{21} & -M_{22} \end{pmatrix}.$$

Consequently $-I \in Z(SL(2,R))$, and so $\{I,-I\} \subseteq Z(SL(2,R))$. Conversely suppose $M \in Z(SL(2,R))$. It must be shown that $M = I$ or $M = -I$. Consider the matrix $\Theta_2(1,2,1)$. Since $\Theta_2(1,2,1)$ has determinant 1 it belongs to $SL(2,R)$ and so $M\Theta_2(1,2,1) = \Theta_2(1,2,1)M$. But

$$M\Theta_2(1,2,1) = \begin{pmatrix} M_{11} & M_{11}+M_{12} \\ M_{21} & M_{21}+M_{22} \end{pmatrix}$$

while

$$\Theta_2(1,2,1)M = \begin{pmatrix} M_{11}+M_{21} & M_{12}+M_{22} \\ M_{21} & M_{22} \end{pmatrix}.$$

This implies $M_{11} = M_{11} + M_{21}$ so $M_{21} = 0$. It also implies $M_{11} + M_{12} = M_{12} + M_{22}$ so $M_{11} = M_{22}$. A similar demonstration using the matrix $\Theta_2(2,1,1)$ shows $M_{12} = 0$. Then $1 = \det(M) = M_{11}M_{22} - M_{12}M_{21} = M_{11}{}^2$, so $M_{11} = \pm 1$. Thus M is either I or $-I$.

NOTE: If F is a field of characteristic 2, then $-1 = 1$ so $-I = I$. Thus $Z(SL(2,F))$ has only one element.

THEOREM 3.2.21 *If F is a finite field, then* $\text{card}(SL(2,F)) = q(q-1)(q+1)$ *where* $q = \text{card}(F)$.

Proof: Matrices $M \in SL(2,F)$ can be chosen with either $M_{11} = 0$ or $M_{11} \neq 0$. In the first case there is, of course, 1 choice for M_{11}. Then regardless of what M_{22} is, $\det(M) = -M_{21}M_{12}$. So there are q choices for M_{22}. If M_{12} is anything but 0, it will satisfy the equation $1 = -M_{21}M_{12}$ provided M_{21} is suitably chosen. Hence there are $q - 1$ choices for M_{12}. Finally note that the choice for M_{12} determines M_{21} as $M_{21} = 1/-M_{12}$. Thus there is 1 choice for M_{21}, and consequently there is a total of $1\cdot(q-1)\cdot 1\cdot q = q(q-1)$ matrices in the first case.

Now assume the second case, that is, $M_{11} \neq 0$. There are then $q - 1$ choices for M_{11}. Let M_{21} and M_{12} be arbitrary (q choices for each). Then the equation $1 = \det(M) = M_{11}M_{22} - M_{12}M_{21}$ can still be satisfied if $M_{22} = (1 + M_{12}M_{21})/M_{11}$ (thus 1 choice for M_{22}). Hence there is a total of $(q-1)\cdot q\cdot q\cdot 1 = q^2(q-1)$ matrices in the second case. Altogether, therefore, there are $q(q-1) + q^2(q-1) = q(q-1)(q+1)$ matrices in $SL(2,F)$.

NOTATION Let $PSL(2,F) = SL(2,F)/Z(SL(2,F))$.

NOTE: When F is a finite field of cardinality q, it is traditional to use the notation $SL(2,q)$ and $PSL(2,q)$ instead of $SL(2,F)$ and $PSL(2,F)$ respectively. This can be done because any two fields of cardinality q are isomorphic.

On the other hand, before any results involving particular matrices in $SL(2,q)$ or particular cosets in $PSL(2,q)$ can be considered it will be necessary to have a definite finite field in mind. In the remainder of this book, whenever p is a prime, $SL(2,p)$ will mean $SL(2,Z(p))$ and $PSL(2,p)$ will mean $PSL(2,Z(p))$ where $Z(p)$ is the field of integers mod p. When q is not a prime, a special explanation will be given for what the underlying field is for elements of $SL(2,q)$ and $PSL(2,q)$.

The group $PSL(2,F)$ is called the **projective special linear** group or the **linear fractional** group (of 2 by 2 matrices with entries in F). The most important property of $PSL(2,F)$ is given in the next theorem. First, however, a preliminary definition is needed.

DEFINITION AND NOTATION If M is a 2 by 2 matrix, then $\operatorname{tr}(M)$ is the **trace** of M if and only if $\operatorname{tr}(M) = M_{11} + M_{22}$.

THEOREM 3.2.22 *If F is a field such that* $\operatorname{char}(F) \neq 2$ *and* $5 \leqslant \operatorname{card}(F)$, *then* $PSL(2,F)$ *is simple.*

Proof: For convenience let Z denote $Z(SL(2,F))$. Suppose H/Z is a nontrivial normal subgroup of $PSL(2,F)$. It must be shown that $H/Z = PSL(2,F)$. That is, it must be shown that $H = SL(2,F)$. And to accomplish this, it suffices by Theorem 3.2.19 to exhibit $a \neq 0$ such that $\Theta_2(2,1,a) \in H$.

Since $H/Z \neq \{1\}$, Z must be a proper subgroup of H. Hence by Theorem 3.2.20 there is some matrix $M \in H$ such that $M \neq I$ and $M \neq -I$. First note that $H \lhd SL(2,F)$ because $H/Z \lhd SL(2,F)/Z$. Therefore

$$\begin{pmatrix} 0 & -1 \\ 1 & 0 \end{pmatrix}\begin{pmatrix} M_{11} & M_{12} \\ M_{21} & M_{22} \end{pmatrix}\begin{pmatrix} 0 & -1 \\ 1 & 0 \end{pmatrix}^{-1} = \begin{pmatrix} -M_{21} & -M_{22} \\ M_{11} & M_{12} \end{pmatrix}\begin{pmatrix} 0 & 1 \\ -1 & 0 \end{pmatrix}$$

$$= \begin{pmatrix} M_{22} & -M_{21} \\ -M_{12} & M_{11} \end{pmatrix} \in H, \tag{1}$$

and

$$\begin{pmatrix} 0 & 1 \\ -1 & 1 \end{pmatrix}\begin{pmatrix} M_{11} & M_{12} \\ M_{21} & M_{22} \end{pmatrix}\begin{pmatrix} 0 & 1 \\ -1 & 1 \end{pmatrix}^{-1} = \begin{pmatrix} M_{21} & M_{22} \\ M_{21} - M_{11} & M_{22} - M_{12} \end{pmatrix}\begin{pmatrix} 1 & -1 \\ 1 & 0 \end{pmatrix}$$

$$= \begin{pmatrix} M_{21} + M_{22} & -M_{21} \\ M_{21} - M_{11} + M_{22} - M_{12} & M_{11} - M_{21} \end{pmatrix}$$

$$\in H, \tag{2}$$

and for any $k \in F$,

$$\begin{pmatrix} 1 & k \\ 0 & 1 \end{pmatrix}\begin{pmatrix} M_{11} & M_{12} \\ M_{21} & M_{22} \end{pmatrix}\begin{pmatrix} 1 & k \\ 0 & 1 \end{pmatrix}^{-1} = \begin{pmatrix} M_{11} + kM_{21} & M_{12} + kM_{22} \\ M_{21} & M_{22} \end{pmatrix}\begin{pmatrix} 1 & -k \\ 0 & 1 \end{pmatrix}$$

$$= \begin{pmatrix} M_{11} + kM_{21} & -kM_{11} - k^2M_{21} + M_{12} + kM_{22} \\ M_{21} & -kM_{21} + M_{22} \end{pmatrix}$$

$$\in H. \tag{3}$$

Several cases will now be distinguished.

CASE ONE, $M_{11} = 1$ and $M_{21} = 0$. Then $1 = \det(M) = M_{11}M_{22} - M_{12}M_{21} = M_{22}$. By Line (1),

$$\begin{pmatrix} 1 & 0 \\ -M_{12} & 1 \end{pmatrix} \in H.$$

That is, $\Theta_2(2,1,-M_{12}) \in H$. If $-M_{12} = 0$, then $M = I$ which is false, so $-M_{12} \neq 0$. Thus $\Theta_2(2,1,-M_{12})$ is the required transvection in H.

CASE TWO, $M_{11} = -1$ and $M_{21} = 0$. Then $1 = \det(M) = -M_{22}$ so $M_{22} = -1$. By Line (1),

$$\begin{pmatrix} -1 & 0 \\ -M_{12} & -1 \end{pmatrix} \in H.$$

But then also $-I \in Z \subseteq H$. Therefore

$$\begin{pmatrix} -1 & 0 \\ -M_{12} & -1 \end{pmatrix}\begin{pmatrix} -1 & 0 \\ 0 & -1 \end{pmatrix} = \begin{pmatrix} 1 & 0 \\ M_{12} & 1 \end{pmatrix} = \Theta_2(2,1,M_{12}) \in H.$$

If $M_{12} = 0$, then $M = -I$ which is false, so $M_{12} \neq 0$. Thus $\Theta_2(2,1,M_{12})$ is the required transvection in H.

CASE THREE, $M_{11} \neq 1, -1$ and $M_{21} = 0$. By Line (2),

$$\begin{pmatrix} M_{22} & 0 \\ -M_{11} + M_{22} - M_{12} & M_{11} \end{pmatrix} \in H.$$

By Line 1,

$$\begin{pmatrix} M_{22} & 0 \\ -M_{12} & M_{11} \end{pmatrix} \in H$$

hence inverting yields

$$\begin{pmatrix} M_{11} & 0 \\ M_{12} & M_{22} \end{pmatrix} \in H.$$

Therefore the product

$$\begin{pmatrix} M_{22} & 0 \\ M_{22} - M_{11} - M_{12} & M_{11} \end{pmatrix}\begin{pmatrix} M_{11} & 0 \\ M_{12} & M_{22} \end{pmatrix} = \begin{pmatrix} M_{22}M_{11} & 0 \\ M_{22}M_{11} - M_{11}{}^2 & M_{11}M_{22} \end{pmatrix} \in H.$$

This and $1 = \det(M) = M_{11}M_{22}$ show $\Theta_2(2,1,1 - M_{11}{}^2) \in H$. If $1 - M_{11}{}^2 = 0$, then $(1 - M_{11})(1 + M_{11}) = 0$ which would imply $M_{11} = \pm 1$ which is false, so $1 - M_{11}{}^2$ is nonzero. Thus $\Theta_2(2,1,1 - M_{11}{}^2)$ is the required transvection in H.

CASE FOUR, $M_{21} \neq 0$ and $\operatorname{tr}(M) \neq 0$. Since $M_{21} \neq 0$, $-M_{11}/M_{21} \in F$. Set $k = -M_{11}/M_{21}$. Then $M_{11} + kM_{21} = 0$, $-kM_{21} + M_{22} = M_{11} + M_{22}$, and $-kM_{11} - k^2M_{21} + M_{12} + kM_{22} = (M_{11}{}^2/M_{21}) - (M_{11}{}^2/M_{21}) + M_{12} - (M_{11}M_{22}/M_{21}) = M_{12} - (M_{11}M_{22}/M_{21}) = (M_{12}M_{21}/M_{21}) - (M_{11}M_{22}/M_{21}) = (M_{12}M_{21} - M_{11}M_{22})/M_{21} = -\det(M)/M_{21} = -1/M_{21}$. Therefore

$$\begin{pmatrix} 0 & -1/M_{21} \\ M_{21} & M_{11} + M_{22} \end{pmatrix} \in H \tag{4}$$

by Line (3).

If $x \in F$ such that $x^2 = 1$, then $0 = x^2 - 1 = (x - 1)(x + 1)$ implying $x = \pm 1$. Thus $1, -1$ are the only two field elements whose squares are 1. Consequently $\operatorname{card}(\{c \mid c \in F \text{ and } c^2 = 1\} \cup \{0\}) = 3 < \operatorname{card}(F)$. Thus there exists $a \in F$ such that $a \neq 0$ and $a^2 \neq 1$. Since $a \neq 0$, $1/a \in F$. Therefore

$$\begin{pmatrix}1/a & 0\\ 0 & a\end{pmatrix}\begin{pmatrix}0 & -1/M_{21}\\ M_{21} & M_{11}+M_{22}\end{pmatrix}\begin{pmatrix}1/a & 0\\ 0 & a\end{pmatrix}^{-1} = \begin{pmatrix}0 & -1/M_{21}\\ aM_{21} & aM_{11}+aM_{22}\end{pmatrix}\begin{pmatrix}a & 0\\ 0 & 1/a\end{pmatrix}$$

$$= \begin{pmatrix}0 & -1/a^2M_{21}\\ a^2M_{21} & M_{11}+M_{22}\end{pmatrix} \in H.$$

Inverting the matrix in Line (4) yields

$$\begin{pmatrix}M_{11}+M_{22} & 1/M_{21}\\ -M_{21} & 0\end{pmatrix} \in H.$$

Therefore the product

$$\begin{pmatrix}0 & -1/a^2M_{21}\\ a^2M_{21} & M_{11}+M_{22}\end{pmatrix}\begin{pmatrix}M_{11}+M_{22} & 1/M_{21}\\ -M_{21} & 0\end{pmatrix} = \begin{pmatrix}1/a^2 & 0\\ e & a^2\end{pmatrix} \in H \qquad (5)$$

where $e = (a^2 - 1)M_{21}(M_{11} + M_{22})$. Using this last matrix instead of M in Line (1) yields

$$\begin{pmatrix}a^2 & -e\\ 0 & 1/a^2\end{pmatrix} \in H.$$

Call this last matrix $\bar{M}$. Now by case hypothesis, $M_{21} \neq 0$ and $M_{11} + M_{22} \neq 0$, and by choice of a, $a^2 - 1 \neq 0$. Therefore $e \neq 0$ and so $-e \neq 0$. If $a^2 = -1$, then $\bar{M} \neq -I$ because $-e \neq 0$. Therefore Case two can be reworked using $\bar{M} \in H$ instead of $M \in H$ to get the required transvection in H. If $a^2 \neq -1$, then Case three can be reworked using $\bar{M} \in H$ instead of $M \in H$.

CASE FIVE, $M_{21} \neq 0$ and $\mathrm{tr}(M) = 0$. First assume $5 < \mathrm{card}(F)$. A well-known theorem in algebra (see, for example, Herstein [18], Lemma 5.2) states that a degree-n polynomial over a field can have at most n roots. Applying this to the polynomial $x^4 - 1 = 0$ yields $\mathrm{card}(\{c \mid c \in F \text{ and } c^4 = 1\} \cup \{0\}) \leqslant 5 < \mathrm{card}(F)$ and so there exists $a_1 \in F$ such that $a_1 \neq 0$ and $a_1^4 \neq 1$. Then, as in Case four (Line (5)),

$$\begin{pmatrix}1/a_1^2 & 0\\ e & a_1^2\end{pmatrix} \in H$$

where $e = (a_1^2 - 1)M_{21}(M_{11} + M_{22})$. In this case, however, $M_{11} + M_{22} = 0$ so $e = 0$. Thus

$$\begin{pmatrix}1/a_1^2 & 0\\ 0 & a_1^2\end{pmatrix} \in H.$$

Using this last matrix instead of M in Line (2) yields

$$\begin{pmatrix}a_1^2 & 0\\ a_1^2 - (1/a_1^2) & 1/a_1^2\end{pmatrix} \in H.$$

This further implies

$$\begin{pmatrix} a_1^2 & 0 \\ a_1^2 - (1/a_1^2) & 1/a_1^2 \end{pmatrix} \begin{pmatrix} 1/a_1^2 & 0 \\ 0 & a_1^2 \end{pmatrix} = \begin{pmatrix} 1 & 0 \\ 1-(1/a_1^4) & 1 \end{pmatrix} \in H.$$

Now if $1 - (1/a_1^4) = 0$, then $a_1^4 = 1$ which is false, so $1 - (1/a_1^4) \neq 0$ and hence $\Theta_2(2,1,1 - (1/a_1^4))$ is the required transvection in H.

Finally suppose card(F) = 5. As in Case four (Line (4)),

$$\begin{pmatrix} 0 & -1/M_{21} \\ M_{21} & M_{11} + M_{22} \end{pmatrix} \in H.$$

Here, however, $M_{11} + M_{22} = 0$ so

$$\begin{pmatrix} 0 & -1/M_{21} \\ M_{21} & 0 \end{pmatrix} \in H.$$

Call this last matrix D. Then using $D \in H$ instead of $M \in H$ in Line (3) yields

$$\begin{pmatrix} D_{11} + kD_{21} & -kD_{11} - k^2 D_{21} + D_{12} + kD_{22} \\ D_{21} & -kD_{21} + D_{22} \end{pmatrix} \in H$$

for any $k \in F$. In particular when $k = 1/M_{21}$, $D_{11} + kD_{21} = 0 + (1/M_{21})M_{21} = 1$, $-kD_{11} - k^2 D_{21} + D_{12} + kD_{22} = -(1/M_{21})0 - (1/M_{21}^2)M_{21} - (1/M_{21}) + (1/M_{21})0 = -2/M_{21}$, and $-kD_{21} + D_{22} = -(1/M_{21})M_{21} + 0 = -1$. Thus

$$\begin{pmatrix} 1 & -2/M_{21} \\ M_{21} & -1 \end{pmatrix} \in H.$$

Inverting D yields

$$\begin{pmatrix} 0 & 1/M_{21} \\ -M_{21} & 0 \end{pmatrix} \in H.$$

Therefore the product

$$\begin{pmatrix} 1 & -2/M_{21} \\ M_{21} & -1 \end{pmatrix} \begin{pmatrix} 0 & 1/M_{21} \\ -M_{21} & 0 \end{pmatrix} = \begin{pmatrix} 2 & 1/M_{21} \\ M_{21} & 1 \end{pmatrix} \in H.$$

Call this last matrix $\bar{M}$. Then $\bar{M}_{21} = M_{21} \neq 0$ and $\operatorname{tr}(\bar{M}) = 3 \neq 0$ (because card(F) = 5). Therefore Case three can be reworked with $\bar{M} \in H$ instead of $M \in H$ to get the required transvection. This completes the proof.

A few words are in order here as to whether or not the requirements on F in the hypothesis of Theorem 3.2.22 are really necessary. First the cardinality condition. The only fields that fail to satisfy this condition have cardinality 2, 3, or 4. The group $PSL(2,4)$ is indeed simple and in fact is isomorphic to $PSL(2,5)$ which has been shown to be simple by the theorem. On the other hand, the groups $PSL(2,2)$ and $PSL(2,3)$ are nonsimple, being in fact solvable (by the note after Theorem 3.2.20, $PSL(2,2)$ is isomorphic to $SL(2,2)$, hence is solvable by Result 4.2.9 and Exercise 7 of Example 4.2. The group $PSL(2,3)$ is solvable by Result 4.6.7).

The other requirement (char(F) $\neq 2$) is unnecessary. Except for the case when card(F) = 2 (which is eliminated by the cardinality requirement anyway), $PSL(2,F)$ is simple even when char(F) = 2. The proof of Theorem 3.2.22, however, relying as it does on Theorem 3.2.19 and its lemma, is just insufficient to show this. For a proof of Theorem 3.2.22 that does not assume the restriction on char(F), see Rotman [31], Lemma 8.22.

We conclude the chapter with some further results about $PSL(2,F)$ when F is a finite field.

THEOREM 3.2.23 *If F is a finite field of cardinality q, then*

$$\mathrm{card}(PSL(2,F)) = \begin{cases} q(q-1)(q+1) & \textit{if } \mathrm{char}(F) = 2, \\ q(q-1)(q+1)/2 & \textit{if } \mathrm{char}(F) \neq 2. \end{cases}$$

Proof: By Theorem 3.2.21, card($SL(2,F)$) = $q(q-1)(q+1)$, and by Theorem 3.2.20 and the note after it, Z($SL(2,F)$) has cardinality 1 if char(F) = 2 and 2 otherwise. This and card($PSL(2,F)$) = card($SL(2,F)$)/card(Z($SL(2,F)$)) complete the proof.

The following notational convention will be used in the remainder of this section and also later in the book.

NOTATION If $M \in SL(2,F)$, let $[M]$ denote the coset of M relative to the center Z($SL(2,F)$). Thus $[M] \in PSL(2,F)$.

THEOREM 3.2.24 *Let F be any field. Suppose* $M \in SL(2,F)$. *Then* tr(M) = 0 *implies* $[M]^2 = 1$.

Proof: Consider the product M^2 =

$$\begin{pmatrix} M_{11} & M_{12} \\ M_{21} & M_{22} \end{pmatrix}\begin{pmatrix} M_{11} & M_{12} \\ M_{21} & M_{22} \end{pmatrix} = \begin{pmatrix} M_{11}^2 + M_{12}M_{21} & M_{12}(M_{11}+M_{22}) \\ M_{21}(M_{11}+M_{22}) & M_{21}M_{12}+M_{22}^2 \end{pmatrix}. \quad (1)$$

Now

$$M_{11} + M_{22} = \mathrm{tr}(M) = 0. \quad (2)$$

This further implies $0 = M_{11}(M_{11} + M_{22}) = M_{11}^2 + M_{11}M_{22}$ so $M_{11}^2 = -M_{11}M_{22}$ and hence

$$M_{11}^2 + M_{12}M_{21} = M_{12}M_{21} - M_{11}M_{22} = -\det(M) = -1. \quad (3)$$

Again from Equation 1, $0 = M_{22}(M_{11} + M_{22}) = M_{11}M_{22} + M_{22}^2$ so that $M_{22}^2 = -M_{11}M_{22}$ and hence

$$M_{22}^2 + M_{21}M_{12} = M_{21}M_{12} - M_{11}M_{22} = -\det(M) = -1. \quad (4)$$

Substituting Equations 2,3, and 4 into Equation 1 yields $M^2 = \begin{pmatrix} -1 & 0 \\ 0 & -1 \end{pmatrix} = -I$. Therefore $[M]^2 = [M^2] = [-I] = 1$ since $-I \in$ Z($SL(2,F)$) by Theorem 3.2.20.

NOTE: The converse of Theorem 3.2.24 is also true provided the underlying field has characteristic $\neq 2$. See Exercise IV.5.

THEOREM 3.2.25 *Let F be a field of characteristic $\neq 3$. Suppose $M \in SL(2,F)$. Then* $o([M]) = 3$ *(in $PSL(2,F)$) if and only if* $\operatorname{tr}(M) = \pm 1$.

Proof: For convenience denote $Z(SL(2,F))$ by Z. First suppose $\operatorname{tr}(M) = -1$. Since $\operatorname{tr}(I) = 2$ and $\operatorname{tr}(-I) = -2$, it follows that $M \neq I, -I$ (the equation $-1 = 2$ implies $0 = 3$ which is false by the hypothesis on char(F), while the equation $-1 = -2$ implies $1 = 0$, false in any field). Thus $[M]$ is nontrivial by Theorem 3.2.20. To show $o([M]) = 3$ then, it suffices to show $M^3 \in Z$. Since $M_{11} + M_{22} = -1$, $M_{11}^2 + M_{11}M_{22} = -M_{11}$. But $1 = \det(M) = M_{11}M_{22} - M_{12}M_{21}$. Combining the last two equations yields $M_{11}^2 + 1 + M_{12}M_{21} = -M_{11}$. Thus $M_{11}^2 + M_{12}M_{21} = -M_{11} - 1 = -M_{11} + \operatorname{tr}(M) = -M_{11} + M_{11} + M_{22}$, so

$$M_{11}^2 + M_{12}M_{21} = M_{22}. \tag{1}$$

Also $M_{11} + M_{22} = -1$ implies $M_{11}M_{22} + M_{22}^2 = -M_{22}$. This and $1 = \det(M) = M_{11}M_{22} - M_{12}M_{21}$ yields $1 + M_{12}M_{21} + M_{22}^2 = -M_{22}$. Thus $M_{22}^2 + M_{12}M_{21} = -M_{22} - 1 = -M_{22} + M_{11} + M_{22}$, so

$$M_{22}^2 + M_{12}M_{21} = M_{11}. \tag{2}$$

Entries of the matrix M^2 are given by

$$\begin{pmatrix} M_{11} & M_{12} \\ M_{21} & M_{22} \end{pmatrix}\begin{pmatrix} M_{11} & M_{12} \\ M_{21} & M_{22} \end{pmatrix} = \begin{pmatrix} M_{11}^2 + M_{12}M_{21} & M_{12}(M_{11} + M_{22}) \\ M_{21}(M_{11} + M_{22}) & M_{21}M_{12} + M_{22}^2 \end{pmatrix}. \tag{3}$$

After substitution of Equations 1 and 2 and $M_{11} + M_{22} = -1$, this becomes

$$M^2 = \begin{pmatrix} M_{22} & -M_{12} \\ -M_{21} & M_{11} \end{pmatrix}.$$

Therefore $M^3 = M^2M =$

$$\begin{pmatrix} M_{22} & -M_{12} \\ -M_{21} & M_{11} \end{pmatrix}\begin{pmatrix} M_{11} & M_{12} \\ M_{21} & M_{22} \end{pmatrix} = \begin{pmatrix} M_{22}M_{11} - M_{12}M_{21} & 0 \\ 0 & -M_{21}M_{12} + M_{11}M_{22} \end{pmatrix} =$$

$\begin{pmatrix} 1 & 0 \\ 0 & 1 \end{pmatrix} = I \in Z$.

Next suppose $\operatorname{tr}(M) = 1$. Again $M \neq I$ and $M \neq -I$ so again it suffices to show $M^3 \in Z$. Similarly to the first case

$$M^3 = \begin{pmatrix} -M_{22}M_{11} + M_{12}M_{21} & 0 \\ 0 & M_{21}M_{12} - M_{11}M_{22} \end{pmatrix} = -I \in Z.$$

To complete the proof, it must be shown that $o([M]) = 3$ implies $\operatorname{tr}(M) = \pm 1$. So suppose $M^3 \in Z$ and $M \notin Z$. Then either $M^3 = I$ or $M^3 = -I$. Assume $M^3 = I$, the other case being similar. From $M^3 = I$ it follows that $M^2 = M^{-1}$. Entries of the matrix M^2 are given by Equation 3 while

$$M^{-1} = \begin{pmatrix} M_{22} & -M_{12} \\ -M_{21} & M_{11} \end{pmatrix}.$$

Since $M \notin Z$, $M \neq I$ so M and I differ in (at least) one entry.
CASE ONE, $M_{12} \neq 0$. Equating $(M^2)_{12}$ and $(M^{-1})_{12}$ yields $M_{12}(M_{11} + M_{22}) = -M_{12}$. Dividing by M_{12} gives $\mathrm{tr}(M) = -1$.
CASE TWO, $M_{21} \neq 0$. Equating $(M^2)_{21}$ and $(M^{-1})_{21}$ and dividing by M_{21} gives $\mathrm{tr}(M) = -1$.
CASE THREE, $M_{12} = M_{21} = 0$. Then $1 = \det(M) = M_{11}M_{22}$. Equating $(M^2)_{11}$ and $(M^{-1})_{11}$ yields $M_{11}{}^2 = M_{22}$, hence $M_{11}{}^2 = 1/M_{11}$ implying $M_{11}{}^3 = 1$. Now since $M \neq I$, $M_{11} \neq 1$ (the other possibility, $M_{22} \neq 1$, is quickly seen to imply $M_{11} \neq 1$ because $1 = M_{11}M_{22}$). Factoring $M_{11}{}^3 - 1 = 0$ yields $(M_{11} - 1)(M_{11}{}^2 + M_{11} + 1) = 0$. This and $M_{11} \neq 1$ imply $M_{11}{}^2 + M_{11} + 1 = 0$. Finally $M_{11}{}^2 = M_{22}$ so $M_{22} + M_{11} + 1 = 0$ and hence $\mathrm{tr}(M) = -1$.

THEOREM 3.2.26 *Let F be a finite field of cardinality q such that* $\mathrm{char}(F) \neq 2$ *or* 3. *Then $PSL(2,F)$ has either $q^2 + q$ or $q^2 - q$ elements of order* 3 *depending on whether or not F has a square root of* -3.
Proof: The major portion of the proof will be concerned with counting the number of matrices M in $SL(2,F)$ having trace ± 1. Three cases are distinguished.
CASE ONE, $M_{11} = 0$. There is, of course, 1 choice for M_{11}. Moreover the trace of M will equal M_{22} so there are exactly 2 choices for M_{22} (namely ± 1). The entry M_{12} may be anything except 0 ($M_{12} = 0$ implies $1 = \det(M) = 0 - 0 = 0$), so there are $q - 1$ choices for M_{12}. Finally $1 = \det(M) = 0 - M_{12}M_{21}$ implies $M_{21} = -1/M_{12}$. Thus there is 1 choice for M_{21}. Altogether there are $1 \cdot 2 \cdot q - 1 \cdot 1 = 2q - 2$ matrices.
CASE TWO, $M_{12} = 0$. There is, of course, 1 choice for M_{12}. Since the choice of M_{21} affects neither the determinant nor the trace of M it is arbitrary; thus q choices. The remaining entries, M_{11} and M_{22}, must satisfy $M_{11}M_{22} = 1$ (since $1 = \det(M)$) and $M_{11} + M_{22} = \pm 1$. Consequently M_{11} must satisfy $M_{11}{}^2 \pm M_{11} + 1 = 0$. The quadratic formula shows that if F has a square root of -3, call it t, then $x^2 + x + 1 = 0$ has the two solutions $x = (-1 \pm t)/2$ (note these two solutions are distinct for their equality would imply $-1 + t = -1 - t$ and hence $2t = 0$ so $t = 0$ implying $-3 = t^2 = 0$ which is false) and $x^2 - x + 1 = 0$ has the two solutions $x = (1 \pm t)/2$. Arguments similar to the one in the last parentheses show the four numbers $(-1 + t)/2$, $(-1 - t)/2$, $(1 + t)/2$, and $(1 - t)/2$ are distinct. On the other hand if a square root of -3 does not exist in F, then the equations $x^2 \pm x + 1 = 0$ have no solutions (if a were such a solution, then $(2a \pm 1)^2 = -3$). Thus the number of choices for M_{11} is either 4 or 0 depending on whether or not F has a square root of -3. Finally $1 = \det(M) = M_{11}M_{22}$ so $M_{22} = 1/M_{11}$ (note $M_{11} \neq 0$ lest $1 = \det(M) = 0$). Hence there is 1 choice for M_{22}. Altogether the number of matrices in this case is

$$\begin{cases} 1 \cdot q \cdot 4 \cdot 1 = 4q & \text{if } F \text{ has a square root of } -3, \\ 0 & \text{otherwise.} \end{cases}$$

CASE THREE, $M_{11} \neq 0$ and $M_{12} \neq 0$. The entry M_{11} cannot be 0 but is otherwise arbitrary, so there are $q - 1$ choices for M_{11}. Then $\pm 1 = M_{11} + M_{22}$ hence there are two choices for M_{22} (namely $\pm 1 - M_{11}$). Now the choice of M_{12} does not affect the trace of M, and it also does not affect the determinant of M since no matter what M_{12} is, some value of M_{21} (namely $M_{21} = (M_{11}M_{22} - 1)/M_{12}$) can always be found to satisfy the equation $1 = M_{11}M_{22} - M_{12}M_{21}$ (recall $M_{12} \neq 0$ by case hypothesis). Thus M_{12} is nonzero but otherwise arbitrary. Thus there are $q - 1$ choices for M_{12}. Finally, as has been noted, M_{21} is determined as $M_{21} = (M_{11}M_{22} - 1)/M_{12}$, hence 1 choice for M_{21}. Altogether there are $(q - 1)\cdot 2\cdot (q - 1)\cdot 1 = 2(q - 1)^2$ matrices.

It is easily seen that Cases one, two, and three are mutually exclusive so the total number of matrices having trace ± 1 is

$$\begin{cases}(2q - 2) + 4q + 2(q - 1)^2 = 2q^2 + 2q & \text{if } \sqrt{-3} \in F,\\ (2q - 2) + 0 + 2(q - 1)^2 = 2q^2 - 2q & \text{if } \sqrt{-3} \notin F.\end{cases}$$

It is easily seen that if $[A] = [B]$ then either $A = B$ or

$$A = \begin{pmatrix} -B_{11} & -B_{12} \\ -B_{21} & -B_{22} \end{pmatrix}.$$

Therefore if $[A] = [B]$, then $\operatorname{tr}(A) = \pm 1$ if and only if $\operatorname{tr}(B) = \pm 1$. Consequently there are exactly one-half as many cosets $[A]$ in $PSL(2,F)$ such that $\operatorname{tr}(A) = \pm 1$ as there are matrices in $SL(2,F)$ with this property. Conclude from this and the previous paragraph that there are

$$\begin{cases} q^2 + q & \text{if } \sqrt{-3} \in F \\ q^2 - q & \text{if } \sqrt{-3} \notin F \end{cases}$$

cosets $[A]$ such that $\operatorname{tr}(A) = \pm 1$. Finally apply Theorem 3.2.25 to conclude the result.

EXERCISES

I Unitriangular Groups

I.1. (a) For $x \in R$ and $1 \leqslant j < i \leqslant n$ show $\Theta_n(i,j,-x)$ is the inverse of $\Theta_n(i,j,x)$ in the group $UT(n,R)$.

(b) Use Part (a) and Theorem 3.1.9 to get an alternate proof of the existence of inverses in $UT(n,R)$.

I.2. (a) Let $H = \{A \mid A \in UT(n,R) \text{ and } A_{i+1,i} = A_{j+1,j} \text{ for } i,j < n\}$. Show $H \lhd UT(n,R)$.

(b) Show $UT(n,R)' \subseteq H$ and $H/UT(n,R)' \cong R$.

(c) Find card(H) in terms of card(R) and n when R is finite.

I.3. Suppose R and S are commutative rings with identity and $g:R \to S$ is an identity-preserving homomorphism. Define $\phi_g:UT(n,R) \to UT(n,S)$ by $\phi_g(A)_{ij} = g(A_{ij})$ for $1 \leqslant i,j \leqslant n$. Show

(a) ϕ_g is a homomorphism.

(b) ϕ_g is injective if and only if g is injective.

(c) ϕ_g is surjective if and only if g is surjective.

I.4. If J is an ideal of R, let $C(J) = \{A \mid A \in UT(n,R)$ and $A_{ij} \in J$ for all $j < i\}$. Show
(a) $C(J) \lhd UT(n,R)$.
(b) $UT(n,R)/C(J) \cong UT(n,R/J)$.

I.5. (a) Let $K = \{A \mid A \in UT(n,R)$ and $A_{21} = 0\}$. Show K is a subgroup of $UT(n,R)$.
(b) Find card(K) in terms of card(R) and n when R is finite.

I.6. For $1 \leqslant i \leqslant n - 1$, show $Z_i(UT(n,R))/Z_{i-1}(UT(n,R)) \cong R \times R \times \ldots \times R$ (i times).

I.7. Suppose the additive group of R is torsion-free. Show $UT(n,R)$ is torsion-free.

I.8. (a) Show central extensions of divisible groups by divisible groups are divisible.
(b) Suppose the additive group of R is divisible. Show $UT(n,R)$ is divisible.

II $UT(\infty,R)$

II.1. Suppose the additive group of R is torsion-free. Show $UT(\infty,R)$ is torsion-free.

II.2. Suppose the additive group of R is divisible. Show $UT(\infty,R)$ is divisible.

III Triangular Groups

III.1. Find card($T(n,F)$) in terms of card(F) and n when F is a finite field.

III.2. For any field F, show $UT(n,F) \lhd T(n,F)$.

III.3. If R is a commutative ring with identity, let $T(n,R)$ be the set of n by n triangular matrices M, with entries in R, such that $1 \leqslant i \leqslant n$ implies M_{ii} is a unit of R. Show $T(n,R)$ is a group and show that when F is a field this definition of $T(n,F)$ and the one given in the text coincide.

IV Special Linear and Projective Special Linear Groups

IV.1. If F is a field of characteristic 0, show $SL(2,F)$ is semidivisible.

IV.2. Show
(a) The additive group of R can be embedded in $SL(2,R)$.
(b) The multiplicative group of units of R can be embedded in $SL(2,R)$.

IV.3. Show the following commutator formula for $SL(2,F)$ when F is a field. For all $t \in F$ and all $y \in F \backslash \{0\}$,

$$\begin{pmatrix} 1 & t(y^2 - 1) \\ 0 & 1 \end{pmatrix} = A^{-1}B^{-1}AB$$

where

$$A = \begin{pmatrix} 1 & t \\ 0 & 1 \end{pmatrix} \quad \text{and} \quad B = \begin{pmatrix} 1/y & 0 \\ 0 & y \end{pmatrix}.$$

IV.4. If p is a prime such that $p \equiv 3 \bmod 8$ or $p \equiv 5 \bmod 8$, show
(a) 8 does not divide card($PSL(2,p)$).

(b) The 2-Sylow subgroups of $PSL(2,p)$ are abelian.

IV.5. Suppose F is a field such that $\text{char}(F) \neq 2$, and let $[M] \in PSL(2,F)$. Show $o([M]) = 2$ if and only if $\text{tr}(M) = 0$.

IV.6. Suppose F is a finite field of cardinality q such that $\text{char}(F) \neq 2$. Show the number of elements of order 2 in $PSL(2,F)$ is either $(q^2 + q)/2$ or $(q^2 - q)/2$ depending on whether or not F has a square root of -1.

IV.7. Suppose F is a field such that $\text{char}(F) \neq 2$, and let $M \in SL(2,F)$. Show $\text{tr}(M^2) = \text{tr}(M)^2 - 2$.

IV.8. Suppose F is a field such that $\text{char}(F) \neq 2$.

(a) Show $[M] \in PSL(2,F)$ has order 4 if and only if $\text{tr}(M) = \pm\sqrt{2}$.

(b) If F is finite of cardinality q, show the number of elements of order 4 in $PSL(2,F)$ is

$$\begin{cases} 0 & \text{if } \sqrt{2} \notin F, \\ q^2 + q & \text{if } \sqrt{2} \in F \text{ and } \sqrt{-2} \in F, \\ q^2 - q & \text{if } \sqrt{2} \in F \text{ and } \sqrt{-2} \notin F. \end{cases}$$

IV.9. Suppose F is a field such that $\text{char}(F) \neq 2, 3$.

(a) Show $[M] \in PSL(2,F)$ has order 6 if and only if $\text{tr}(M) = \pm\sqrt{3}$.

(b) If F is finite of cardinality q, show the number of elements of order 6 in $PSL(2,F)$ is

$$\begin{cases} 0 & \text{if } \sqrt{3} \notin F, \\ q^2 + q & \text{if } \sqrt{3} \in F \text{ and } \sqrt{-1} \in F, \\ q^2 - q & \text{if } \sqrt{3} \in F \text{ and } \sqrt{-1} \notin F. \end{cases}$$

MISCELLAENOUS EXERCISES

NOTATION Let R be a commutative ring with identity. Then $GL(2,R)$ will denote the set of 2 by 2 matrices M, with entries in R, such that $\det(M)$ is a unit of R.

1. Show $GL(2,R)$ is a group and give an explicit formula for inverses. This group is called the **general linear group** (of 2 by 2 matrices with entries in R).
2. Show
   (a) $SL(2,R) \triangleleft GL(2,R)$.
   (b) $GL(2,R)/SL(2,R)$ is isomorphic to the (multiplicative) group of units of R.
   (c) If F is a finite field, then $\text{card}(GL(2,F)) = q(q-1)^2(q+1)$ where q is the cardinality of F.

DEFINITION AND NOTATION If A is a 2 by 2 matrix, then A^t is the **transpose** of A if and only if $(A^t)_{ij} = A_{ji}$ for $1 \leqslant i,j \leqslant 2$.

3. (a) Let $A, B \in GL(2,R)$. Show $(AB)^t = B^t A^t$.
   (b) Let $A \in GL(2,R)$. Show $(A^t)^{-1} = (A^{-1})^t$.
4. Let $A \in GL(2,R)$ and let $H_A = \{M \mid M \in GL(2,R) \text{ and } M^t AM = A\}$. Show H_A is a subgroup of $GL(2,R)$.
5. Show
   (a) $Z(GL(2,R)) = \{\left(\begin{smallmatrix} k & 0 \\ 0 & k \end{smallmatrix}\right) \mid k \text{ is a unit of } R\}$.
   (b) $SL(2,R) \cap Z(GL(2,R)) = Z(SL(2,R))$.

DEFINITION AND NOTATION Let F be a field. Then $PGL(2,F)$ is the **projective general linear group** (of 2 by 2 matrices with entries in F) if and only if $PGL(2,F) = GL(2,F)/Z(GL(2,F))$.

6. Let F be a field such that $\text{char}(F) \neq 2$. Let $H = SL(2,F)Z(GL(2,F))$. Show
   (a) $H/Z(GL(2,F)) \cong PSL(2,F)$.
   (b) $PGL(2,F)$ is an extension of (a group isomorphic to) $PSL(2,F)$ by $Z(2)$.
   (c) $H = \{A \mid A \in GL(2,F) \text{ and } \det(A) \text{ is a square in } F\}$.

SUGGESTIONS FOR FURTHER READING

The groups $GL(m,F)$, $SL(m,F)$, and $PSL(m,F)$ where $3 \leqslant m$ are discussed in Rotman [31] and simplicity is proven for the latter class. Reference Newman [28] is concerned with various aspects of the theory of matrix groups, and Dickson [8] concerns finite matrix groups.

PART TWO
EXAMPLES

In the three chapters that follow, various examples of finite, infinite abelian, and infinite nonabelian groups will be presented and their properties examined.

In general the properties of a group G we will be most concerned with are simply the propositions that G does or does not belong to a particular class of groups $\mathcal{H}$. While the classes of groups to be considered will vary from chapter to chapter, certain classes will be worthy of consideration in all three chapters as they are meaningful in all three branches of group theory. Most notable of these are the "indecomposability" classes which give some sort of an indication whether or not a group is "fundamental" or has been built up in some fashion from other, more fundamental groups. The indecomposability classes to be considered are the classes of *indecomposable* groups, *subdirectly irreducible* groups, and *simple* groups. (Note the following inclusions: Subdirectly irreducible groups are indecomposable by Theorem A.10.10, and it is easily seen that simple groups are subdirectly irreducible.)

Though not as important as the question of which indecomposability classes a group belongs to, a second area of investigation which is meaningful in all three branches of group theory is the question of the existence and behavior of maximal subgroups. A number of conditions on groups pertaining to maximal subgroups will be considered. Among these are the classes of coatomic groups, MN groups, groups with no maximal subgroups, groups with a unique maximal subgroup, and groups in which maximal subgroups have prime-power index.

Most of the examples will be followed by exercises and notes. The exercises will be concerned with further properties the group may or may not possess, alternate constructions, automorphisms of the group, and characterizations.

The idea of a characterization is as follows. After noting that a group G has certain properties, it is natural to ask whether or not G is the only group in which all of these properties hold simultaneously. In considerations of this type, isomorphic copies of G will not be counted as distinct from G. Thus the questions becomes, if a group G belongs to classes $\mathcal{H}_1, \mathcal{H}_2, \ldots, \mathcal{H}_n$, can it be shown that $A \in \mathcal{H}_1 \cap \mathcal{H}_2 \cap \ldots \cap \mathcal{H}_n$ implies $A \cong G$? A result of this type is called a **characterization** of G.

The notes following each example will be used for a variety of purposes. First the notes to a group G will attempt to point out what is so important about G that merits its inclusion in the book. In many cases, however, this will be done implicitly by listing the uses of G as a counterexample. The notes will also state and reference proofs of properties or characterizations of G considered too deep for inclusion in the text. A third function of the notes will be to give comparative references to other examples in the book with properties similar to those of G, and cross-reference further results on G found in different sections.

When G has a commonly accepted name, this will be given at the beginning of the section. Many of these names (such as *the tetrahedral group*) are geometric in origin. They owe their existence to the fact that the group can be constructed as a group of some kind of geometric transformations of some geometric object. No attempt will be made to justify the name by describing the object or its transformations.

4

Finite Groups

Groups in this chapter are arranged in a natural order, namely that of increasing cardinality. The reader, however, should keep in mind a second and more important order; the progression in stages away from the concept of commutativity. These stages are the group classes *abelian, nilpotent, supersolvable*, solvable,* and *arbitrary,* each class being a subclass of its successor.

Outside of some of the exercises, abelian groups are not examined in this chapter. The famous basis theorem of Frobenius and Stickelberger (see Rotman [31], Theorem 4.6) states that a finite abelian group is isomorphic to the direct product of a finite number of cyclic groups of prime-power cardinalities. As these cyclic groups are of a very uncomplicated structure, so too are finite abelian groups. Consequently (excepting the aforementioned exercises) finite abelian groups will be dispensed with in this chapter.

Four examples of groups that are nilpotent but not abelian are presented in the text and several others in the exercises. All of these groups are p-groups, and indeed another structure theorem (see Rotman [31], Theorem 6.26) constrains a finite nilpotent group to being (isomorphic to) a direct product of a finite number of p-groups, various primes p allowed.

Two examples of groups that are supersolvable but not nilpotent are presented (Examples 4.2 and 4.5).

Next come three examples of solvable groups that are not supersolvable (Examples 4.6, 4.9, and 4.10). In examining these groups particular attention will be

---

*In this chapter the word *supersolvable* will be used to mean finite and hypercyclic. It is not difficult to show that supersolvable groups are solvable.

paid to determining whether or not they belong to the following classes. All finite supersolvable groups belong to each of these classes, but each class also has finite groups in it that are not supersolvable. Thus the classes give a measure of how close a finite nonsupersolvable group comes to being supersolvable.

(a) The class of converse lagrange groups. See Deskins [7], Theorem 1, for a proof that finite supersolvable groups are converse lagrange. (This theorem actually proves more, namely that a finite group is supersolvable if and only if each subgroup is converse lagrange.)
(b) The class of groups G such that G' is nilpotent. See Scott [33], Theorem 7.2.13 for a proof that supersolvable groups belong to this class.
(c) The class of groups G such that minimal normal subgroups of G are cyclic. (A normal subgroup H of G is **minimal normal** if and only if $B \lhd G$ and $\{1\} \subseteq B \subseteq H$ imply $B = \{1\}$ or $B = H$.) That all supersolvable groups belong to this class follows immediately from Theorem A.7.3.

Finally four examples of nonsolvable groups are presented (Examples 4.11–4.14). All four of these groups happen to be simple, while some examples of nonsimple, nonsolvable groups are treated in the exercises. With regards to finite nonsolvable groups, we will be interested in knowing whether or not these groups belong to the following classes. All finite solvable groups belong to each of these classes, but each class also has finite groups in it that are not solvable. Thus the classes give a measure of how close a finite nonsolvable group comes to being solvable.

(a) The class of finite groups satisfying extended Sylow properties $E(\pi)$, $C(\pi)$, and $D(\pi)$ for various sets of primes π. See Scott [33], Theorem 9.3.10 for a proof that finite solvable groups satisfy extended Sylow property $D(\pi)$ (and hence $E(\pi)$ and $C(\pi)$ also) for any set of primes π.
(b) The class of finite groups G such that maximal subgroups of G have prime-power index (different primes allowed).
(c) The class of finite groups in which Carter subgroups exist. See Schenkman [32], Theorem VII.4.a for a proof that finite solvable groups belong to this class.

It is interesting to note that while the examples 4.11–4.14 all belong to the same family of simple groups (they are all $PSL(2,F)$ for various fields F) they have such diverse structure as measured by classes (a), (b), and (c) of the last list. Two additional classes will demonstrate the diversity of Examples 4.11–4.14 even further; the class of minimal simple groups, and the class of groups in which all nontrivial elements have prime-power order (different primes allowed).

This chapter (compared with the two that follow) is rich is characterizations. Most of these depend on the cardinality of the group being characterized. Those that do not, while being spectacular results, are in many cases much too difficult to be included as exercises. Some will be listed in the notes following their groups.

Example 4.1 (The Dihedral Groups)

For $n \in N$ such that $3 \leqslant n$, let $D_n = G(X, Y \mid X^n = 1, Y^2 = 1, (XY)^2 = 1)$. (See Note (1).)

RESULT 4.1.1
(a) $D_n \cong Dih(Z(n))$.
(b) $\text{card}(D_n) = 2n$.
(c) $\langle x \rangle \lhd D_n$.
(d) *Every element of* D_n *can be expressed as* $x^i y^e$ *where* $0 \leqslant i \leqslant n-1$ *and* $e =$ 0 *or* 1.
(e) $o(x) = n$.

Let $H = \langle x \rangle$. Clearly $xxx^{-1} \in H$. From the third relation $yxy = x^{-1}$. But $y = y^{-1}$ by the second relation, so $yxy^{-1} = x^{-1} \in H$. Thus $H \lhd D_n$ by Theorem A.10.3, proving Part (c). By Theorem A.10.4, $D_n/H = \langle Hx, Hy \rangle$. But $x \in H$ so $Hx = 1$ and thus $D_n/H = \langle Hy \rangle$. This and $(Hy)^2 = 1$ (because of the second relation) imply $\text{card}(D_n/H) \leqslant 2$. Also $\text{card}(H) \leqslant n$ by the first relation. Consequently $\text{card}(D_n) = \text{card}(D_n/H) \cdot \text{card}(H) \leqslant 2n$. Next note that $Dih(Z(n)) = Z(n) \times_g Z(2)$ which implies $\text{card}(Dih(Z(n))) = 2n$. Thus $\text{card}(D_n) \leqslant 2n \leqslant \text{card}(Dih(Z(n)))$.

Let $a = (1,0)$ and $b = (0,1)$ in $Dih(Z(n))$. It is claimed that $a^m = (m,0)$ for all positive integers m. This is shown by induction, noting the induction begins at $m = 1$. Assuming the induction hypothesis $a^{m-1} = (m-1,0)$, then $a^m = aa^{m-1} = (1,0)(m-1,0) = (1+g_0(m-1),0+0) = (1+m-1,0) = (m,0)$ as required. Now it can be shown that a and b generate $Dih(Z(n))$. Let $(i,j) \in Dih(Z(n))$. Then $(i,j) = (i,0)(0,j)$, so by the claim, $(i,j) = (1,0)^i(0,j)$ which equals a^i if $j = 0$ and which equals $a^i b$ if $j = 1$. Note also that $0 \leqslant i \leqslant n-1$ since $i \in Z(n)$.

Define the function $f: \{X, Y\} \to \{a, b\}$ by $f(X) = a$ and $f(Y) = b$. In order to apply Theorem 2.2.6, it must be shown that the three relations of D_n are "satisfied" in $Dih(Z(n))$. Thus it must be shown that $a^n = 1$, $b^2 = 1$, and $(ab)^2 = 1$. This is done as follows: by the previous paragraph, $a^n = (n,0) = (0,0) = 1$; $b^2 = (0,1)(0,1) = (0 + g_1(0), 1+1) = (0,0) = 1$; $ab = (1,0)(0,1) = (1+g_0(0),0+1) = (1,1)$ so $(ab)^2 = (1,1)(1,1) = (1+g_1(1),1+1) = (1-1,0) = (0,0) = 1$. Theorem 2.2.6 may now be applied to yield an isomorphism $d: D_n \to Dih(Z(n))$ such that $d(x) = a$ and $d(y) = b$. Theorem 2.2.6 further states $\text{card}(D_n) = 2n$. This proves Parts (a) and (b).

To show Part (d), let $w \in D_n$. Then $d(w) \in Dih(Z(n))$ so as seen in the second paragraph, $d(w)$ is either a^i or $a^i b$ where $0 \leqslant i \leqslant n-1$. Thus $d(w)$ is either $d(x^i)$ or $d(x^i y)$. Injectivity of d now yields $w = x^i$ or $x^i y$ proving Part (d).

Since $a^n = 1$ and $a^j \neq 1$ for $1 \leqslant j \leqslant n-1$ (by the second paragraph $a^j = (j,0) \neq (0,0)$), the order of a is n. But d is an isomorphism and $d(x) = a$ so $o(x) = n$ proving Part (e).

RESULT 4.1.2 *If* $0 \leqslant i \leqslant n-1$, *then*
(a) $x^i y = yx^{n-i}$.

(b) $o(x^iy) = 2$.
The third relation implies $yxy = x^{-1}$. Thus $x^{-i} = (yxy)^i = (yxy^{-1})^i$ ($y = y^{-1}$ because $y^2 = 1$) $= yx^iy^{-1} = yx^iy$. Consequently $x^iy = y^{-1}x^{-i} = yx^{n-i}$ since $o(x) = n$ by Result 4.1.1, Part (e). This proves Part (a).

By Part (a), $(x^iy)^2 = x^iyx^iy = x^i(x^{n-i}y)y = x^ny^2 = 1$. Thus x^iy has order 2 or 1. But it cannot have order 1 for then $x^i = y$ implying $y \in \langle x\rangle$ and hence $D_n = \langle x\rangle$ so $\text{card}(D_n) = n$ contradicting Result 4.1.1, Part (b).

RESULT 4.1.3

$$Z(D_n) = \begin{cases} \{1, x^{n/2}\} & \textit{if } n \textit{ is even,} \\ \{1\} & \textit{if } n \textit{ is odd.} \end{cases}$$

If n is odd, then $Z(n)$ has no elements of order 2. That is, $Z(n)[2]$ is trivial. Thus $Z(D_n)$ is trivial by Result 4.1.1, Part (a) and Corollary 3 to Theorem 1.2.5.

Assume now that n is even, say $n = 2k$. It is first claimed that $Z(n)[2]$ has just the two elements 0 and k. Let $t \in Z(n)[2]$. Thus $2t = 0$ meaning $2t \equiv 0 \bmod n$ and so $n \mid 2t$. That is, $2k \mid 2t$ and so $k \mid t$. Hence $kc = t$ for some integer c. If c is even, say $c = 2e$, then $t = 2ke = ne$ so $n \mid t$ and hence $t = 0$ in $Z(n)$. If c is odd, say $c = 2e + 1$, then $t - k = kc - k = k(c - 1) = k2e = ne$ so $n \mid t - k$ and hence $t = k$ in $Z(n)$ establishing the claim. Thus $Z(n)[2]$ has cardinality 2, hence (again by Result 4.1.1, Part (a) and Corollary 3 to Theorem 1.2.5), so does $Z(D_n)$.

Thus to complete the proof it suffices to show $x^{n/2} \in Z(D_n)$. Let w be any element of D_n. Then $w = x^i$ or $w = x^iy$ for some i by Result 4.1.1, Part (d). In the first case $x^{n/2}$ and w certainly commute. In the second case $(x^iy)x^{n/2} = x^i(yx^{n/2}) = x^i(x^{n-(n/2)}y)$ (Result 4.1.2) $= x^{i-(n/2)}y = x^{n/2}(x^iy)$.

RESULT 4.1.4 $D_{2n}/Z(D_{2n}) \cong D_n$.
First consider the function $e{:}Z(n) \to 2Z(2n)$ defined by $e(i) = 2i$. It is easily seen that e is an isomorphism (injectivity proof: if $e(i) = 0$, then $2i = 0$ in $Z(2n)$ which means $2n \mid 2i$. Hence $n \mid i$ so $i = 0$ in $Z(n)$).

Now by Result 4.1.1, Part (a), $D_{2n}/Z(D_{2n}) \cong Dih(Z(2n))/Z(Dih(Z(2n))) = Dih(Z(2n))/Z(2n)[2]^*$ (Corollary 3 to Theorem 1.2.5) $\cong Dih(Z(2n)/Z(2n)[2])$ (Theorem 1.2.9) $\cong Dih(2Z(2n))$ (Theorem A.4.1) $\cong Dih(Z(n))$ (because of the isomorphism e) $\cong D_n$ by Result 4.1.1 again.

RESULT 4.1.5 *For* $2 \leqslant n$, D_{2^n} *is class-n nilpotent.*
This will be shown by induction on n. To show that induction begins let $G = D_4$. Then $\text{card}(Z(G)) = 2$ by Result 4.1.3 so $\text{card}(G/Z(G)) = 8/2 = 4$ and so $G/Z(G)$ is abelian by Theorem A.1.2. Therefore $G' \subseteq Z(G)$ by Theorem A.6.4 which further implies $G = Z_2(G)$. This and $G \neq Z(G)$ (cardinality difference) imply G is class-2 nilpotent, so induction begins.

Assume now the induction hypothesis, $D_{2^{n-1}}$ is class-$(n - 1)$ nilpotent. Then $D_{2^n}/Z(D_{2^n})$ is class-$(n - 1)$ nilpotent because it is isomorphic to $D_{2^{n-1}}$ by Result 4.1.4. Then D_{2^n} is nilpotent by Theorem A.5.3, and furthermore must be class-n nilpotent by Theorem A.5.5.

RESULT 4.1.6 *For $2 \leqslant n$, D_{2^n} is subdirectly irreducible and*

$$H(D_{2^n}) = Z(D_{2^n}) = \{1, x^{2^{n-1}}\}.$$

For notational convenience let G denote D_{2^n}. By Result 4.1.5, G is nilpotent, and by Result 4.1.3,

$$Z(G) = \{1, x^{2^{n-1}}\}.$$

Hence $Z(G) \cong Z(2)$ so G is subdirectly irreducible by Theorem A.5.10. Thus $H(G)$ is nontrivial. Also $H(G) \subseteq Z(G)$ because $Z(G)$ is a nontrivial normal subgroup of G. Then $\{1\} \neq H(G) \subseteq Z(G)$ implies $H(G) = Z(G)$ because the center has only two elements.

RESULT 4.1.7 *For $2 \leqslant n$, D_{2^n} is indecomposable.*
By Result 4.1.6 and Theorem A.10.10.

RESULT 4.1.8 *For $3 \leqslant n$,*
(a) *D_n is supersolvable.*
(b) *D_n is solvable.*
Part (a) by Theorem 1.2.11, Part (b) by Theorem 1.2.10.

EXERCISES

1. If $n \mid m$, show D_n can be embedded in D_m.
2. If $n \mid m$, show there is a normal subgroup K of D_m such that $D_m/K \cong D_n$.
3. Show $D_n{}' = \langle x^2 \rangle$.

NOTES

1. Some authors use the notation D_{2n} instead of D_n.
2 Particular dihedral groups are examined in Examples 4.2 and 4.3, and in Miscellaneous Exercises IV.1 and IV.2.

□□□□□

Example 4.2

Let $G = D_3 = G(X, Y \mid X^3 = 1, Y^2 = 1, (XY)^2 = 1)$.

RESULT 4.2.1 $\text{card}(G) = 6$.
By Result 4.1.1, Part (b).

RESULT 4.2.2 *The subgroups of G are $\{1\}$, $\langle x \rangle$, $\langle y \rangle$, $\langle xy \rangle$, $\langle x^2y \rangle$, and G.*
Clearly all entries in the above list are subgroups. Moreover since $\langle x^2 \rangle = \langle x \rangle$ (x^2

belongs to $\langle x \rangle$ and $o(x^2) = 3 = o(x)$), all cyclic subgroups of G have been accounted for in the list.

Now let B be any subgroup of G. If $B = G$, then B appears in the list, so assume B is proper. Then card$(B) = 3, 2$, or 1 by Lagrange's theorem. In any case B is cyclic, so has been accounted for.

The subgroups of G are illustrated below. Note that the subgroup lattice L(G) is self-dual.

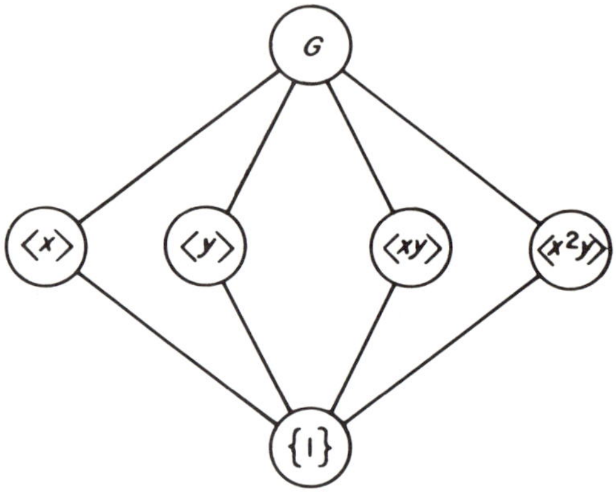

RESULT 4.2.3 *The normal subgroups of G are* $\{1\}$, $\langle x \rangle$, *and G.*
By Result 4.2.2, $\langle x \rangle$ is the only subgroup of cardinality 3 ($o(y) = o(xy) = o(x^2y) = 2$ by Result 4.1.2, Part (b) so card$(\langle y \rangle)$ = card$(\langle xy \rangle)$ = card$(\langle x^2y \rangle) = 2$) hence $\langle x \rangle$ is the only 3-Sylow subgroup of G so must be normal (Theorem A.10.15, Part (b)).

Next note that $n_2(G) = 3$ because $\langle y \rangle$, $\langle xy \rangle$, and $\langle x^2y \rangle$ are all 2-Sylow subgroups of G. Therefore the normalizer of each 2-Sylow subgroup of G has index 3 in G by Theorem A.10.15, Part (a). Thus for $S = \langle y \rangle, \langle xy \rangle$, or $\langle x^2y \rangle$, $N_G(S) = S$ and hence S is not a normal subgroup.

RESULT 4.2.4 *G is subdirectly irreducible.*
By Result 4.2.3, $H(G) = \langle x \rangle \cap G = \langle x \rangle$, so $H(G)$ is nontrivial and hence G is subdirectly irreducible.

RESULT 4.2.5 *G is indecomposable.*
By Result 4.2.4 and Theorem A.10.10.

RESULT 4.2.6 $Z(G) = \{1\}$.
By Result 4.1.3.

RESULT 4.2.7 *G is not nilpotent.*
By Result 4.2.6 and the Corollary to Theorem A.5.6.

RESULT 4.2.8 *G is supersolvable.*
By Result 4.1.8.

RESULT 4.2.9 *G is solvable.*
By Result 4.1.8.

RESULT 4.2.10 *G is not an* MN *group.*
By Result 4.2.2 there exists no subgroup B meeting the conditions $\langle y \rangle \subset B \subset G$. Thus $\langle y \rangle$ is a maximal subgroup of G. But by Result 4.2.3 $\langle y \rangle$ is not normal.

EXERCISES

1. Show all normal subgroups of G are characteristic.
2. Identify G' and $\Phi(G)$.
3. Show $\langle y \rangle$, $\langle xy \rangle$, and $\langle x^2 y \rangle$ are Carter sugroups of G.
4. Describe the decomposition of G into conjugacy classes and write the class equation.
5. Let A be a nonabelian group of cardinality 6. Show $A \cong G$.
6. Show $S_3 \cong G$.
7. Show $SL(2,2) \cong G$.
8. Show G is a complete group.
9. If A is any group and B is a complete normal subgroup of A, show $A = B \oplus C_A(B)$. Conclude that if G can be embedded in any group A as a normal subgroup, then the image of this embedding is a direct factor of A.
10. If A is a nonperfect complete group, show there exists no group C such that $C' \cong A$. Conclude there exists no group C such that $C' \cong G$.
11. Show $\mathrm{L}(G) \cong \mathrm{L}(Z(3) \times Z(3))$.

NOTES

1. Most commonly known as S_3 (see Exercise 6), G is one of the most well-known and also one of the most important finite groups.

   As a counterexample G shows that the class of nilpotent groups is not extension closed, and even that extensions of abelian groups by abelian groups need not be nilpotent. This is because $\langle x \rangle$ and $G/\langle x \rangle$ are both abelian while G is not nilpotent.
2. The MN property is actually an equivalent to nilpotency for finite groups (see Scott [33], Theorem 6.4.14) so Results 4.2.7 and 4.2.10 are more closely related than they appear to be.
3. A characterization of G is given in Exercise 5. On the other hand, the purpose of Exercise 11 is to show that G cannot be characterized solely by its subgroup lattice. Note Exercise 11 also shows that there exist no characterizations, solely in terms of subgroup lattices, for the classes of abelian or nilpotent groups.

□□□□□

Example 4.3

Let $G = D_4 = G(X, Y \mid X^4 = 1, Y^2 = 1, (XY)^2 = 1)$.

RESULT 4.3.1 $\text{card}(G) = 8$.
By Result 4.1.1, Part (b).

RESULT 4.3.2 *The subgroups of G are* $\{1\}$, $\langle x^2\rangle$, $\langle y\rangle$, $\langle xy\rangle$, $\langle x^2y\rangle$, $\langle x^3y\rangle$, $\{1,x^2,y,x^2y\}$, $\{1,x^2,xy,x^3y\}$, $\langle x\rangle$, *and G.*
First note that $\langle x^2\rangle = Z(G)$ by Result 4.1.3 so $\langle x^2\rangle \lhd G$. Then $\langle x^2\rangle\langle y\rangle$ and $\langle x^2\rangle\langle xy\rangle$ are both subgroups of G. That is, $\{1,x^2,y,x^2y\}$ and $\{1,x^2,xy,x^3y\}$ are both subgroups. Consequently every entry in the above list is in fact a subgroup of G.

It is clear that all cyclic subgroups have been accounted for in the above list (the only possible omission is $\langle x^3\rangle$ but $x^3 \in \langle x\rangle$ and $o(x^3) = 4 = o(x)$ so $\langle x^3\rangle = \langle x\rangle$). Next suppose B is any subgroup of G. It must be shown that B appears in the list. If $B = \{1\}$ or $B = G$, B certainly appears so assume B is both nontrivial and proper. Therefore by Lagrange's theorem $\text{card}(B) = 2$ or 4. In the first case B is cyclic and hence has been listed. In the second case $[G{:}B] = 8/4 = 2$ so B is normal by Theorem A.10.2. Thus $G/B \cong Z(2)$ and hence $(Bx)^2 = 1$ so $x^2 \in B$. Since $\text{card}(\langle x^2\rangle) = 2$, $B\setminus\langle x^2\rangle$ is nonempty. Let $b \in B\setminus\langle x^2\rangle$. If $b = x$, then $\langle x\rangle \subseteq B$. But $\text{card}(\langle x\rangle) = 4$ so $\langle x\rangle = B$ and hence B is in the list. If $b = y$, then $x^2 \in B$ and $y \in B$ imply $x^2y \in B$. Hence $\{1,x^2,y,x^2y\} \subseteq B$ so again by a cardinality consideration $\{1,x^2,y,x^2y\} = B$ and B has been listed. If $b = x^3$, then $B = \langle x\rangle$. If $b = xy$ or $b = x^3y$, then $B = \{1,x^2,xy,x^3y\}$. If $b = x^2y$, then $B = \{1,x^2, y,x^2y\}$. So in all cases B is in the list, proving the result.

The subgroups of G are illustrated below.

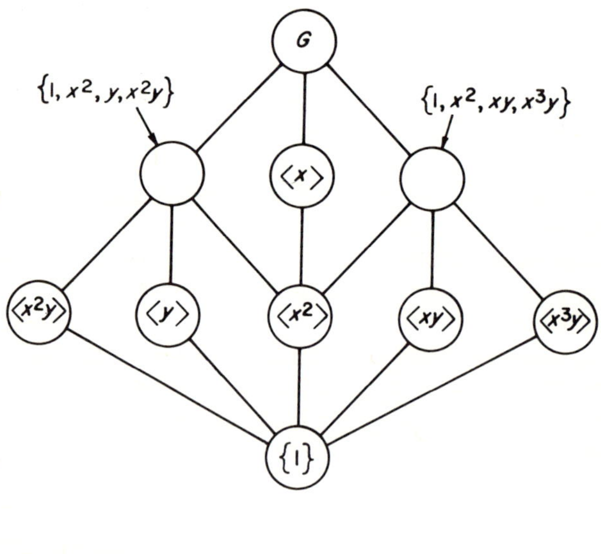

RESULT 4.3.3 *The normal subgroups of G are* $\{1\}$, $\langle x^2\rangle$, $\langle x\rangle$, $\{1,x^2,y,x^2y\}$, $\{1,x^2,xy,x^3y\}$, *and G.*
Clearly the subgroups $\{1\}$ and G are normal. The subgroups $\langle x\rangle$, $\{1,x^2,y,x^2y\}$, and $\{1,x^2,xy,x^3y\}$ all have cardinality 4, hence index 2. Thus they are normal in G by Theorem A.10.2. Also then, $\langle x^2\rangle$ is the intersection of the normal subgroups $\{1,x^2,y,x^2y\}$ and $\{1,x^2,xy,x^3y\}$ so $\langle x^2\rangle \triangleleft G$.

To complete the proof it must be shown that all other subgroups of G are nonnormal. By Result 4.3.2, the subgroups not listed above are the four subgroups $\langle x^k y\rangle$ where $k = 0, 1, 2, 3$. By Result 4.1.2, $x^{-1}(x^k y)x = x^{-1}x^k x^3 y = x^{k+2}y$. But $\langle x^k y\rangle = \{1,x^k y\}$ since $o(x^k y) = 2$. It follows then that $x^{k+2}y \notin \langle x^k y\rangle$ because $x^{k+2}y = 1$ would imply $x^{k+2} = y$ and hence that $y \in \langle x\rangle$ so $G = \langle x,y\rangle = \langle x\rangle$, while $x^{k+2}y = x^k y$ would imply $x^2 = 1$. Thus $\langle x^k y\rangle$ is nonnormal.

RESULT 4.3.4 *G is nilpotent.*
By Result 4.1.5.

RESULT 4.3.5 *G is subdirectly irreducible.*
By Result 4.1.6.

RESULT 4.3.6 *G is indecomposable.*
By Result 4.1.7.

RESULT 4.3.7
(a) *G is supersolvable.*
(b) *G is solvable.*
Both parts by Result 4.1.8. Alternate proof: Part (a) by Result 4.3.4 and the Corollary to Theorem A.7.2, Part (b) by Result 4.3.4 and Theorem A.6.7.

EXERCISES

1. Identify G' and $\Phi(G)$.
2. Describe the decomposition of G into conjugacy classes and write the class equation.
3. Show
   (a) There is an automorphism d such that $d(x) = x$ and $d(y) = xy$.
   (b) $\text{Aut}(G) = \{1,d,d^2,d^3,g_y,d\circ g_y,d^2\circ g_y,d^3\circ g_y\}$ where g_y is the inner automorphism determined by y.
   (c) $\text{Aut}(G) \cong G$.
4. Show $\{1\}$, $\langle x^2\rangle$, $\langle x\rangle$, and G are the only characteristic subgroups of G.
5. Let A be a group with the following properties:
   (a) $\text{card}(A) = 8$.
   (b) A is nonabelian.
   (c) A has 2 or more elements of order 2.
   Show $A \cong G$.

6. Show $G \cong Z(2)$ wr $Z(2)$.
7. Show $G \cong UT(3,2)$.
8. Let A be a group such that $L(A) \cong L(G)$. Show $A \cong G$.

NOTES

1. As a counterexample G shows
   (a) Subgroups of subdirectly irreducible groups (even nilpotent subdirectly irreducible groups) need not be subdirectly irreducible. This is because of Result 4.3.5 and the fact that the subgroup $\{1,x^2,y,x^2y\}$ is clearly not subdirectly irreducible.
   (b) Quotient groups of subdirectly irreducible groups (even nilpotent subdirectly irreducible groups) need not be subdirectly irreducible. This is because of Result 4.3.5 and the fact that $G/\langle x^2\rangle$ is clearly not subdirectly irreducible.
   (c) Subgroups of indecomposable groups need not be indecomposable ($\{1,x^2,y,x^2y\}$ is clearly decomposable).
   (d) Quotient groups of indecomposable groups need not be indecomposable ($G/\langle x^2\rangle$ is clearly decomposable).

   Both (a) and (b) are in contrast to the abelian case where subgroups and quotient groups of subdirectly irreducible groups must be subdirectly irreducible.
2. Characterizations of G are given in Exercises 5 and 8. Note that Exercise 5 and Exercise 5 of Example 4.4 show that Examples 4.3 and 4.4 are the only (up to isomorphism) nonabelian groups of cardinality 8.
3. See Exercise 9 of Example 4.4.

□□□□□

Example 4.4 (The Quaternion Group)

Let $G = G(X, Y \mid X^4 = 1, X^2Y^{-2} = 1, Y^{-1}XYX = 1)$.

RESULT 4.4.1

(a) *G is isomorphic to the subgroup of $SL(2,3)$ generated by the matrices $a = \begin{pmatrix} 1 & 1 \\ 1 & 2 \end{pmatrix}$ and $b = \begin{pmatrix} 0 & 1 \\ 2 & 0 \end{pmatrix}$.*
(b) $\mathrm{card}(G) = 8$.
(c) *The elements of G are x, x^2, x^3, y, xy, x^2y, x^3y, and* 1.
(d) $o(x^2) = 2$. *All other nontrivial elements have order* 4.

Let $H = \langle x\rangle$. Clearly $xxx^{-1} \in H$. From the third relation, $y^{-1}xy = x^{-1}$. Inverting yields $y^{-1}x^{-1}y = x$, therefore $x^{-1} = yxy^{-1}$ so $yxy^{-1} \in H$. Applying Theorem A.10.3 now shows $H \lhd G$.

By Theorem A.10.4, $G/H = \langle Hx, Hy \rangle$. But $x \in H$ so $Hx = 1$. Thus $G/H = \langle Hy \rangle$. From the second relation, $y^2 = x^2$. Therefore $(Hy)^2 = Hx^2 = 1$ implying that $\text{card}(G/H) \leqslant 2$. From the first relation, $\text{card}(H) \leqslant 4$. But $\text{card}(G) = \text{card}(G/H) \cdot \text{card}(H)$ so $\text{card}(G) \leqslant 8$. Furthermore note that since $\text{card}(G/H) = 1$ or 2 and since $\text{card}(H) = o(x) = 1, 2$, or 4 ($o(x) | 4$ by the first relation), $\text{card}(G) = 1, 2, 4$, or 8.

Consider the matrices a and b in $SL(2,3)$. (It is easily seen that $\det(a) = \det(b) = 1$ in $Z(3)$ so both matrices actually belong to $SL(2,3)$.) Let $T = \langle a, b \rangle$. Define the function $f: \{X, Y\} \to \{a, b\}$ by $f(X) = a$ and $f(Y) = b$. Then the relations of G are "satisfied" in T as seen as follows: $a^2 = \begin{pmatrix} 1 & 1 \\ 1 & 2 \end{pmatrix}\begin{pmatrix} 1 & 1 \\ 1 & 2 \end{pmatrix} = \begin{pmatrix} 2 & 0 \\ 0 & 2 \end{pmatrix}$ so $a^4 = \begin{pmatrix} 2 & 0 \\ 0 & 2 \end{pmatrix}\begin{pmatrix} 2 & 0 \\ 0 & 2 \end{pmatrix} = I$; $b^2 = \begin{pmatrix} 0 & 1 \\ 2 & 0 \end{pmatrix}\begin{pmatrix} 0 & 1 \\ 2 & 0 \end{pmatrix} = \begin{pmatrix} 2 & 0 \\ 0 & 2 \end{pmatrix} = a^2$ so $a^2 b^{-2} = I$; and $b^{-1}aba = \begin{pmatrix} 0 & 2 \\ 1 & 0 \end{pmatrix}\begin{pmatrix} 1 & 1 \\ 1 & 2 \end{pmatrix}\begin{pmatrix} 0 & 1 \\ 2 & 0 \end{pmatrix}\begin{pmatrix} 1 & 1 \\ 1 & 2 \end{pmatrix} = \begin{pmatrix} 2 & 1 \\ 1 & 1 \end{pmatrix}\begin{pmatrix} 1 & 2 \\ 2 & 2 \end{pmatrix} = I$. Theorem 2.2.5 may now be applied to yield existence of a surjective homomorphism $d: G \to T$ such that $d(x) = a$ and $d(y) = b$. Then $G/\ker d \cong T$. This and $\text{card}(G) = 1, 2, 4$, or 8 imply $\text{card}(T) = 1, 2, 4$, or 8. In the first three cases T would be abelian (third case by Theorem A.1.2). This, however, is not true since it is easily seen that $ab \neq ba$. Thus $\text{card}(T) = 8$. This and $\text{card}(G) \leqslant 8$ imply $\ker d = \{1\}$ so d is an isomorphism. This proves Parts (a) and (b).

Now note that $1, x, x^2, x^3, y, xy, x^2y, x^3y$ are elements of G and are eight in number. Part (c) then will be completed if these eight elements can be shown to be distinct. First note that $a^4 = I$ and $a^2 \neq I$ as shown in the previous paragraph. Thus $o(a) = 4$ which in turn implies $o(x) = 4$ since d is an isomorphism. Thus the elements $1, x, x^2, x^3$ are distinct, hence so too are the elements y, xy, x^2y, x^3y (if $x^i y = x^k y$ for $i \neq k$, then $x^i = x^k$). Finally if $x^i = x^k y$ for some i and k, then $x^{i-k} = y$ so $y \in \langle x \rangle$ and hence $G = \langle x \rangle$ implying G, and hence T, is abelian – a contradiction. This completes the proof of Part (c).

To show Part (d) first note that if g is any nontrivial element of G, then $\text{card}(G) = 8$ implies $o(g) = 2, 4$, or 8. But $o(g) = 8$ would imply G is cyclic, hence abelian, a contradiction. Thus all nontrivial elements have order either 2 or 4. Now $(x^2)^2 = x^4 = 1$ so $o(x^2) = 2$. It suffices then to show $t^2 \neq 1$ for $t \neq 1, x^2$.

It is clear that $x^2 \neq 1$ and $(x^3)^2 = x^2$ so $(x^3)^2 \neq 1$. By the second relation, $y^2 = x^2 \neq 1$ and $(x^2y)^2 = (y^3)^2 = y^2 \neq 1$. By the third relation, $xyx = y$, hence $(xy)^2 = y^2 \neq 1$. Finally $(x^3y)^2 = x^3yx^3y = x^2(xyx)x^2y = x^2yx^2y = (x^2y)^2 \neq 1$.

RESULT 4.4.2 *The subgroups of G are:* $\{1\}$, $\langle x^2 \rangle$, $\langle x \rangle$, $\langle y \rangle$, $\langle xy \rangle$, *and G.*
All of the above are clearly subgroups of G. Preliminary to showing all subgroups of G appear in the above list, it will be shown that all cyclic subgroups do. Since $x^3 \in \langle x \rangle$, $\langle x^3 \rangle \subseteq \langle x \rangle$. But both x^3 and x have order 4 by Result 4.4.1, Part (d). Hence $\text{card}(\langle x^3 \rangle) = \text{card}(\langle x \rangle)$ implying $\langle x^3 \rangle = \langle x \rangle$. Since $x^2y = y^2y \in \langle y \rangle$ and since $o(x^2y) = o(y)$, $\langle x^2y \rangle = \langle y \rangle$. Invert the third relation, then premultiply by x and postmultiply by x^{-1} to get $y^{-1}x^{-1}yx^{-1} = 1$. But $x^{-1} = x^3$ so $y^{-1}x^3yx^3 = 1$. Premultiply and postmultiply by y to get $x^3yx^3y = y^2 = x^2$. Thus $(x^3y)^3 = x^2(x^3y) = xy$ implying $xy \in \langle x^3y \rangle$ and hence $\langle x^3y \rangle = \langle xy \rangle$ by a comparison of orders. Thus all cyclic subgroups of G have been accounted for.

Now let B be any subgroup of G. If $B = \{1\}$ or if $B = G$, then B appears in the list so assume B is both proper and nontrivial. By Lagrange's theorem then card(B) = 2 or 4. In either case B has an element of order 2 by Cauchy's theorem. Applying Result 4.4.1, Part (d) yields $x^2 \in B$. If card(B) = 2, then $B = \langle x^2 \rangle$ so B appears in the list. Finally suppose card(B) = 4. Then $\langle x^2 \rangle \subset B$. Let $b \in B \backslash \langle x^2 \rangle$. By Result 4.4.1, Part (d) again, $o(b) = 4$. Thus $B = \langle b \rangle$ so B is cyclic and consequently has been listed

The subgroups of G are illustrated below.

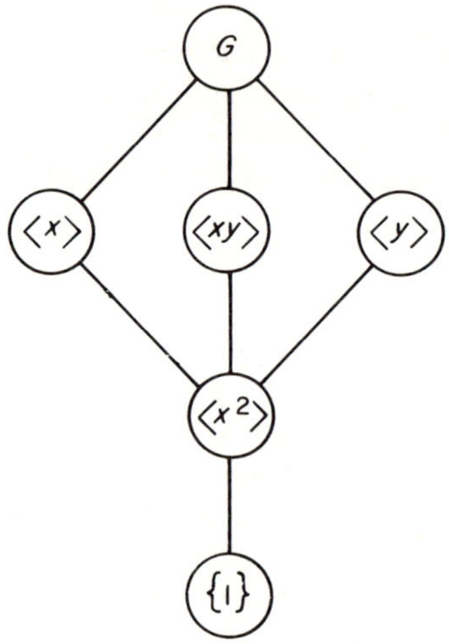

RESULT 4.4.3 *G is hamiltonian.*
The subgroups $\langle x \rangle$, $\langle xy \rangle$, and $\langle y \rangle$ each have cardinality 4 by Result 4.4.1, Part (d). This and Result 4.4.1, Part (b) imply these three subgroups have index 2, hence are normal by Theorem A.10.2. By Result 4.4.2, the remaining proper nontrivial subgroup of G is $\langle x^2 \rangle$. But $\langle x^2 \rangle$ is the intersection of the normal subgroups $\langle x \rangle$ and $\langle y \rangle$, hence is a normal subgroup itself. Thus all subgroups are normal. It remains to show that G is nonabelian. But this is clear for if G is abelian, then the third relation implies $x^2 = 1$ which is false.

RESULT 4.4.4 *G is nilpotent.*
By Result 4.4.1, Part (b) and Theorem A.5.7.

RESULT 4.4.5
(a) *G is solvable.*
(b) *G is supersolvable.*
Part (a) by Result 4.4.4 and Theorem A.6.7. Part (b) by Result 4.4.4 and the Corollary to Theorem A.7.2.

RESULT 4.4.6 $Z(G) = \langle x^2 \rangle$.
From the second relation $x^2 = y^2$ and so x^2 commutes with y. It also commutes with x, hence it must belong to the center (since it commutes with all generators). Thus $\langle x^2 \rangle \subseteq Z(G)$. But if this inclusion were proper, then by Result 4.4.2, $Z(G) = \langle x \rangle$, $\langle xy \rangle$, $\langle y \rangle$, or G. The last case is ruled out because G is nonabelian. In any of the other cases $G/Z(G)$ has cardinality 2, and so must be cyclic. Thus these three cases are ruled out by Theorem A.10.5. Conclude $Z(G) = \langle x^2 \rangle$.

RESULT 4.4.7 *G is subdirectly irreducible.*
By Results 4.4.4 and 4.4.6, and Theorem A.5.10, or by inspection of the subgroup-lattice diagram.

RESULT 4.4.8 *G is indecomposable.*
By Result 4.4.7 and Theorem A.10.10, or by inspection of the subgroup-lattice diagram.

EXERCISES

1. Identify G' and $\Phi(G)$.
2. Describe the decomposition of G into conjugacy classes and write the class equation.
3. Show G has automorphisms h and d such that $h(x) = xy$ and $h(y) = x$; and $d(x) = y$ and $d(y) = x$.
4. Show $\{1\}$, $Z(G)$, and G are the only characteristic subgroups of G.
5. Let A be a group with the following properties:
   (a) $\text{card}(A) = 8$.
   (b) A is nonabelian.
   (c) A has a unique element of order 2.
   Show $A \cong G$.
6. Let A be a hamiltonian group of cardinality 8. Show $A \cong G$.
7. Let A be a group such that $L(A) \cong L(G)$. Show $A \cong G$.
8. Let R be the division ring of quaternions (it makes no difference whether rational quaternions or real quaternions are considered here) and let T be the subgroup of the multiplicative group of R generated by i and j. Show $T \cong G$.
9. Show $N(G) \cong N(D_4)$.

NOTES

1. The group G is often denoted by Q, and is called the quaternion group because of Exercise 8. It is one of the infinite family of groups $G(X, Y \mid X^{2n} = 1, X^nY^{-2} = 1, Y^{-1}XYX = 1)$ which are known as the **dicyclic groups**. Those dicyclic groups in which n is a power of 2 are called **generalized quaternion groups**.
2. Characterizations of G are given in Exercises 5, 6, and 7. The Exercise 5 char-

acterization and Exercise 5 of Example 4.3 show that Examples 4.3 and 4.4 are the only (up to isomorphism) nonabelian groups of cardinality 8.

Note that the Exercise 7 characterization does not depend on cardinality. A fourth characterization, also cardinality-independent, is the following: G is (up to isomorphism) the only indecomposable hamiltonian group. This is an immediate corollary of the theorem: *Every hamiltonian group is isomorphic to a direct product* $G \times A_1 \times A_2$ *where* A_1 *is an abelian group in which every element is of finite and odd order and* A_2 *is an abelian group of exponent* 2. This theorem is proved in Hall [14], Theorem 12.5.4.

Finally the following is a characterization semidependent on cardinality. If A is a finite p-group with (at least) two different cyclic subgroups of index p but only one subgroup of cardinality p, then $A \cong G$. See Zassenhaus [36], page 148 for a proof.

3. The automorphism group of G is described in Example 4.9, Exercise 15.

□□□□□

Example 4.5

Let $G = \mathrm{G}(X, Y \mid X^6 = 1, X^3Y^{-2} = 1, Y^{-1}XYX = 1)$.

RESULT 4.5.1

(a) *G is isomorphic to* $T = Z(3) \times_g Z(4)$ *where* $g{:}Z(4) \to \mathrm{Aut}(Z(3))$ *is defined by* $g_k = \theta^k$, *θ being the automorphism of* $Z(3)$ *defined by* $\theta(t) = -t$. (θ is an automorphism by Theorem 1.2.4 and hence $(Z(4), g, Z(3))$ is a semidirect system by Theorem 1.2.3.)

(b) $\mathrm{card}(G) = 12$.

(c) $G = \{x^k y^e \mid 0 \leqslant k \leqslant 5$ *and* $e = 0$ *or* $1\}$.

Let $H = \langle x \rangle$. The third relation implies $y^{-1}xy = x^{-1}$. Taking inverses yields $x = y^{-1}x^{-1}y$. Hence $yxy^{-1} = x^{-1}$. Thus $H \lhd G$ by Theorem A.10.3. By Theorem A.10.4, $G/H = \langle Hx, Hy \rangle$. But $Hx = 1$ so $G/H = \langle Hy \rangle$. Since $(Hy)^2 = Hy^2 = Hx^3 = 1$, $\mathrm{card}(G/H) \leqslant 2$. Also $x^6 = 1$ implies $\mathrm{card}(H) \leqslant 6$. Consequently $\mathrm{card}(G) = \mathrm{card}(G/H)\cdot\mathrm{card}(H) \leqslant 12$.

Next note that $\mathrm{card}(T) = 12$ so $\mathrm{card}(G) \leqslant 12 = \mathrm{card}(T)$.

Let $a = (1,2)$ and $b = (0,1)$ in the group T. It is claimed that $T = \{a^k b^e \mid 0 \leqslant k \leqslant 5$ and $e = 0$ or $1\}$, and consequently that a and b generate T. Since the twelve elements in the last set of braces belong to T and since $\mathrm{card}(T) = 12$, it suffices to show that these twelve elements are distinct. First consider the powers of a: $a = (1,2)$; $a^2 = (1,2)(1,2) = (1+g_2(1), 2+2) = (2,0)$; $a^3 = (1,2)(2,0) = (1+g_2(2), 2+0) = (0,2)$; $a^4 = a^2a^2 = (2,0)(2,0) = (2+g_0(2), 0+0) = (1,0)$; $a^5 = aa^4 = (1,2)(1,0) = (1+g_2(1), 2+0) = (2,2)$; $a^6 = a^4a^2 = (1,0)(2,0) = (1+g_0(2), 0+0) = (0,0) = 1$. These calculations show $o(a) = 6$ and hence the six elements

a^k, $0 \leqslant k \leqslant 5$ are distinct. This further implies the six elements a^kb, $0 \leqslant k \leqslant 5$ are distinct since $a^kb = a^ib$ implies $a^k = a^i$. Finally if some a^k and some a^ib were equal, then $a^{k-i} = b$ which is false since it was seen that $b = (0,1)$ is not a power of a. This establishes the claim.

Define the function $f: \{X, Y\} \to \{a, b\}$ by $f(X) = a$ and $f(Y) = b$. In order to apply Theorem 2.2.6 it must be shown that the relations of G are "satisfied" in T. The first relation is satisfied because $a^6 = 1$ as shown above. The other calculations are: $b^2 = (0,1)(0,1) = (0+g_1(0), 1+1) = (0,2) = a^3$ so $a^3b^{-2} = 1$; and $aba = (1,2)(0,1)(1,2) = (1+g_2(0), 2+1)(1,2) = (1,3)(1,2) = (1+g_3(1), 3+2) = (1-1,1) = (0,1) = b$ so $b^{-1}aba = 1$. Theorem 2.2.6 may now be applied to yield an isomorphism $d: G \to T$ such that $d(x) = a$ and $d(y) = b$. Theorem 2.2.6 further states $\text{card}(G) = \text{card}(T) = 12$ and $G = \{d^{-1}(t) \mid t \in T\}$. The last equation and $T = \{a^kb^e \mid 0 \leqslant k \leqslant 5 \text{ and } e = 0 \text{ or } 1\}$ imply $G = \{x^ky^e \mid 0 \leqslant k \leqslant 5 \text{ and } e = 0 \text{ or } 1\}$.

RESULT 4.5.2 *The elements x^2 and x^4 have order* 3; *x^3 has order* 2; *x^5 and x have order* 6; *all other nontrivial elements have order* 4.

That $o(x) = 6$ follows from the relation $x^6 = 1$ and the facts that $x^2 \neq 1$ and $x^3 \neq 1$. It is immediate from $o(x) = 6$ that $o(x^2) = o(x^4) = 3$, $o(x^3) = 2$, and $o(x^5) = 6$.

Now $y^4 = (y^2)^2 = (x^3)^2 = x^6 = 1$. This and $y^2 \neq 1$ imply $o(y) = 4$. The last relation implies $xyx = y$. Let $1 \leqslant i \leqslant 5$. Then $x^{i-1}(xyx) = x^{i-1}y$ so $x^iyx = x^{i-1}y$ and hence $x^iyx(x^{i-1}y) = x^{i-1}yx^{i-1}y$. That is $(x^iy)^2 = (x^{i-1}y)^2$ for $1 \leqslant i \leqslant 5$. Thus $y^2 = (xy)^2 = \ldots = (x^5y)^2$. This, $y^4 = 1$, and $y^2 \neq 1$ imply $o(x^iy) = 4$ for $1 \leqslant i \leqslant 5$.

RESULT 4.5.3 *The subgroups of G are:* $\{1\}$, $\langle x^3\rangle$, $\langle x^2\rangle$, $\langle y\rangle$, $\langle xy\rangle$, $\langle x^2y\rangle$, $\langle x\rangle$, *and G.*

Clearly all of the above are subgroups of G. It will first be shown that all cyclic subgroups of G are in the above list. Thus it must be shown that the subgroups generated by x^3y, x^4y, x^4, x^5, and x^5y appear in the list. Since $x^4 = (x^2)^2$ and $o(x^4) = 3 = o(x^2)$ (Result 4.5.2), $\langle x^4\rangle = \langle x^2\rangle$. Similarly $x^5 = x^{-1}$ and $o(x^5) = o(x)$ imply $\langle x^5\rangle = \langle x\rangle$. Now $y^3 = y^2y = x^3y$ by the second relation. This and $o(x^3y) = o(y)$ (Result 4.5.2) imply $\langle x^3y\rangle = \langle y\rangle$. The third relation implies $xyx = y$. Thus $x(xyx)xy = xyxy = y^2$. That is, $(x^2y)^2 = y^2$ so $(x^2y)^3 = (x^2y)y^2 = x^2y^2y = x^2x^3y = x^5y$. This and $o(x^5y) = o(x^2y)$ (Result 4.5.2) imply $\langle x^5y\rangle = \langle x^2y\rangle$. The demonstration that $\langle x^4y\rangle = \langle xy\rangle$ is similar.

Now let B be any subgroup of G. If $B = \{1\}$ or $B = G$, B appears in the list, so it can be assumed that $\text{card}(B) \neq 1, 12$. By Lagrange's theorem then, $\text{card}(B) = 2, 3, 4$, or 6. If $\text{card}(B) = 2$ or 3, then B is cyclic so has been accounted for. If $\text{card}(B) = 6$, then by Cauchy's theorem (Theorem A.1.5), there is an element of order 2 in B and an element of order 3. Thus by Result 4.5.2, $x^3 \in B$ and $x^2 \in B$ (the other possibility, $x^4 \in B$, implies $x^2 = (x^4)^{-1} \in B$). Consequently $x = x^3(x^2)^{-1} \in B$, hence $\langle x\rangle \subseteq B$. But $\text{card}(\langle x\rangle) = o(x) = 6$ by Result 4.5.2 so $B = \langle x\rangle$ and hence B has been accounted for. Finally assume $\text{card}(B) = 4$. As in the previous case, $x^3 \in B$. Since $\text{card}(\langle x^3 \rangle) = 2 < 4$, $\langle x^3 \rangle \subset B$. Let $b \in B \setminus \langle x^3 \rangle$. Then $o(b) = 4$, hence $B = \langle b \rangle$ so B has been accounted for.

The subgroups of G are illustrated below.

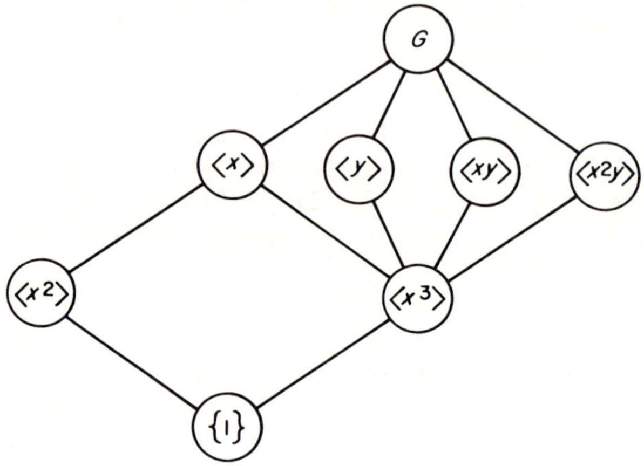

RESULT 4.5.4 $Z(G) = \langle x^3 \rangle$.
By the second relation x^3 commutes with y. Hence x^3 commutes with both generators of G, and so must belong to the center. Consequently $\langle x^3 \rangle \subseteq Z(G)$. Thus $\text{card}(Z(G)) = 2, 4, 6$, or 12, implying $\text{card}(G/Z(G)) = 6, 3, 2$, or 1. In the last three cases, $G/Z(G)$ is cyclic which implies G is abelian by Theorem A.10.15. This, however, is not true (if G were abelian, the last relation would yield $x^2 = 1$ which is false). Conclude $\text{card}(Z(G)) = 2$ and so $Z(G) = \langle x^3 \rangle$.

RESULT 4.5.5 *G is indecomposable.*
By Result 4.5.3, every proper subgroup of G is cyclic, hence abelian. By Result 4.5.4, however, G is nonabelian. Conclude G cannot be expressed as $G = A \oplus B$ for proper subgroups A and B by the Corollary to Theorem A.2.1.

RESULT 4.5.6 *G is not subdirectly irreducible.*
By Result 4.5.4, $\langle x^3 \rangle \lhd G$. By Result 4.5.3, $\langle x^2 \rangle$ is the only 3-Sylow subgroup of G hence is normal (Theorem A.10.15, Part (b)). Thus $H(G) \subseteq \langle x^3 \rangle \cap \langle x^2 \rangle = \{1\}$ proving G is not subdirectly irreducible.

EXERCISES

1. Identify G' and $\Phi(G)$.
2. Show $G/Z(G) \cong D_3$.

3. Show
   (a) G is solvable.
   (b) G is supersolvable.
   (c) G is not nilpotent.
4. Identify the normal subgroups of G.
5. Show all normal subgroups of G are characteristic.
6. Show there exist $d,h \in \mathrm{Aut}(G)$ such that $d(x) = x$ and $d(y) = xy$; and $h(x) = x^5$ and $h(y) = y$.
7. Show that $\langle y \rangle$, $\langle xy \rangle$, and $\langle x^2y \rangle$ are Carter subgroups of G.
8. Describe the decomposition of G into conjugacy classes and write the class equation.
9. Let A be a group with the following properties:
   (a) $\mathrm{card}(A) = 12$.
   (b) A is nonabelian.
   (c) A has a unique element of order 2.
   Show $A \cong G$.
10. Let A be a group with the following properties:
   (a) $\mathrm{card}(A) = 12$.
   (b) A is indecomposable.
   (c) $Z(A)$ is nontrivial.
   Show $A \cong G$.

NOTES

1. The group G is the smallest example of an indecomposable group that is not subdirectly irreducible.
2. Characterizations of G are given in Exercises 9 and 10. The Exercise 10 characterization and Exercise 9 of Example 4.6 show that Examples 4.5 and 4.6 are the only (up to isomorphism) indecomposable groups of cardinality 12.
3. The group G is a dicyclic group (see Example 4.4, Note 1).
4. The argument in the last paragraph of Result 4.5.2 generalizes to show that $o(x^iy) = 4$ in any dicyclic group.

□□□□□

Example 4.6 (The Tetrahedral Group)

Let $G = G(X, Y \mid X^3 = 1, Y^3 = 1, (XY)^2 = 1)$.

AUXILIARY NOTATION Let $H = \langle xy, yx \rangle$.

RESULT 4.6.1

(a) $G \cong PSL(2,3)$.

(b) $\text{card}(G) = 12$.

(c) *The elements of G are* x, x^2, y, y^2, xy, yx, x^2y, xy^2, xy^2x, yx^2, y^2x, *and* 1.

(d) $H \lhd G$ *and* $H = \{1, xy, yx, xy^2x\}$.

(e) xy, yx, *and* xy^2x *have order* 2. *All other nontrivial elements have order* 3.

The third relation implies $xyx = y^{-1}$, hence $(xyx)(xyx) = y^{-2}$ from which $xyx^2yx = y$ since $y^3 = 1$. Therefore $yx^2y = x^{-1}yx^{-1}$, that is $(yx)(xy) = x^{-1}yx^{-1}$. Similarly $(xy)(yx) = y^{-1}xy^{-1}$. Thus $x(xy)x^{-1} = x^2yx^{-1} = x^{-1}yx^{-1} = (yx)(xy) \in H$; $x(yx)x^{-1} = xy \in H$; $y(xy)y^{-1} = yx \in H$; and $y(yx)y^{-1} = y^2xy^{-1} = y^{-1}x \cdot y^{-1} = (xy)(yx) \in H$. These four calculations and Theorem A.10.3 show $H \lhd G$.

Next note that $(yx)(xy) = x^{-1}yx^{-1} = x^2yx^2 = x(xyx)x = x(xyxy)y^{-1}x = xy^{-1}x = xy^2x = (xy)(yx)$. Hence the generators of H commute and so H is abelian. Let $K = \langle xy \rangle$. Since H is abelian, $K \lhd H$. By the third relation, $\text{card}(K) \leqslant 2$. By Theorem A.10.4, $H/K = \langle Kyx \rangle$. But $(yx)^2 = 1$ (premultiply the third relation by y, then postmultiply by y^{-1}), so $\text{card}(H/K) \leqslant 2$. Hence $\text{card}(H) \leqslant 4$.

A second application of Theorem A.10.4 yields $G/H = \langle Hx, Hy \rangle$. But $x^2y^{-1} = x^2(xyx) = yx \in H$ so $Hx^2 = Hy$ and hence $G/H = \langle Hx \rangle$. Thus $\text{card}(G/H) \leqslant 3$. Consequently $\text{card}(G) \leqslant 12$. An application of Theorem 3.2.23 shows that $\text{card}(PSL(2,3)) = 3(3-1)(3+1)/2 = 12$. Thus $\text{card}(G) \leqslant 12 = \text{card}(PSL(2,3))$.

Let $a = [\Theta_2(1,2,1)]$ and $b = [\Theta_2(2,1,1)]$ in $PSL(2,3)$. Then $a^2 = [\Theta_2(1,2,2)]$ and $b^2 = [\Theta_2(2,1,2)]$ (Corollary to Theorem 3.1.3). Since the four matrices $\Theta_2(1,2,1)$, $\Theta_2(2,1,1)$, $\Theta_2(1,2,2)$, and $\Theta_2(2,1,2)$ generate $SL(2,3)$ (Theorem 3.2.18), it follows that a, b, a^2, and b^2 generate $PSL(2,3)$, that is, $PSL(2,3) = \langle a, b \rangle$.

Now define the function $f: \{X, Y\} \to \{a, b\}$ by $f(X) = a$ and $f(Y) = b$. In order to apply Theorem 2.2.6 it must be shown that the relations of G are "satisfied" in $PSL(2,3)$. Thus it must be shown that $a^3 = 1$, $b^3 = 1$, and $(ab)^2 = 1$. This is done as follows: $a^3 = [\Theta_2(1,2,3)] = [\Theta_2(1,2,0)] = 1$ by the Corollary to Theorem 3.1.3; also by this corollary, $b^3 = [\Theta_2(2,1,3)] = 1$; next note that $ab = [\begin{pmatrix} 1 & 1 \\ 0 & 1 \end{pmatrix}][\begin{pmatrix} 1 & 0 \\ 1 & 1 \end{pmatrix}] = [\begin{pmatrix} 2 & 1 \\ 1 & 1 \end{pmatrix}]$ so $(ab)^2 = 1$ since the last matrix in brackets has trace 0 (Theorem 3.2.24). Theorem 2.2.6 may now be applied to yield existence of an isomorphism $d: G \to PSL(2,3)$ such that $d(x) = a$ and $d(y) = b$; hence $\text{card}(G) = \text{card}(PSL(2,3)) = 12$. Thus the proofs of Parts (a) and (b) are complete.

Now $xy, yx \in H$ so $\{1, xy, yx, xy^2x\} \subseteq H$ since $xy^2x = (xy)(yx)$. But as noted before, $\text{card}(H) \leqslant 4$. Thus to show $H = \{1, xy, yx, xy^2x\}$, it suffices to show the elements 1, xy, yx, and xy^2x are distinct. But since d is an isomorphism, it suffices to show 1, ab, ba, and ab^2a are distinct in $PSL(2,3)$. This is seen as follows: $ab = [\begin{pmatrix} 2 & 1 \\ 1 & 1 \end{pmatrix}] = \{\begin{pmatrix} 2 & 1 \\ 1 & 1 \end{pmatrix}, \begin{pmatrix} 1 & 2 \\ 2 & 2 \end{pmatrix}\}$, $ba = [\begin{pmatrix} 1 & 0 \\ 1 & 1 \end{pmatrix}][\begin{pmatrix} 1 & 1 \\ 0 & 1 \end{pmatrix}] = \{\begin{pmatrix} 1 & 1 \\ 1 & 2 \end{pmatrix}, \begin{pmatrix} 2 & 2 \\ 2 & 1 \end{pmatrix}\}$, and $ab^2a = (ab)(ba) = [\begin{pmatrix} 2 & 1 \\ 1 & 1 \end{pmatrix}][\begin{pmatrix} 1 & 1 \\ 1 & 2 \end{pmatrix}] = \{\begin{pmatrix} 0 & 1 \\ 2 & 0 \end{pmatrix}, \begin{pmatrix} 0 & 2 \\ 1 & 0 \end{pmatrix}\}$. This completes the proof of Part (d).

Now since $\text{card}(H) = 4$ by the previous paragraph, and since $\text{card}(\langle x \rangle) = o(x) = 3$ (because $x^3 = 1$ and $x \neq 1$ lest $1 = d(x) = a$) it follows that $\text{card}(H\langle x \rangle) = 12$ and hence $G = H\langle x \rangle$. Therefore the elements of G are $1 \cdot 1 = 1$; $1 \cdot x = x$; $1 \cdot x^2 = x^2$;

$xy \cdot 1 = xy$; $xy \cdot x = xyx = y^2$ (by the second and third relations); $xy \cdot x^2 = xyx^2 = (xyx)x = y^2x$; $yx \cdot 1 = yx$; $yx \cdot x = yx^2$; $yx \cdot x^2 = y$; $xy^2x \cdot 1 = xy^2x$; $xy^2x \cdot x = xy^2x^2 = x(xyx)x^2 = x^2y$; and $xy^2x \cdot x^2 = xy^2$. This completes the proof of Part (c).

The proof of Part (e) is as follows. That $o(xy) = 2$ follows from the third relation. It has been previously seen that $(yx)^2 = 1$, so $o(yx) = 2$. The third relation implies $xyx = y^{-1} = y^2$ hence $xy^2x = x^2yx^2$. Consequently $(xy^2x)^2 = (xy^2x)x^2yx^2 = 1$ and thus $o(xy^2x) = 2$. That $o(x) = o(y) = 3$ is obvious. This in turn implies $o(x^2) = o(y^2) = 3$. Since $(yx^2)^3 = yx^2yx^2yx^2 = yx(xyx)(xyx)x = yxy^2y^2x = yxyx = 1$, $o(yx^2) = 3$. Similarly $o(y^2x) = 3$ because $y^2xy^2xy^2x = (xyx)x(xyx)x(xyx)x = xyx^3yx^3yx^2 = xy^3x^2 = 1$. Finally it is easily seen that $x^2y = (y^2x)^{-1}$ and $xy^2 = (yx^2)^{-1}$. Therefore $o(x^2y) = o(yx^2) = 3$.

RESULT 4.6.2 *The subgroups of G are* $\{1\}$, $\langle x\rangle$, $\langle x^2y\rangle$, $\langle yx^2\rangle$, $\langle xy\rangle$, $\langle yx\rangle$, $\langle xy^2x\rangle$, $\langle y\rangle$, *H, and G.*

All of the above are clearly subgroups. Preliminary to proving all subgroups of G appear in the above list, it will first be shown that H is the only subgroup of cardinality 4. By Result 4.6.1, Part (d), $H \lhd G$. Therefore $N_G(H) = G$, so by Theorem A.10.15 $n_2(G) = 1$ as claimed.

It will next be shown that G has no subgroups of cardinality 6. Suppose F is such a subgroup. By Sylow's theorem (Theorem A.1.6), $n_3(F) \equiv 1 \bmod 3$ and $n_3(F) \mid 6$. Hence $n_3(F) = 1$. Let C denote this unique 3-Sylow subgroup of F. Then as $g[C]$ is also a 3-Sylow subgroup of F for all $g \in \mathrm{Aut}(F)$, C must be a characteristic subgroup of F. But $F \lhd G$ by Theorem A.10.2 so $C \lhd G$ by Theorem A.10.6. It follows then that $G = H \oplus C$. But H is abelian (Theorem A.1.2) as is C (card$(C) = 3$). Therefore G must be abelian by the Corollary to Theorem A.2.1. This contradiction (the elements xy and yx are distinct by Result 4.6.1, Part (c)) shows G has no subgroups of cardinality 6.

Next it will be shown that all cyclic subgroups of G appear in the list. By the second and third relations, $xyx = y^2$. Therefore $(x^2y)^2 = x^2yx^2y = x(xyx)xy = xy^2xy = xy^2(xyx)x^{-1} = xy^2y^2x^{-1} = xyx^{-1} = xyx^2$. Thus $xyx^2 \in \langle x^2y\rangle$. But $o(xyx^2) = 3 = o(x^2y)$ by Result 4.6.1, Part (e). Consequently $\langle xyx^2\rangle = \langle x^2y\rangle$. Next note that $(yx^2)(xy^2) = yx^3y^2 = yy^2 = 1$ so $xy^2 = (yx^2)^{-1}$ and hence $\langle xy^2\rangle = \langle yx^2\rangle$. As it is easily seen that the subgroups $\langle x^2\rangle$ and $\langle y^2\rangle$ are equal to $\langle x\rangle$ and $\langle y\rangle$ respectively, all cyclic subgroups of G have been accounted for.

Finally let B be a subgroup of G. By Lagrange's theorem and the second paragraph, card$(B) = 1, 2, 3, 4$, or 12. If card$(B) = 1, 2$, or 3, B is cyclic and hence has been accounted for. If card$(B) = 4$, then $B = H$ as has been shown previously. If card$(B) = 12$, then $B = G$.

The subgroups of G are illustrated below.

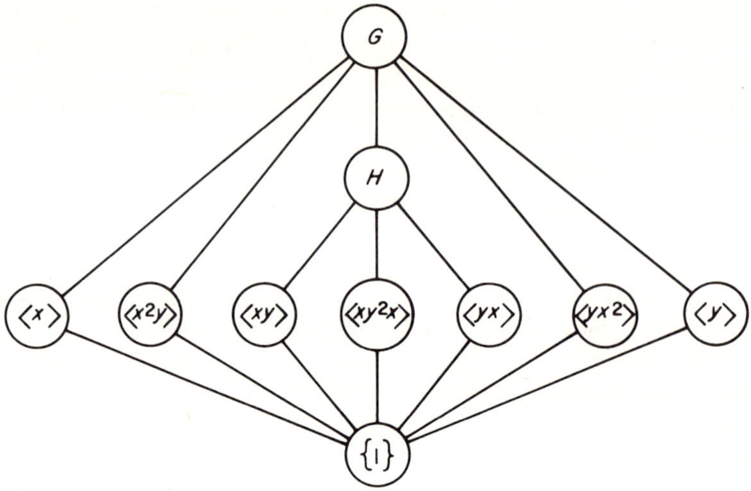

RESULT 4.6.3 *The normal subgroups of G are* {1}, *H, and G.*
The three subgroups listed above are normal (H by Result 4.6.1, Part (d)) so it suffices to show all other subgroups are nonnormal. If K were a normal subgroup of G of cardinality 2, then $\text{card}(K\langle x\rangle) = 6$ which is a contradiction (G has no subgroups of cardinality 6 by Result 4.6.2). Hence the subgroups $\langle xy\rangle$, $\langle yx\rangle$, and $\langle xy^2x\rangle$ are nonnormal. Again if K were a normal subgroup of cardinality 3, then $\text{card}(K\langle xy\rangle) = 6$ which is again false. Hence the subgroups $\langle x\rangle$, $\langle y\rangle$, $\langle x^2y\rangle$, and $\langle yx^2\rangle$ are nonnormal.

RESULT 4.6.4 *G is subdirectly irreducible.*
By Result 4.6.3, $\text{H}(G) = H \cap G = H \neq \{1\}$ so G is subdirectly irreducible.

RESULT 4.6.5 *G is indecomposable.*
By Result 4.6.4 and Theorem A.10.10.

RESULT 4.6.6 *G is not converse lagrange.*
By Result 4.6.2, G has no subgroup of cardinality 6.

RESULT 4.6.7 *G is solvable.*
The subgroup H is normal by Result 4.6.3. This subgroup is abelian (hence solvable) and its factor G/H has cardinality 3 so it too is abelian (hence solvable). Conclude G is solvable by Theorem A.6.6.

RESULT 4.6.8 *G is not supersolvable.*
If G were supersolvable, $G/\{1\}$ would have a nontrivial normal cyclic subgroup.

But $G/\{1\} \cong G$ and G has no nontrivial normal cyclic subgroups by Result 4.6.3.

RESULT 4.6.9 *G is not nilpotent.*
By Result 4.6.8 and the Corollary to Theorem A.7.2.

RESULT 4.6.10 *G has maximal subgroups of nonprime index.*
By Result 4.6.2, $\langle x \rangle$ is a maximal subgroup of index 4.

RESULT 4.6.11
(a) $G' = H$.
(b) $\text{card}(G') = 4$.
(c) G' *is abelian (hence nilpotent) but noncyclic.*
(d) G *has a noncyclic minimal normal subgroup.*
Since $G' \lhd G$, $G' = G$, H, or $\{1\}$ by Result 4.6.3. In the first case, $G = G' = G'' = \ldots$ so G is nonsolvable contradicting Result 4.6.7. In the third case G is abelian which is false ($xy \neq yx$). Thus $G' = H$ proving Part (a). Part (b) follows from Part (a) and Result 4.6.1, Part (d).

Since $\text{card}(G') = 4$, G' is abelian by Theorem A.1.2. Suppose G' is cyclic. Then by Part (b) again, $G' \cong Z(4)$. But G has no elements of order 4 by Result 4.6.1, Part (e). This contradiction proves Part (c).

By Result 4.6.3, there is no nontrivial normal subgroup underneath H so H is a minimal normal subgroup. By Parts (a) and (c), H is noncyclic. This proves Part (d).

EXERCISES

1. Show $Z(G)$ is trivial.
2. Show all normal subgroups of G are fully invariant.
3. Identify $\Phi(G)$.
4. Describe the decomposition of G into conjugacy classes and write the class equation.
5. Show the 3-Sylow subgroups of G are Carter subgroups.
6. Show there exist $f, g \in \text{Aut}(G)$ such that $f(x) = x^2y$ and $f(y) = y$; and $g(x) = y$ and $g(y) = x$.
7. Let A be a group such that $\text{card}(A) = 12$ and A is not supersolvable. Show $A \cong G$.
8. Let A be a group such that $\text{card}(A) = 2^2p$ for some prime p and A is not supersolvable. Show $A \cong G$.
9. Let A be a group such that $\text{card}(A) = 12$, A is indecomposable, and $Z(A)$ is trivial. Show $A \cong G$.
10. Let A be a group of cardinality 12 having no normal subgroups of cardinality 3. Show $A \cong G$.

11. Let A be a group such that $L(A) \cong L(G)$. Show $A \cong G$.
12. Show G is isomorphic to the subgroup of S_4 generated by $f = \binom{1\ 2\ 3\ 4}{2\ 3\ 1\ 4}$ and $g = \binom{1\ 2\ 3\ 4}{2\ 4\ 3\ 1}$. (This subgroup, $\langle f,g \rangle$, is the alternating group A_4.)

NOTES

1. The group G is isomorphic to, and is most popularly known as, the alternating group A_4. See Exercise 12.
2. As a counterexample, G shows that none of the following classes of groups are extension closed:
   (a) supersolvable and hypercyclic groups
   (b) converse lagrange groups
   (c) $\{A \mid$ minimal normal subgroups of A are cyclic$\}$
   This is because H and G/H satisfy (a), (b), and (c) above, while G does not.
3. For a finite group A, the conditions *A is supersolvable* and *maximal subgroups of A have prime index* (different primes allowed) are equivalent (see Hall [14], Corollary 10.5.1 and Theorem 10.5.8). Thus Results 4.6.8 and 4.6.10 are related more closely than they appear to be.
4. Consider the two chains of subgroups $\{1\} \subseteq \langle x \rangle \subseteq G$ and $\{1\} \subseteq \langle xy \rangle \subseteq H \subseteq G$. First note that both chains are "maximal" in the sense that no further subgroups can be added to either chain (thus $\{1\} \subseteq H \subseteq G$ would be an example of a nonmaximal chain). Secondly note that the two chains have different lengths. In lattice-theoretic terminology, the requirement that any two maximal chains between the same points have the same length is called the **Jordan–Dedekind chain condition.** Thus $L(G)$ fails to satisfy this condition. In general, the subgroup lattice of a finite group A satisfies the Jordan–Dedekind chain condition if and only if A is supersolvable (see Hall [14], Theorem 19.3.1).
5. Characterizations of G are given in Exercises 7–11 with Exercise 8 being a generalization of Exercise 7. The Exercise 9 characterization and Exercise 10 of Example 4.5 show Examples 4.5 and 4.6 are the only (up to isomorphism) indecomposable groups of cardinality 12.
6. The automorphism group of G is described in Example 4.9, Exercise 16.

□□□□□

Example 4.7

Let $G = G(X, Y \mid X^8 = 1, Y^2 = 1, YXYX^3 = 1)$.

RESULT 4.7.1

(a) *Define $\phi: Z(8) \to Z(8)$ by $\phi(i) = 5i$. Then $\phi \in \mathrm{Aut}(Z(8))$ and $o(\phi) = 2$. Thus $(Z(2), g, Z(8))$ is a semidirect system when g is defined by $g_0 = 1$ and $g_1 = \phi$. Finally $G \cong Z(8) \times_g Z(2)$.*

(b) card(G) = 16.
(c) $G = \{x^k y^e \mid 0 \leq k \leq 7 \text{ and } e = 0 \text{ or } 1\}$.
Let $H = \langle x \rangle$. Since $y = y^{-1}$ by the second relation and $yxy = x^5$ by the first and third, $yxy^{-1} \in H$. Clearly $xxx^{-1} \in H$ as well so $H \lhd G$ by Theorem A.10.3. Application of Theorem A.10.4 yields $G/H = \langle Hx, Hy \rangle$. But $Hx = 1$ so $G/H = \langle Hy \rangle$ and hence card(G/H) ≤ 2. Also $H = \langle x \rangle$ implies card(H) ≤ 8. Thus card(G) = card(G/H)·card(H) ≤ 16.

Now let ϕ and g be defined as above. Then for $i,j \in Z(8)$, $\phi(i+j) = 5(i+j) = 5i + 5j = \phi(i) + \phi(j)$ so ϕ is a homomorphism. If $\phi(i) = 0$, then $5i = 0$. But 5 and 8 are relatively prime so there exist integers m and n such that $5m + 8n = 1$. Then $i = (5m + 8n)i = 5mi + 8ni = 0$, proving ϕ is injective. Application of the pigeonhole principle, or an easy surjectivity proof, shows $\phi \in \text{Aut}(Z(8))$.

For all $i \in Z(8)$, $\phi^2(i) = 25i = i$ so ϕ has order 2. Thus $(Z(2),g,Z(8))$ is a semidirect system by Theorem 1.2.3. Let $T = Z(8) \times_g Z(2)$. Let $a = (1,0)$ and $b = (0,1)$. First it is claimed that $\langle a,b \rangle = T$. Since card(T) = 16, to show this it suffices to exhibit 16 distinct elements of $\langle a,b \rangle$. Clearly $a^k = (k,0)$ for $0 \leq k \leq 7$. Also for such k, $a^k b = (k,0)(0,1) = (k,1)$ so $T = \langle a,b \rangle$ as claimed. At this stage of the proof card(G) ≤ 16 = card(T).

Define the function $f: \{X, Y\} \to \{a,b\}$ by $f(X) = a$ and $f(Y) = b$. In order to apply Theorem 2.2.6 it must be shown that the relations of G are "satisfied" in T. This is done as follows: $a^8 = 1$ is clear; $b^2 = (0,1)(0,1) = (0+g_1(0), 1+1) = (0,0) = 1$; and $baba^3 = (0,1)(1,0)(0,1)(3,0) = (0+g_1(1), 1+0)(0+g_1(3), 1+0) = (5,1)(7,1) = (5+g_1(7), 1+1) = (0,0) = 1$. Theorem 2.2.6 may now be applied to yield an isomorphism $d: G \to T$ such that $d(x) = a$ and $d(y) = b$. Theorem 2.2.6 also states card(G) = card(T) = 16 and $G = \{d^{-1}(t) \mid t \in T\}$. This proves Parts (a) and (b). Since 16 elements of T have been accounted for in the previous paragraph, card(T) = 16 implies these are the only elements of T. That is, $T = \{a^k b^e \mid 0 \leq k \leq 7 \text{ and } e = 0 \text{ or } 1\}$. Thus $G = \{d^{-1}(t) \mid t \in T\} = \{x^k y^e \mid 0 \leq k \leq 7$ and $e = 0$ or $1\}$ proving Part (c).

RESULT 4.7.2 $Z(G) = \{1, x^2, x^4, x^6\}$.
The relations imply $yxy^{-1} = x^5$ so $x^2 = x^{10} = (yxy^{-1})^2 = yx^2y^{-1}$. Hence $x^2y = yx^2$. Consequently x^2 commutes with both generators of G so belongs to the center. This implies $\{1, x^2, x^4, x^6\} \subseteq Z(G)$. By Lagrange's theorem card($Z(G)$) = 4, 8, or 16. If card($Z(G)$) = 8 or 16, then $G/Z(G)$ is cyclic and hence $G = Z(G)$ by Theorem A.10.5. The third relation then becomes $y^2x^4 = 1$ and so $x^4 = 1$ which is false. Conclude card($Z(G)$) = 4 so $Z(G) = \{1, x^2, x^4, x^6\}$.

RESULT 4.7.3 *x^4, y, and x^4y have order 2; x^2, x^6, x^2y, and x^6y have order 4; all other nontrivial elements of G have order 8.*
Clearly $o(x) = 8$ which further implies $o(x^4) = 2$, $o(x^2) = o(x^6) = 4$, and $o(x^i) = 8$ for $i = 3$, 5, and 7. That $o(y) = 2$ is immediate from the second relation. Since $x^4 \in Z(G)$ by Result 4.7.2, $(x^4y)^2 = x^8y^2 = 1$ so $o(x^4y) = 2$. Also $x^2 \in Z(G)$ so $(x^2y)^4 = x^8y^4 = 1$, while $(x^2y)^2 = x^4y^2 = x^4 \neq 1$ so $o(x^2y) = 4$. That $o(x^6y) = 4$ is similar.

Now $(xy)^2 = x(yxy) = xx^5$ (by the first and third relations) $= x^6$ so $(xy)^8 = (x^6)^4 = 1$, while $(xy)^4 = (x^6)^2 = x^4 \neq 1$. Thus $o(xy) = 8$. The proofs that $o(x^k y) = 8$ for $k = 3, 5$, and 7 are similar.

RESULT 4.7.4 *The subgroups of G are* $\{1\}$, $\langle y\rangle$, $\langle x^4\rangle$, $\langle x^4y\rangle$, $\langle x^2y\rangle$, $\langle x^2\rangle$, $\langle x\rangle$, $\langle xy\rangle$, $\{1,x^4,y,x^4y\}$, $\{1,x^2,x^4,x^6,y,x^2y,x^4y,x^6y\}$, *and G.*

The first eight entries in the above list are clearly subgroups of G, as is the last. To show the remaining two entries are subgroups use will be made of the fact that if $N \lhd G$, then NB is a subgroup of G whenever B is. By Result 4.7.2, $Z(G) = \{1,x^2,x^4,x^6\}$. Thus $Z(G)\langle y\rangle = \{1,x^2,x^4,x^6,y,x^2y,x^4y,x^6y\}$ is a subgroup (the tenth entry). Also $\langle x^4\rangle \subseteq Z(G)$ so $\langle x^4\rangle \lhd G$ and hence $\langle x^4\rangle\langle y\rangle = \{1, x^4,y,x^4y\}$ is a subgroup (the ninth entry).

It will next be shown that all cyclic subgroups have been accounted for. Since $x^2 \in Z(G)$, $(x^2y)^3 = x^6y^3 = x^6y$ so $x^6y \in \langle x^2y\rangle$. But $o(x^6y) = o(x^2y)$ (Result 4.7.3) hence $\langle x^6y\rangle = \langle x^2y\rangle$. Also $(xy)^3 = x^7y$, $(xy)^5 = x^5y$, and $(xy)^7 = x^3y$ so $\langle x^ky\rangle = \langle xy\rangle$ for $k = 3,5$, or 7. Finally $\langle x^6\rangle = \langle x^2\rangle$ and $\langle x^k\rangle = \langle x\rangle$ for $k = 3,5$, or 7.

Next suppose B is any subgroup of G. Then card(B) = 1,2,4,8, or 16 by Lagrange's theorem. If card(B) = 1 or 16, $B = \{1\}$ or G and so B appears in the list. If card(B) = 2, then B is cyclic and hence has been accounted for. Suppose card(B) = 4. By Cauchy's theorem (Theorem A.1.5) there exists $b \in B$ such that $o(b) = 2$. Let $b' \in B\backslash\langle b\rangle$. If $o(b') = 4$, then $B = \langle b'\rangle$ implying B is cyclic and hence has been listed. Assume then that $o(b') = 2$. Therefore by Result 4.7.3, two of the elements x^4, y, x^4y belong to B. But then so too does the third ($x^4 = (x^4y)y$, $y = x^4(x^4y)$, and $x^4y = (x^4)(y)$) consequently $\{1,x^4,y,x^4y\} \subseteq B$ so $\{1,x^4,y,x^4y\} = B$ because card(B) = 4.

Finally suppose card(B) = 8. Then $B \lhd G$ by Theorem A.10.2 so $G/B \cong Z(2)$. Hence $(Bx)^2 = 1$ implying $x^2 \in B$ and so $Z(G) \subseteq B$ by Result 4.7.2. Let $b \in B\backslash Z(G)$. If $o(b) = 8$, then $B = \langle b\rangle$ so B is cyclic and thus in the list. If $o(b) \neq 8$, then by Result 4.7.3, $b = y$, x^2y, x^4y, or x^6y (recall $b \notin Z(G)$). In all four cases $y \in B$ (if $b = x^2y$, then $y = x^6b \in B$, etcetera). Hence $Z(G)\langle y\rangle \subseteq B$. That is $\{1,x^2,x^4,x^6,y,x^2y,x^4y,x^6y\} \subseteq B$. Then card$(B)$ = 8 implies $B = \{1,x^2,x^4,x^6, y,x^2y,x^4y,x^6y\}$.

The subgroups of G are illustrated on page 123. Note that the subgroup lattice $L(G)$ is self-dual.

EXERCISES

1. Identify G' and $\Phi(G)$.
2. Identify the normal subgroups of G.
3. Show G is nilpotent.
4. Show G is subdirectly irreducible and hence is indecomposable.
5. Show all proper subgroups of G are abelian.
6. Describe the decomposition of G into conjugacy classes and write the class equation.

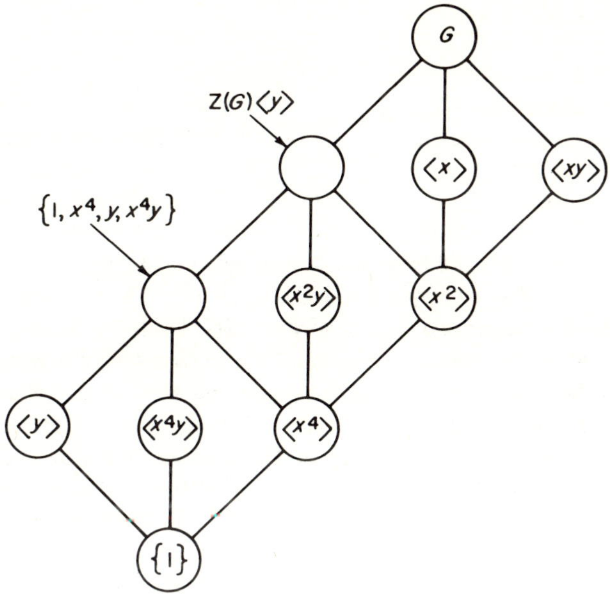

7. Show $\mathrm{Aut}(G) \cong D_4 \times Z(2)$.
8. Identify the claracteristic subgroups of G.
9. Show $\mathrm{L}(G) \cong \mathrm{L}(Z(8) \times Z(2))$.

NOTE

1. Exercise 9 shows G cannot be characterized by its subgroup lattice. Moreover it shows G cannot be characterized by its subgroup lattice even among groups of cardinality 16.

□□□□□

Example 4.8

Let $G = G(X, Y \mid X^4 = 1, Y^4 = 1, XYXY^{-1} = 1)$.

RESULT 4.8.1

(a) *Let* $Q = G(S, T \mid S^4 = 1, S^2T^{-2} = 1, T^{-1}STS = 1)$* *be the quaternion group. Then G is isomorphic to the subgroup of* $Q \times Z(4)$ *generated by* $(s,0)$ *and* $(t,1)$.

(b) $\text{card}(G) = 16$.

(c) $G = \{x^iy^k \mid 0 \leqslant i,k \leqslant 3\}$.

Let $H = \langle x^2 \rangle$. Note that $yx^2 = yx^3x^{-1}$, but by the first and third relations $xy = yx^3$ so $yx^2 = xyx^{-1} = xyx^3 = x(yx^3) = x(xy) = x^2y$. Thus x^2 commutes with y, hence with both generators of G and so $x^2 \in Z(G)$ implying $H \lhd G$.

By Theorem A.10.4, $G/H = \langle Hx, Hy \rangle$. Now $HxHy = Hxy = Hyx^3 = HyHx^3 = HyHx$. Thus the generators of G/H commute and so G/H is abelian. Let $B/H = \langle Hx \rangle$. Then $(Hx)^2 = 1$ implies $\text{card}(B/H) \leqslant 2$. Since G/H is abelian, its subgroup B/H is normal. A second application of Theorem A.10.4 shows $(G/H)/(B/H) = \langle (B/H)Hy \rangle$. This and $y^4 = 1$ imply $\text{card}((G/H)/(B/H)) \leqslant 4$, hence $\text{card}(G/B) \leqslant 4$. Also $x^4 = 1$ so $\text{card}(H) \leqslant 2$. Then $\text{card}(G) = \text{card}(G/B)\cdot\text{card}(B/H)\cdot\text{card}(H) \leqslant 4\cdot2\cdot2 = 16$.

Let $a = (s,0)$ and $b = (t,1)$ in the group $Q \times Z(4)$, and let $T = \langle a, b \rangle$. It is claimed that $16 \leqslant \text{card}(T)$. To show this it suffices to exhibit sixteen distinct elements of T. Consider the elements $\{a^ib^k \mid 0 \leqslant i,k \leqslant 3\}$. Since there are indeed sixteen of these elements, it suffices to show they are distinct. First consider the powers of a and b: $a = (s,0)$; $a^2 = (s^2,0)$; $a^3 = (s^3,0)$; $a^4 = (1,0) = 1$; $b = (t,1)$; $b^2 = (t^2,2) = (s^2,2)$; $b^3 = (t^3,3)$; and $b^4 = b^2b^2 = (s^2,2)(s^2,2) = (1,0) = 1$. Thus $o(a) = o(b) = 4$. To demonstrate distinctness of the 16 elements, suppose $a^ib^k = a^ub^v$ where $0 \leqslant i,k,u,v \leqslant 3$. Then $a^{i-u} = b^{v-k}$. But as seen above, only the identity element is simultaneously a power of both a and b. Therefore $a^{i-u} = b^{v-k} = 1$ hence $4 \mid i - u$ and $4 \mid v - k$. This and $0 \leqslant i,k,u,v \leqslant 3$ imply $i = u$ and $v = k$ as required. Thus $16 \leqslant \text{card}(T)$ as claimed, and so $\text{card}(G) \leqslant 16 \leqslant \text{card}(T)$.

Define the function $f: \{X, Y\} \to \{a, b\}$ by $f(X) = a$ and $f(Y) = b$. In order to apply Theorem 2.2.6 it must be shown that the relations of G are "satisfied" in T. That $a^4 = b^4 = 1$ has been noted in the previous paragraph. Next note that $aba = (s,0)(t,1)(s,0) = (sts,1) = (t,1) = b$ so $abab^{-1} = 1$. Theorem 2.2.6 may now be applied to yield an isomorphism $d: G \to T$ such that $d(x) = a$ and $d(y) = b$. Theorem 2.2.6 further states $\text{card}(G) = \text{card}(T) = 16$ and $G = \{d^{-1}(t) \mid t \in T\}$. Since sixteen elements of T have been accounted for in the previous paragraph,

---

*The quaternion group, Example 4.4, was defined in terms of generators called X and Y, not S and T. It should be obvious, however, that the "two" quaternion groups defined here and in Example 4.4 are isomorphic. Substitutions of this nature will be made several times later in this chapter wihout comment.

$\text{card}(T) = 16$ implies these are the only elements of T. That is, $T = \{a^i b^k \mid 0 \leq i,k \leq 3\}$. Therefore $G = \{d^{-1}(t) \mid t \in T\} = \{x^i y^k \mid 0 \leq i,k \leq 3\}$.

RESULT 4.8.2 $Z(G) = \{1, x^2, y^2, x^2y^2\}$.
From the third relation $xyx = y$ and hence $xyx^{-1} = yx^{-2} = yx^2$. Square both sides to get $xy^2x^{-1} = yx^2yx^2 = yx(xyx)x = yxyx = y(xyx) = y^2$. Postmultiplying by x yields $xy^2 = y^2x$ and so y^2 commutes with both generators and hence belongs to the center. Postmultiplying the third relation by yx^{-1} yields $xy = yx^{-1}$ and so $y^{-1}xy = x^{-1}$. Square both sides to get $y^{-1}x^2y = x^{-2} = x^2$. Premultiplying by y yields $x^2y = yx^2$. Thus $x^2 \in Z(G)$. Also x^2y^2 must belong to the center since x^2 and y^2 do.

Hence $\{1, x^2, y^2, x^2y^2\} \subseteq Z(G)$. If this inclusion is proper, however, then by Lagrange's theorem $\text{card}(Z(G)) = 8$ or 16. In either case $G = Z(G)$ by Theorem A.10.5 which is false (if G were abelian, the third relation would imply $x^2 = 1$). Conclude $\{1, x^2, y^2, x^2y^2\} = Z(G)$.

RESULT 4.8.3 *x^2, y^2, and x^2y^2 have order 2. All other nontrivial elements of G have order 4.*
By the first relation $(x^2)^2 = 1$, implying $o(x^2) = 2$. Similarly $o(y^2) = 2$. Since $x^2 \in Z(G)$ by Result 4.8.2, $(x^2y^2)^2 = x^4y^4 = 1$. Thus $o(x^2y^2) = 2$. To complete the proof it must be shown that if t is any of the remaining nontrivial elements of G, then $t^4 = 1$ while $t^2 \neq 1$. First note that the equation $p^4 = 1$ holds for every element p of the quaternion group Q (Result 4.4.1, Part (d)) and hence $w^4 = 1$ holds for every element w of $Q \times Z(4)$. Conclude that $t^4 = 1$ holds for all $t \in G$ since G can be embedded in $Q \times Z(4)$ by Result 4.8.1. Thus it suffices to show $t^2 \neq 1$ for all $t \in G$, $t \neq 1, x^2, y^2, x^2y^2$. In showing this the facts $x^2, y^2 \in Z(G)$ (Result 4.8.2) will be used freely. The calculations are: $(xy)^2 = xyxy = y^2 \neq 1$; $(xy^2)^2 = x^2y^4 = x^2 \neq 1$; $(xy^3)^2 = xy^3xy^3 = xyxy^2y^3 = yy^2y^3 = y^2 \neq 1$; $(x^2y)^2 = x^4y^2 = y^2 \neq 1$; $(x^2y^3)^2 = x^4y^6 = y^2 \neq 1$; $(x^3y)^2 = x^3yx^3y = x^3x^2yxy = xyxy = y^2 \neq 1$; $(x^3y^2)^2 = x^6y^4 = x^2 \neq 1$; $(x^3y^3)^2 = x^3y^3x^3y^3 = x^3x^2y^3xy^3 = x^3x^2yxy^2 \cdot y^3 = xyxy = y^2 \neq 1$. That $t^2 \neq 1$ when $t = x, x^3, y$, or y^3 is obvious.

RESULT 4.8.4 *Let $K = \langle y^2 \rangle$. Then $K \lhd G$ and $G/K \cong D_4$.*
By Result 4.8.2, $K \subseteq Z(G)$ so $K \lhd G$. By Theorem A.10.4, $G/K = \langle Kx, Ky \rangle$. Note that $(Kx)^4 = 1$, $(Ky)^2 = 1$, and $(KxKy)^2 = Kxyxy = Ky^2 = 1$. Therefore G/K "satisfies" the relations of D_4. Also $\text{card}(D_4) = 8$ (Result 4.3.1) and $\text{card}(G/K) = \text{card}(G)/\text{card}(K) = \text{card}(G)/o(y^2) = 16/2$ (Results 4.8.1, Part (b) and 4.8.3) $= 8$. Thus $\text{card}(D_4) \leq 8 \leq \text{card}(G/K)$. Hence Theorem 2.2.6 may be applied to yield $D_4 \cong G/K$.

RESULT 4.8.5 *The subgroups of G are $\{1\}$, $\langle x \rangle$, $\langle x^2 \rangle$, $\langle y \rangle$, $\langle xy^2 \rangle$, $\langle xy \rangle$, $\langle x^2y \rangle$, $\langle y^2 \rangle$, $\langle x^2y^2 \rangle$, $\langle x^3y \rangle$, $\{1, x^2, y^2, x^2y^2\}$, $\{1, x, x^2, x^3, y^2, xy^2, x^2y^2, x^3y^2\}$, $\{1, y, x^2, x^2y, y^2, y^3, x^2y^2, x^2y^3\}$, $\{1, xy, x^2, x^3y, y^2, xy^3, x^2y^2, x^3y^3\}$, and G.*
The first ten entries in the above list are clearly subgroups, as is the last. Further-

more the eleventh entry is a subgroup by Result 4.8.2. Since $Z(G) \lhd G$, $Z(G)B$ is a subgroup of G whenever B is. Thus $Z(G)\langle x\rangle = \{1, x, x^2, x^3, y^2, xy^2, x^2y^2, x^3y^2\}$ is a subgroup (the twelfth entry), $Z(G)\langle y\rangle = \{1, y, x^2, x^2y, y^2, y^3, x^2y^2, x^2y^3\}$ is a subgroup (the thirteenth entry), and $Z(G)\langle xy\rangle = \{1, xy, x^2, x^3y, y^2, xy^3, x^2y^2, x^3y^3\}$ is a subgroup (the fourteenth entry).

The next step of the proof will be to show all cyclic subgroups of G appear in the list. To do this the cyclic subgroups generated by x^3, y^3, xy^3, x^2y^3, x^3y^3, and x^3y^2 must be accounted for. Result 4.8.3 will be used repeatedly. Since $o(x^3) = 4 = o(x)$ and $x^3 \in \langle x\rangle$, $\langle x^3\rangle = \langle x\rangle$ and similarly $\langle y^3\rangle = \langle y\rangle$. From the third relation $xyx = y$ so $(xy)^3 = xyxyxy = xy(xyx)y = xyyy = xy^3$. Thus $xy^3 \in \langle xy\rangle$ hence $\langle xy^3\rangle = \langle xy\rangle$ by a comparison of orders. Next note that $(x^2y)^3 = x^6y^3$ (because $x^2 \in Z(G)$ by Result 4.8.2) $= x^2y^3$. This and $o(x^2y) = 4 = o(x^2y^3)$ imply $\langle x^2y^3\rangle = \langle x^2y\rangle$. The demonstrations that $\langle x^3y^3\rangle = \langle x^3y\rangle$ and $\langle x^3y^2\rangle = \langle xy^2\rangle$ are similar.

It will next be shown that the above list accounts for all subgroups B such that $y^2 \in B$. By Result 4.8.4, $G/\langle y^2\rangle$ is isomorphic to D_4 and consequently has ten subgroups (Result 4.3.2). Therefore the set $\{C \mid C$ is a subgroup of G and $\langle y^2\rangle \subseteq C\}$ has ten elements. But note there are ten subgroups in the list having $\langle y^2\rangle$ as a subset (the fourth, fifth, seventh, eighth, and tenth through fifteenth entries). Thus every subgroup B with $y^2 \in B$ is in the list.

It can now be shown that every subgroup of G has been accounted for. Let B be a subgroup of G. By Lagrange's theorem, card(B) = 1, 2, 4, 8, or 16. If card(B) = 1 or 16, then $B = \{1\}$ or $B = G$ and B is in the list. Assume card(B) = 8. Then $B \lhd G$ by Theorem A.10.2 and hence $G/B \cong Z(2)$. Thus By has order 2 in G/B so $(By)^2 = 1$ implying $y^2 \in B$, so as shown in the preceding paragraph, B is in the list. Next suppose card(B) = 4. By Cauchy's theorem (Theorem A.1.5), there exists $b \in B$ of order 2. By Result 4.8.3, $b = y^2$, x^2, or x^2y^2. In the first case B appears in the list as shown above. Assume $b = x^2$. Then card$(\langle x^2\rangle)$ = 2 so $B\backslash\langle x^2\rangle$ is nonempty. Let $c \in B\backslash\langle x^2\rangle$. Since card$(B)$ = 4, $o(c)$ = 2 or 4. If $o(c)$ = 2, then by Result 4.8.3 again $c = y^2$ or $c = x^2y^2$. In either case $y^2 \in B$ (if $c = x^2y^2$, then $y^2 = b^{-1}c \in B$) so B has been accounted for. On the other hand if $o(c)$ = 4, then $B = \langle c\rangle$, hence B is cyclic and so here too B has been accounted for. The remaining case, $b = x^2y^2$, is similar. Finally note that if card(B) = 2, then B is cyclic and hence has been listed.

The subgroups of G are illustrated on page 127.

RESULT 4.8.6 *G is nilpotent.*
By Result 4.8.1, Part (b) and Theorem A.5.7.

LEMMA *If A is a group and $a \in A$, then the subgroup $Z(A)\langle a\rangle$ is abelian.*
Let $ca^n, c^*a^m \in Z(A)\langle a\rangle$. Then $ca^nc^*a^m = cc^*a^{n+m} = c^*ca^{n+m} = c^*a^mca^n$.

RESULT 4.8.7 *Proper subgroups of G are abelian.*

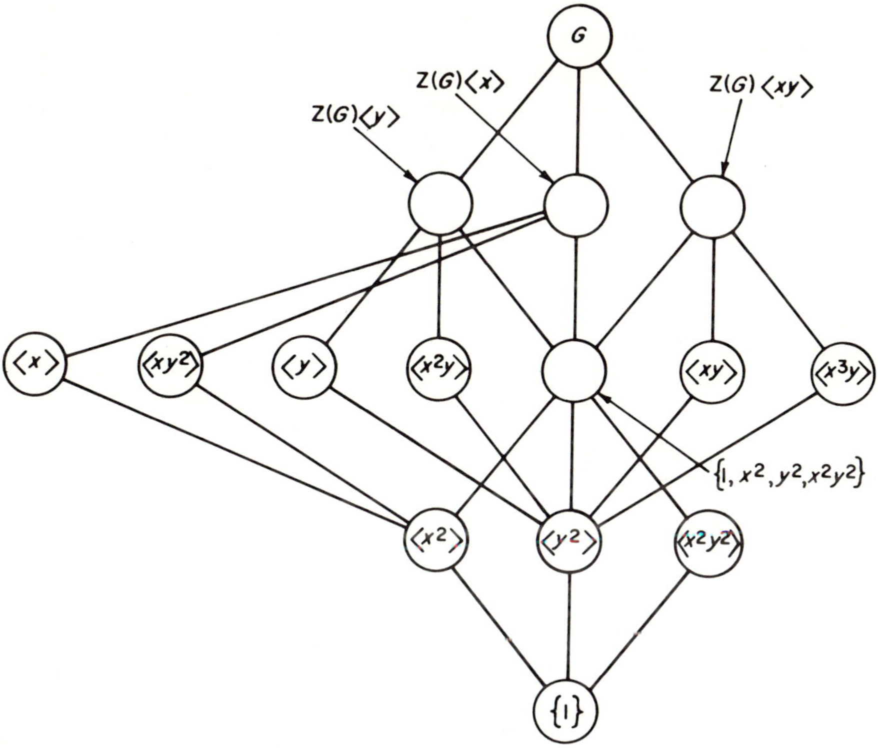

All entries in the list of subgroups (Result 4.8.5) are cyclic with the exception of entries 11–14. By Result 4.8.2, $\{1, x^2, y^2, x^2y^2\} = Z(G)$ so the eleventh entry is abelian. The twelfth, thirteenth, and fourteenth entries are $Z(G)\langle x\rangle$, $Z(G)\langle y\rangle$, and $Z(G)\langle xy\rangle$ respectively. Hence these subgroups are abelian by the lemma.

RESULT 4.8.8 *G is indecomposable.*
Since proper subgroups of G are abelian (Result 4.8.7), while G itself is not abelian (Result 4.8.2), the Corollary to Theorem A.2.1 shows that G cannot be expressed as $G = A \oplus B$ where A and B are proper subgroups.

EXERCISES

1. Show G is not subdirectly irreducible.
2. Identify G' and $\Phi(G)$.
3. Identify the normal subgroups of G and draw the normal subgroup diagram.
4. Identify the characteristic subgroups of G.
5. Describe the decomposition of G into conjugacy classes and write the class equation.

NOTE

1. As a counterexample G shows
   (a) The converse of Theorem A.5.9 is false. That is, there exist indecomposable nilpotent groups with decomposable centers ($Z(G) = \langle x^2 \rangle \oplus \langle y^2 \rangle$).
   (b) In contrast to the abelian case, torsion indecomposable nilpotent groups need not be subdirectly irreducible.

□□□□□

Example 4.9 (The Octohedral Group)

Let $G = \mathrm{G}(X, Y \mid X^3 = 1, Y^2 = 1, (XY)^4 = 1)$.

RESULT 4.9.1

(a) $G \cong S_4$.
(b) $\mathrm{card}(G) = 24$.
(c) *$PSL(2,3)$ can be embedded in G.*

Let $H = \langle x, yxy \rangle$. It is easily seen that xxx^{-1}, $x(yxy)x^{-1}$, yxy^{-1}, and $y(yxy)y^{-1}$ all belong to H so $H \lhd G$ by Theorem A.10.3. By Theorem A.10.4 then, $G/H = \langle Hx, Hy \rangle = \langle Hy \rangle$. Thus from the second relation, $\mathrm{card}(G/H) \leqslant 2$. Let A be the tetrahedral group (Example 4.6), $A = \mathrm{G}(S, T \mid S^3 = 1, T^3 = 1, (ST)^2 = 1)$. Define the function $h: \{S, T\} \to \{x, yxy\}$ by $h(S) = x$ and $h(T) = yxy$. It is obvious that the relations of A are "satisfied" in H (namely that $x^3 = 1$, $(yxy)^3 = 1$, and $(xyxy)^2 = 1$). Theorem 2.2.5 may now be applied to yield a surjective homomorphism $e: A \to H$. Thus $\mathrm{card}(H) \leqslant \mathrm{card}(A)$. But $\mathrm{card}(A) = 12$ by Result 4.6.1, Part (b), so $\mathrm{card}(H) \leqslant 12$. Thus $\mathrm{card}(G) = \mathrm{card}(G/H) \cdot \mathrm{card}(H) \leqslant 2 \cdot 12 = 24$. Also $\mathrm{card}(S_4) = 4! = 24$ so $\mathrm{card}(G) \leqslant 24 = \mathrm{card}(S_4)$.

Let j and k be the permutations $j = \left(\begin{smallmatrix}1&2&3&4\\2&4&3&1\end{smallmatrix}\right)$ and $k = \left(\begin{smallmatrix}1&2&3&4\\3&2&1&4\end{smallmatrix}\right)$. To show that j and k generate S_4 it suffices, because of Lagrange's theorem to show $12 < \mathrm{card}(\langle j, k \rangle)$. This will be done by exhibiting 13 distinct elements of $\langle j, k \rangle$ as follows: $j = \left(\begin{smallmatrix}1&2&3&4\\2&4&3&1\end{smallmatrix}\right)$; $k = \left(\begin{smallmatrix}1&2&3&4\\3&2&1&4\end{smallmatrix}\right)$; $jk = \left(\begin{smallmatrix}1&2&3&4\\2&4&3&1\end{smallmatrix}\right)\left(\begin{smallmatrix}1&2&3&4\\3&2&1&4\end{smallmatrix}\right) = \left(\begin{smallmatrix}1&2&3&4\\3&4&2&1\end{smallmatrix}\right)$; $(jk)^2 = \left(\begin{smallmatrix}1&2&3&4\\3&4&2&1\end{smallmatrix}\right)\left(\begin{smallmatrix}1&2&3&4\\3&4&2&1\end{smallmatrix}\right) = \left(\begin{smallmatrix}1&2&3&4\\2&1&4&3\end{smallmatrix}\right)$; $j^2 = \left(\begin{smallmatrix}1&2&3&4\\2&4&3&1\end{smallmatrix}\right)\left(\begin{smallmatrix}1&2&3&4\\2&4&3&1\end{smallmatrix}\right) = \left(\begin{smallmatrix}1&2&3&4\\4&1&3&2\end{smallmatrix}\right)$; $j^2k = \left(\begin{smallmatrix}1&2&3&4\\4&1&3&2\end{smallmatrix}\right) \cdot \left(\begin{smallmatrix}1&2&3&4\\3&2&1&4\end{smallmatrix}\right) = \left(\begin{smallmatrix}1&2&3&4\\3&1&4&2\end{smallmatrix}\right)$; $kj = \left(\begin{smallmatrix}1&2&3&4\\3&2&1&4\end{smallmatrix}\right)\left(\begin{smallmatrix}1&2&3&4\\2&4&3&1\end{smallmatrix}\right) = \left(\begin{smallmatrix}1&2&3&4\\2&4&1&3\end{smallmatrix}\right)$; $kj^2 = \left(\begin{smallmatrix}1&2&3&4\\3&2&1&4\end{smallmatrix}\right)\left(\begin{smallmatrix}1&2&2&4\\4&1&3&2\end{smallmatrix}\right) = \left(\begin{smallmatrix}1&2&3&4\\4&3&1&2\end{smallmatrix}\right)$; $(kj)k = \left(\begin{smallmatrix}1&2&3&4\\2&4&1&3\end{smallmatrix}\right)\left(\begin{smallmatrix}1&2&3&4\\3&2&1&4\end{smallmatrix}\right) = \left(\begin{smallmatrix}1&2&3&4\\1&4&2&3\end{smallmatrix}\right)$; $(kj^2)k = \left(\begin{smallmatrix}1&2&3&4\\4&3&1&2\end{smallmatrix}\right)\left(\begin{smallmatrix}1&2&3&4\\3&2&1&4\end{smallmatrix}\right) = \left(\begin{smallmatrix}1&2&3&4\\1&3&4&2\end{smallmatrix}\right)$; $(kj)^2 = \left(\begin{smallmatrix}1&2&3&4\\2&4&1&3\end{smallmatrix}\right)\left(\begin{smallmatrix}1&2&3&4\\2&4&1&3\end{smallmatrix}\right) = \left(\begin{smallmatrix}1&2&3&4\\4&3&2&1\end{smallmatrix}\right)$; $j(jk)^2 = \left(\begin{smallmatrix}1&2&3&4\\2&4&3&1\end{smallmatrix}\right)\left(\begin{smallmatrix}1&2&3&4\\2&1&4&3\end{smallmatrix}\right) = \left(\begin{smallmatrix}1&2&3&4\\4&2&1&3\end{smallmatrix}\right)$; and the identity element. Conclude $S_4 = \langle j, k \rangle$.

Now define the function $f: \{X, Y\} \to \{j, k\}$ by $f(X) = j$ and $f(Y) = k$. In order to apply Theorem 2.2.6 it must be shown that the relations of G are "satisfied" in S_4. This is done as follows: $j^3 = j^2 j = \left(\begin{smallmatrix}1&2&3&4\\4&1&3&2\end{smallmatrix}\right)\left(\begin{smallmatrix}1&2&3&4\\2&4&3&1\end{smallmatrix}\right) = 1$; $k^2 = \left(\begin{smallmatrix}1&2&3&4\\3&2&1&4\end{smallmatrix}\right)\left(\begin{smallmatrix}1&2&3&4\\3&2&1&4\end{smallmatrix}\right) = 1$; and $(jk)^4 = (jk)^2(jk)^2 = \left(\begin{smallmatrix}1&2&3&4\\2&1&4&3\end{smallmatrix}\right)\left(\begin{smallmatrix}1&2&3&4\\2&1&4&3\end{smallmatrix}\right) = 1$. Theorem

2.2.6 may now be applied to yield an isomorphism $d:G \to S_4$. This proves Part (a). Part (b) follows immediately.

Note again that card(G/H) $\leqslant 2$. This and card(G) = 24 imply $12 \leqslant$ card(H). This and card(H) $\leqslant 12$ imply card(H) = 12, so card(H) = card(A). Since $e:A \to H$ is a surjective homomorphism, $A/\ker e \cong H$. Then card(H) = card(A) implies card(ker e) = 1 and consequently ker $e = \{1\}$. Conclude $A \cong H$. But $PSL(2,3) \cong A$ (Result 4.6.1) so $PSL(2,3) \cong H$ proving Part (c).

RESULT 4.9.2

(a) *S_3 can be embedded in G.*

(b) *D_4 can be embedded in G.*

By Theorem A.10.8, S_3 can be embedded in S_4. This and $S_4 \cong G$ (Result 4.9.1) prove Part (a). Let $H = \langle y \rangle$ in the group D_4, and let C denote the core (in the group D_4) of H. Then card(D_4) = 8 (Result 4.3.1) and card(H) = o(y) = 2 imply $[D_4:H] = 4$. Also card(H) = 2 implies $C = \{1\}$ or H. But in the second case, $\langle y \rangle \lhd D_4$ in contradiction to Result 4.3.3. Thus $C = \{1\}$. Application of Theorem A.1.3 shows $D_4/\{1\}$, and hence D_4, can be embedded in S_4. Finally $S_4 \cong G$ (Result 4.9.1 again) shows D_4 can be embedded in G.

RESULT 4.9.3 *G is a converse lagrange group.*

The divisors of card(G) = 24 are 1, 2, 3, 4, 6, 8, 12, and 24. Clearly G has subgroups of cardinalities 1 and 24. By Result 4.9.1, Part (c), $PSL(2,3)$ can be embedded in G. Since $PSL(2,3)$ has subgroups of cardinalities 2, 3, 4, and 12 (Result 4.6.2), so too does G. Finally G has subgroups of cardinalities 6 and 8 by Result 4.9.2, Parts (a) and (b) respectively.

RESULT 4.9.4

(a) *$G' \cong PSL(2,3)$.*

(b) *G' is not nilpotent.*

By Result 4.9.1, Part (c), G has a subgroup $R \cong PSL(2,3)$. Since $[G:R] = 24/12 = 2$, $R \lhd G$ by Theorem A.10.2 and hence G/R is isomorphic to $Z(2)$ and is therefore abelian which implies $G' \subseteq R$ by Theorem A.6.4. Now $G' \lhd G$ and $G' \subseteq R$ imply $G' \lhd R$. By Theorem 4.6.3, R has three normal subgroups; R, $\{1\}$, and a normal subgroup H of cardinality 4. If it can be shown that 3 divides card(G'), then $G' \neq \{1\}$ and $G' \neq H$ so $G' = R$ and the proof of Part (a) will be complete.

Since $xyx^{-1}y^{-1} \in G'$, $xyx^2y \in G'$ by the first two relations. Then since G' is normal, $y^{-1}xy(xyx^2y)(y^{-1}xy)^{-1} = y^{-1}xyxyxy \in G'$. But by the relations, $(xy)^3 = (xy)^{-1} = y^{-1}x^{-1} = yx^{-1}$. Thus $y^{-1}yx^{-1} \in G'$. Inverting yields $x \in G'$ so o(x)|card(G'). From the first relation, however, o(x) = 3 unless $x = 1$. If $x = 1$, then $G = \langle y \rangle$, hence card(G) $\leqslant 2$ which is false and so o(x) = 3 and therefore 3|card(G') as required. This proves Part (a). Part (b) follows from Part (a) and Result 4.6.9.

RESULT 4.9.5 *G is solvable.*
By Results 4.9.4, 4.6.1, Part (a), and 4.6.7, G' is solvable. Since G/G' is abelian, it too is solvable. Conclude the result by Theorem A.6.6.

RESULT 4.9.6
(a) $G'' \lhd G$.
(b) $\text{card}(G'') = 4$.
(c) *G'' is noncyclic.*
Part (a) by Theorem A.6.2. Parts (b) and (c) by Results 4.9.4, 4.6.1, Part (a), and 4.6.11.

RESULT 4.9.7 *The normal subgroups of G are G, G', G'', and $\{1\}$.*
Since all of the above subgroups are normal, it suffices to show these are the only normal subgroups. Let $N \lhd G$. Then $G' \subseteq NG' \subseteq G$. But $\text{card}(G') = 12$ by Results 4.9.4 and 4.6.1, Part (b). Therefore $NG' = G'$ or G. In the first case, $N \subseteq G'$ and so $N \lhd G' \cong PSL(2,3)$ (the isomorphism by Result 4.9.4). This implies $N = G'$, G'', or $\{1\}$ (Results 4.6.3 and 4.6.11, Part (a)).

In the second case ($NG' = G$), $G/G' \cong N/(N \cap G')$ so $2 \cdot \text{card}(N \cap G') = \text{card}(N)$. But $N \cap G' \lhd G'$ so by a second application of Results 4.6.3 and 4.6.11 conclude that $N \cap G' = G'$, G'', or $\{1\}$. This implies $\text{card}(N \cap G') = 12, 4$, or 1 and so $\text{card}(N) = 2 \cdot \text{card}(N \cap G') = 24, 8$, or 2. If $\text{card}(N) = 24$, then $N = G$. If $\text{card}(N) = 8$, then $\text{card}(G/N) = 3$ so G/N is abelian and hence $G' \subseteq N$ (Theorem A.6.4). Thus the contradiction $12 | 8$ shows the case $\text{card}(N) = 8$ cannot occur. Finally suppose $\text{card}(N) = 2$. Then $\text{card}(N \cap G') = 1$ so $N \cap G' = \{1\}$ and hence $G = G' \oplus N$. Theorem A.6.3 now shows $G' = G'' \oplus N'$, hence $G' = G''$ since N, having cardinality 2, is abelian. Result 4.9.4 and the last equation imply $PSL(2,3)' = PSL(2,3)$ in contradiction to Result 4.6.11. Therefore the case $\text{card}(N) = 2$ cannot occur.

RESULT 4.9.8
(a) *G is subdirectly irreducible.*
(b) *G is indecomposable.*
By Result 4.9.7, $\text{H}(G) = G \cap G' \cap G'' = G''$, so $\text{H}(G)$ is nontrivial ($\text{card}(G'') = 4$ by Result 4.9.6, Part (b)) and hence G is subdirectly irreducible proving Part (a). Part (b) follows from Part (a) and Theorem A.10.10.

EXERCISES

1. Show G has a unique subgroup of cardinality 12.
2. Show $\text{Z}(G)$ is trivial.
3. Find a maximal subgroup of G of nonprime index.
4. Show $\Phi(G)$ is trivial.
5. Show S_4, and hence G, has 9 elements of order 2, 8 elements of order 3, and 6 elements of order 4.

6. (a) Show G has a noncyclic minimal normal subgroup.
   (b) Show G is nonsupersolvable.
7. Show $n_2(G) = 3$ and $n_3(G) = 4$.
8. Show the 2-Sylow subgroups of G are Carter subgroups.
9. Show G is complete.
10. Show
    (a) If G can be embedded in a group A as a normal subgroup, then the image of this embedding is a direct factor of A.
    (b) There exists no group C such that $C' \cong G$.
11. Show each normal subgroup of G is fully invariant.
12. Show G'' is the intersection of the 2-Sylow subgroups of G.
13. Let A be a group with the following properties:
    (a) $\text{card}(A) = 24$.
    (b) A is not supersolvable.
    (c) $Z(A)$ is trivial.
    Show $A \cong G$.
14. Let A be a group with the following properties:
    (a) $\text{card}(A) = 24$.
    (b) A has no normal subgroups of cardinalities 2 or 3.
    Show $A \cong G$.
15. Let Q be the quaternion group (Example 4.4). Show $G \cong \text{Aut}(Q)$.
16. Let A be the tetrahedral group (Example 4.6). Show $G \cong \text{Aut}(A)$.

NOTES

1. As a counterexample G separates the classes of supersolvable and converse lagrange groups and also shows that
   (a) The class of converse lagrange groups is not subgroup closed (G is converse lagrange by Result 4.9.3, but its subgroup G' is not by Results 4.9.4 and 4.6.6).
   (b) The class of groups $\{G \mid G' \text{ is nilpotent}\}$ is not extension closed (G' and G/G' both belong to this class, but G does not by Result 4.9.10).
2. Characterizations of G are given in Exercises 13 and 14. Note that the Exercise 13 characterization and Exercise 8 of Example 4.10 show that Examples 4.9 and 4.10 are the only (up to isomorphism) indecomposable nonsupersolvable groups of cardinality 24.

□□□□□

Example 4.10 (The Binary Tetrahedral Group)

Let $G = \mathrm{G}(X, Y \mid X^3Y^{-3} = 1, (XY)^2Y^{-3} = 1)$.

RESULT 4.10.1
(a) $G \cong SL(2,3)$.
(b) $\mathrm{card}(G) = 24$.
(c) $\mathrm{Z}(G) = \langle y^3 \rangle$ *and* $\mathrm{card}(\mathrm{Z}(G)) = 2$.

By the first relation, $x^3 = y^3$ so y^3 commutes with both x and y, hence with all generators of G. Thus $y^3 \in \mathrm{Z}(G)$. Let $H = \langle y^3 \rangle$. Then $H \subseteq \mathrm{Z}(G)$ so $H \lhd G$. By Theorem A.10.4, $G/H = \langle Hx, Hy \rangle$. Now $y^3 = (xy)^2$ implies $y^2 = xyx$ and hence $y^4 = (xyx)(xyx) = xyx^2yx$. But $x^3 = (xy)^2$ implies $x^2 = yxy$ so $y^4 = xy(yxy)yx = xy^2xy^2x = x(xyx)x(xyx)x = x^2yx^3yx^2 = x^2yy^3yx^2 = x^2y^5x^2$. Thus y^6 is equal to $(x^2y^5x^2)(xyx) = x^2y^5x^3yx = x^2y^5y^3yx = x^2y^9x = x^2x^9x = x^{12} = y^{12}$. Consequently $y^6 = 1$, and so $\mathrm{card}(H) \leqslant 2$. Let A be the tetrahedral group (Example 4.6), $A = \mathrm{G}(S, T \mid S^3 = 1, T^3 = 1, (ST)^2 = 1)$. Define the function $f: \{S, T\} \to \{Hx, Hy\}$ by $f(S) = Hx$ and $f(T) = Hy$. Since $x^3 = y^3$, $(Hx)^3 = Hy^3 = 1$; and since $(xy)^2 = y^3$, $(HxHy)^2 = Hy^3 = 1$. Moreover $(Hy)^3 = 1$. Thus the relations of A are "satisfied" in G/H. Apply Theorem 2.2.5 which yields a surjective homomorphism $d: A \to G/H$. Thus $\mathrm{card}(G/H) \leqslant \mathrm{card}(A)$. But $\mathrm{card}(A) = 12$ by Result 4.6.1, Part (b). Consequently $\mathrm{card}(G) = \mathrm{card}(G/H) \cdot \mathrm{card}(H) \leqslant 12 \cdot 2 = 24$. By Theorem 3.2.21, $\mathrm{card}(SL(2,3)) = 3(3-1)(3+1) = 24$ and so $\mathrm{card}(G) \leqslant 24 = \mathrm{card}(SL(2,3))$.

Let $a = \left(\begin{smallmatrix}2&2\\0&2\end{smallmatrix}\right)$ and $b = \left(\begin{smallmatrix}2&0\\2&2\end{smallmatrix}\right)$ in $SL(2,3)$. Then $a^2 = \left(\begin{smallmatrix}2&2\\0&2\end{smallmatrix}\right)\left(\begin{smallmatrix}2&2\\0&2\end{smallmatrix}\right) = \left(\begin{smallmatrix}1&2\\0&1\end{smallmatrix}\right)$ and $b^2 = \left(\begin{smallmatrix}2&0\\2&2\end{smallmatrix}\right)\left(\begin{smallmatrix}2&0\\2&2\end{smallmatrix}\right) = \left(\begin{smallmatrix}1&0\\2&1\end{smallmatrix}\right)$. Inverting yields $a^{-2} = \left(\begin{smallmatrix}1&1\\0&1\end{smallmatrix}\right)$ and $b^{-2} = \left(\begin{smallmatrix}1&0\\1&1\end{smallmatrix}\right)$ and hence the four matrices a^2, b^2, a^{-2}, and b^{-2} generate $SL(2,3)$ by Theorem 3.2.18. That is, $SL(2,3) = \langle a, b \rangle$. Now define the function $F: \{X, Y\} \to \{a, b\}$ by $F(X) = a$ and $F(Y) = b$. In order to apply Theorem 2.2.6 it must be shown that the relations of G are "satisfied" in $SL(2,3)$. This is done as follows: $a^3 = a^2a = \left(\begin{smallmatrix}1&2\\0&1\end{smallmatrix}\right)\left(\begin{smallmatrix}2&2\\0&2\end{smallmatrix}\right) = \left(\begin{smallmatrix}2&0\\0&2\end{smallmatrix}\right)$ while $b^3 = b^2b = \left(\begin{smallmatrix}1&0\\2&1\end{smallmatrix}\right)\left(\begin{smallmatrix}2&0\\2&2\end{smallmatrix}\right) = \left(\begin{smallmatrix}2&0\\0&2\end{smallmatrix}\right)$ so $a^3b^{-3} = 1$; $ab = \left(\begin{smallmatrix}2&2\\0&2\end{smallmatrix}\right)\left(\begin{smallmatrix}2&0\\2&2\end{smallmatrix}\right) = \left(\begin{smallmatrix}2&1\\1&1\end{smallmatrix}\right)$ so $(ab)^2 = \left(\begin{smallmatrix}2&1\\1&1\end{smallmatrix}\right)\left(\begin{smallmatrix}2&1\\1&1\end{smallmatrix}\right) = \left(\begin{smallmatrix}2&0\\0&2\end{smallmatrix}\right)$ hence $(ab)^2b^{-3} = 1$. Theorem 2.2.6 may now be applied to yield an isomorphism $D: G \to SL(2,3)$. This proves Part (a). Part (b) follows immediately.

Since $\mathrm{card}(H) \leqslant 2$, $\mathrm{card}(G)/2 \leqslant \mathrm{card}(G)/\mathrm{card}(H)$. Hence $12 \leqslant \mathrm{card}(G/H)$. But as noted in the first paragraph, $\mathrm{card}(G/H) \leqslant \mathrm{card}(A) = 12$. Thus $\mathrm{card}(G/H) = 12$ proving $d: A \to G/H$ is an isomorphism and so $G/H \cong A$. But $\mathrm{card}(G) = 24$ and $\mathrm{card}(G/H) = 12$ imply $\mathrm{card}(H) = 2$. Also $\mathrm{card}(\mathrm{Z}(G)) = 2$ by Part (a) and Theorem 3.2.20. This and $H \subseteq \mathrm{Z}(G)$ (as observed in the first paragraph) shows $H = \mathrm{Z}(G)$.

RESULT 4.10.2
(a) $\mathrm{Z}(G) \subseteq G'$.
(b) $\mathrm{card}(G') = 8$.
(c) G' *is nilpotent.*

(d) *G' is the only 2-Sylow subgroup of G.*
(e) *$G' \cong Q$, the quaternion group (Example 4.4).*
The relations imply $x^2 = yxy$ and $y^2 = xyx$. Premultiply $x^2 = yxy$ by x^{-1} then postmultiply by y^{-2} to get $xy^{-2} = x^{-1}yxy^{-1}$. Thus $xy^{-2} \in G'$. Similarly $y^2 = xyx$ implies $yx^{-2} \in G'$. Hence $(yx^{-2})(xy^{-2}) \in G'$, that is, $yx^{-1}y^{-2} \in G'$. Then since G' is normal, $y^{-1}(yx^{-1}y^{-2})y \in G'$, that is, $x^{-1}y^{-1} \in G'$. Inverting yields $yx \in G'$. Square this to get $yxyx \in G'$. But $yxyx = y(xyx) = yy^2 = y^3$. Hence $y^3 \in G'$ implying $Z(G) \subseteq G'$ by Result 4.10.1, Part (c).

Again by Result 4.10.1, $G \cong SL(2,3)$ so $G/Z(G) \cong PSL(2,3)$. Let $B/Z(G) = (G/Z(G))'$. Then card$(B/Z(G))$ = 4 by Results 4.6.1, Part (a) and 4.6.11. This and card$(Z(G))$ = 2 show card(B) = 8. Since $B/Z(G) = (G/Z(G))'$, $G' \subseteq B$. If this inclusion is proper, then card(G') = 1, 2, or 4. If card(G') = 1, then G is abelian, an obvious contradiction. If card(G') = 2, then $G' = Z(G)$ because $Z(G) \subseteq G'$. This implies $G/Z(G)$ and hence $PSL(2,3)$ is abelian, again a contradiction. If card(G') = 4, then card$(G'/Z(G))$ = 2 and so $PSL(2,3)$ has a normal subgroup of cardinality 2 in contradiction to Result 4.6.3. Thus $G' \subset B$ must be false and so $G' = B$ implying card(G') = card(B) = 8, proving Part (b).

Part (c) follows immediately from Part (b) and Theorem A.5.7.

By Theorem A.10.15 $n_2(G) = [G:N_G(G')]$ since G' is a 2-Sylow subgroup of G by Part (b). Thus $n_2(G) = 1$ proving Part (d).

By Results 4.4.1, Part (a) and 4.10.1, Part (a), G has a subgroup $R \simeq Q$. Then card(R) = 8 (Result 4.4.1) so R is a 2-Sylow subgroup of G. Conclude $R = G'$ by Part (d) hence $G' \cong Q$ proving Part (e).

RESULT 4.10.3 *G has no subgroups of cardinality 12 and consequently G is not converse lagrange.*
Suppose K is a subgroup of cardinality 12. Then $K \lhd G$ by Theorem A.10.2 and hence $G/K \cong Z(2)$. Thus $G' \subseteq K$ by Theorem A.6.4. This and Lagrange's theorem imply $8 \mid 12$ by Result 4.10.2, Part (b).

RESULT 4.10.4 *The normal subgroups of G are $\{1\}$, $Z(G)$, G', and G.*
Since all of the above subgroups are normal it suffices to show that these are the only normal subgroups. Let $C \lhd G$. Then normality of C and $Z(G)$ imply that $CZ(G) \lhd G$ and hence $CZ(G)/Z(G) \lhd G/Z(G)$. Since $G \cong SL(2,3)$ (Result 4.10.1), $G/Z(G) \cong PSL(2,3)$ which has only three normal subgroups, one each of cardinalities 12, 4, and 1 (Results 4.6.3 and 4.6.1, Part (d)). Therefore since $CZ(G)/Z(G) \cong C/(C \cap Z(G))$, conclude card$(C/(C \cap Z(G)))$ = 12, 4, or 1. Furthermore, since card$(Z(G))$ = 2 (Result 4.10.1, Part (c)), card$(C \cap Z(G))$ = 2 or 1. Thus card(C) = (12, 4, or 1) $\times$ (2 or 1) = 24, 8, 2, 12, 4, or 1.

If card(C) = 24, $C = G$. If card(C) = 8, $C = G'$ by Result 4.10.2, Part (d). If card(C) = 2, let $C = \langle c \rangle$. Then for all $g \in G$, $g^{-1}cg \in C$ so $g^{-1}cg = c$ (if $g^{-1}cg = 1$ for some g, then $c = 1$ which is false). Thus $cg = gc$ for all $g \in G$ and so $c \in Z(G)$ and hence $C = Z(G)$.

Result 4.10.3 eliminates card(C) = 12. If card(C) = 4, then card(CS) = 12 where S is any 3-Sylow subgroup of G. Application of Result 4.10.3 again therefore eliminates the case card(C) = 4. Finally if card(C) = 1, then $C = \{1\}$.

RESULT 4.10.5
(a) *G is subdirectly irreducible.*
(b) *G is indecomposable.*
By Result 4.10.4, H(G) = Z(G) $\cap G' \cap G$ = Z(G) because Z(G) $\subseteq G'$ by Result 4.10.2, Part (a). Thus H(G) is nontrivial (card(Z(G)) = 2 by Result 4.10.1, Part (c)) and hence G is subdirectly irreducible proving Part (a). Part (b) follows from Part (a) and Theorem A.10.10.

RESULT 4.10.6 *G is solvable.*
By Result 4.10.2, Part (c), G' is nilpotent and hence solvable (Theorem A.6.7). Also G/G' is abelian, hence solvable. Conclude the result by Theorem A.6.6.

RESULT 4.10.7 *G is not supersolvable.*
It is easily seen that quotient groups of supersolvable groups must be supersolvable. But $G \cong SL(2,3)$ (Result 4.10.1), hence $G/\mathrm{Z}(G) \cong PSL(2,3)$, so $G/\mathrm{Z}(G)$ is nonsupersolvable (Results 4.6.1, Part (a) and 4.6.8). Thus G is nonsupersolvable.

RESULT 4.10.8 *Minimal normal subgroups of G are cyclic.*
Since Z(G) $\subseteq G'$ (Result 4.10.2, Part (a)), Result 4.10.4 shows Z(G) is a minimal normal subgroup and is moreover the only minimal normal subgroup. Since card(Z(G)) = 2 (Result 4.10.1, Part (c)), Z(G) is cyclic.

EXERCISES

1. Show G has one element of order 2, eight elements of order 3, six elements of order 4, and eight elements of order 6.
2. Show $G'' = \mathrm{Z}(G)$.
3. Show all normal subgroups of G are fully invariant.
4. Identify the Carter subgroups of G.
5. Show G has 15 subgroups (including itself) and draw its subgroup diagram.
6. Identify $\Phi(G)$.
7. Describe the decomposition of G into conjugacy classes and write the class equation. In particular show that all elements of order 4 are conjugate.
8. Let A be a group such that
   (a) card(A) = 24.
   (b) A is nonsupersolvable.
   (c) Z(A) is nontrivial.
   (d) A is indecomposable.
   Show $A \cong G$.

9. If A is a group such that $L(A) \cong L(G)$ show $A \cong G$.

10. Let Q be the quaternion group (Example 4.4). Let h be the automorphism given in Example 4.4, Exercise 3. Define $g: Z(3) \to \mathrm{Aut}(Q)$ by $g_0 = 1, g_1 = h$, and $g_2 = h^2$. Show $Q \times_g Z(3) \cong G$.

NOTES

1. As a counterexample G
   (a) separates the class of supersolvable groups and $\{H \mid H' \text{ is nilpotent}\}$
   (b) separates the class of supersolvable groups and $\{H \mid$ minimal normal subgroups of H are cyclic$\}$
   (c) shows $\{H \mid$ minimal normal subgroups of H are cyclic$\}$ is not quotient closed (G belongs to this class by Result 4.10.8, but $G/Z(G) \cong PSL(2,3)$ does not by Result 4.6.11).
2. Characterizations of G are given in Exercises 8 and 9. Note that the Exercise 8 characterization and Exercise 13 of Example 4.9 show that Example 4.9 and 4.10 are the only (up to isomorphism) indecomposable nonsupersolvable groups of cardinality 24.

□□□□□

Example 4.11 (The Icosahedral Group)

Let $G = G(X, Y \mid X^5 = 1, Y^5 = 1, (XY)^2 = 1, (Y^{-1}X)^3 = 1)$.

LEMMA *Let A be a (not necessarily finite) group. Let H and B be subgroups with $B = \langle b \rangle$ finite and cyclic. Suppose $bh \in HB$ for all $h \in H$. Then HB is a subgroup of A.*

It is first claimed that $b^n h \in HB$ for all $h \in H$ and all natural numbers n. This is proved by induction noting the induction begins at $n = 0$. Assume now the induction hypothesis $b^{n-1}h \in HB$ for all $h \in H$. Note that $b^n h = b^{n-1}(bh)$. But by hypothesis, $bh = h_2 b^i$ for some $h_2 \in H$, so $b^n h = b^{n-1}h_2 b^i$. By the induction hypothesis, $b^{n-1}h_2 \in HB$ so $b^{n-1}h_2 = h_3 b^j$ for some $h_3 \in H$. Thus $b^n h = h_3 b^j b^i = h_3 b^{i+j} \in HB$ proving the claim.

Now let $x_1, x_2 \in HB$. Thus $x_1 = h_1 b^i$ and $x_2 = h_2 b^j$ for $h_1, h_2 \in H$. Since $o(b)$ is finite it may be assumed that i is nonnegative. Then by the preceding paragraph, $b^i h_2 = h_3 b^k$ for some $h_3 \in H$. Hence $x_1 x_2 = h_1 b^i h_2 b^j = h_1 h_3 b^k b^j = (h_1 h_3) b^{k+j} \in HB$.

Finally it must be shown HB is closed under inverses. Let $x \in HB$ so $x = hb^i$ where $h \in H$. Then $x^{-1} = b^{-i}h^{-1}$. Let $n = o(b)$. Then $b^{-1} = b^{n-1}$ but as shown before $b^{n-1}h^{-1} \in HB$. Thus $x^{-1} \in HB$ as required.

RESULT 4.11.1 *Let* $H = \langle y^{-1}x, xy^{-1}\rangle$ *and let* $B = \langle x\rangle$. *Then*
(a) $G = HB$.
(b) $G \cong PSL(2,5)$.
(c) $\text{card}(G) = 60 = 2^2\cdot 3\cdot 5$.
(d) $H \cong PSL(2,3)$ *and so* $\text{card}(H) = 12$.
For convenience denote $y^{-1}x$ by s and xy^{-1} by t. Thus $H = \langle s,t\rangle$. Let M denote the tetrahedral group (Example 4.6), $M = G(A,B \mid A^3 = 1, B^3 = 1, (AB)^2 = 1)$. Define the function $f:\{A,B\} \to \{s,t\}$ by $f(A) = s$ and $f(B) = t$. In order to apply Theorem 2.2.5 it must be shown that H "satisfies" the relations of M. This is done as follows: $s^3 = 1$ immediately from the fourth relation. Premultiplying the fourth relation by x, then postmultiplying by x^{-1} yields $t^3 = 1$. That $(st)^2 = 1$ will be shown as follows. The third relation implies $xyx = y^{-1}$ and hence $st = y^{-1}x^2y^{-1} = xyxx^2xyx = xyx^{-1}yx$. From the fourth relation $xy^{-1}xy^{-1}x = y$ which when inverted yields $x^{-1}yx^{-1}yx^{-1} = y^{-1}$. Pre- and postmultiplication by x^2 then gives $xyx^{-1}yx = x^2y^{-1}x^2$ so $st = x^2y^{-1}x^2$. Square this to get $(st)^2 = x^2y^{-1}x^4y^{-1}x^2 = x^2y^{-1}x^{-1}y^{-1}x^2$. But $yxy = x^{-1}$ (third relation) implies that $y^{-1}x^{-1}y^{-1} = x$. Therefore $(st)^2 = x^2xx^2 = 1$. Thus Theorem 2.2.5 may be applied to yield existence of a surjective homomorphism $d{:}M \to H$ such that $d(a) = s$ and $d(b) = t$. Applying Result 4.6.1, Part (c) yields $H = \{s, s^2, t, t^2, st, ts, s^2t, st^2, st^2s, ts^2, t^2s, 1\}$. (Although d is in fact an isomorphism, this has not yet been shown. Consequently it has not yet been shown that $\text{card}(H) = 12$ and hence that the elements $s, s^2, \ldots, t^2s, 1$ are distinct. This will not affect the proof.)

The lemma will now be used to show HB is a subgroup of G. By the first relation, B is a finite cyclic group. Thus in order to apply the lemma it must be shown that $xh \in HB$ for all $h \in H$. This is done by separate consideration of the elements of H.

(1) Since $s = x^{-1}yx^{-1}y$ from the fourth relation, $xs = yx^{-1}y = (yx^{-1})\cdot(yx^{-1})x = t^{-2}x \in HB$.

(2) $xs^2 = xs(x^{-1}x)s = (xs)x^{-1}(xs)$ so by (1), $xs^2 = t^{-2}xx^{-1}t^{-2}x = t^{-4}x = t^{-1}x \in HB$ because $t^3 = 1$ as seen in the first paragraph.

(3) From the third relation, $x = y^{-1}x^{-1}y^{-1}$ so $x^2 = xy^{-1}x^{-1}y^{-1}$ implying $xt = x^2y^{-1} = xy^{-1}x^{-1}y^{-2} = xy^{-1}x^{-1}y^3 = ts^{-1}y^2 = ts^{-1}y^2(xx^{-1})$. From the third relation again, $x^{-1} = yxy$ so $y^2x = y^2xyy^{-1} = y(yxy)y^{-1} = yx^{-1}y^{-1}$. Therefore $xt = ts^{-1}y^2xx^{-1} = ts^{-1}yx^{-1}y^{-1}x^{-1} = ts^{-1}t^{-1}y^{-1}xx^{-2} = ts^{-1}t^{-1}sx^{-2} \in HB$.

(4) $xt^2 = xtxy^{-1}$ so by (3), $xt^2 = (ts^{-1}t^{-1}sx^{-2})xy^{-1} = ts^{-1}t^{-1}sx^{-1}y^{-1}$. From $x^{-1} = yxy$ again, $x^{-1}y^{-1} = yx = yx^{-1}x^2 = t^{-1}x^2$. Thus $xt^2 = ts^{-1}t^{-1}s\cdot t^{-1}x^2 \in HB$.

(5) $xst = xs(x^{-1}x)t = (xs)x^{-1}(xt)$ so by (1) and (3), $xst = (t^{-2}x)x^{-1}(ts^{-1}\cdot t^{-1}sx^{-2}) = t^{-1}s^{-1}t^{-1}sx^{-2} \in HB$.

(6) Premultiply $(st)^2 = 1$ (which was shown in the first paragraph) by s^2 and postmultiply by t^2 to get $ts = s^2t^2$. Thus $xts = xs^2t^2 = xs^2(x^{-1}x)t^2 = (xs^2)x^{-1}\cdot(xt^2)$ so by (2) and (4), $xts = (t^{-1}x)x^{-1}(ts^{-1}t^{-1}st^{-1}x^2) = s^{-1}t^{-1}st^{-1}x^2 \in HB$.

(7) By (2), $xs^2t = t^{-1}xt$. Then by (3), $xs^2t = t^{-1}ts^{-1}t^{-1}sx^{-2} = s^{-1}t^{-1}sx^{-2} \in HB$.

(8) By (1), $xst^2 = t^{-2}xt^2$. Then by (4), $xst^2 = t^{-2}ts^{-1}t^{-1}st^{-1}x^2 = t^{-1}s^{-1}t^{-1}\cdot st^{-1}x^2 \in HB$.

(9) $xst^2s = xst^2(x^{-1}x)s = (xst^2)x^{-1}(xs)$ so by (1) and (8), $xst^2s = t^{-1}s^{-1}t^{-1}\cdot st^{-1}x^2x^{-1}t^{-2}x = t^{-1}s^{-1}t^{-1}st^{-1}xt^{-2}x = t^{-1}s^{-1}t^{-1}st^{-1}xtx$ since $t^3 = 1$. Therefore by (3), $xst^2s = t^{-1}s^{-1}t^{-1}st^{-1}ts^{-1}t^{-1}sx^{-2}x = t^{-1}s^{-1}t^{-2}sx^{-1} \in HB$.

(10) Premultiply $(st)^2 = 1$ (which was shown in the first paragraph) by s^2 and postmultiply by t^2s to get $ts^2 = s^2t^2s$. Therefore $xts^2 = xs^2t^2s = (xs)st^2s$ so by (1), $xts^2 = t^{-2}xst^2s$, so by (9), $xts^2 = t^{-2}t^{-1}s^{-1}t^{-2}sx^{-1} = s^{-1}t^{-2}sx^{-1} \in HB$.

(11) $xt^2s = xt^2(x^{-1}x)s = (xt^2)x^{-1}(xs)$. Therefore by (1) and (4), $xt^2s = ts^{-1}t^{-1}st^{-1}x^2x^{-1}t^{-2}x = ts^{-1}t^{-1}st^{-1}xtx$. Therefore by (3), $xt^2s = ts^{-1}t^{-1}st^{-1}t\cdot s^{-1}t^{-1}sx^{-2}x = ts^{-1}t^{-2}sx^{-1} \in HB$.

(12) Finally note that $x = x\cdot 1 \in HB$.

Applying the lemma shows HB is a subgroup of G. Now note that $x \in HB$ and $y = x(x^{-1}y) = xs^{-1} \in HB$. Thus both generators of G belong to HB so $G = HB$ proving Part (a).

Apply Theorem A.10.7, Part (a) to get $[G{:}H] = [B{:}H \cap B]$. Since $d{:}M \to H$ is a surjective homomorphism and since card(M) = 12 by Result 4.6.1, Part (b), card(H)$\,|\,$12. Thus card($H \cap B$)$\,|\,$12. From the first relation, card(B)$\,|\,$5 and so card($H \cap B$)$\,|\,$5. Consequently card($H \cap B$) = 1 thus $[B{:}H \cap B]$ = card(B) $\leqslant 5$, which implies $[G{:}H] \leqslant 5$. This and card(H) $\leqslant 12$ imply card(G) $\leqslant 60$. By Theorem 3.2.23, card($PSL(2,5)$) = $5(5-1)(5+1)/2 = 60$. Thus card(G) $\leqslant 60 =$ card($PSL(2,5)$).

Lct $j - [\left(\begin{smallmatrix}1&0\\3&1\end{smallmatrix}\right)]$ and $k = [\left(\begin{smallmatrix}4&1\\1&3\end{smallmatrix}\right)]$ in $PSL(2,5)$. Now by Theorem 3.2.18, $SL(2,5)$ is generated by its transvections. But any transvections $\Theta_2(1,2,n)$ and $\Theta_2(2,1,m)$ are equal to $\Theta_2(1,2,1)^n$ and $\Theta_2(2,1,1)^m$ respectively by the Corollary to Theorem 3.1.3. Therefore $SL(2,5) = \langle \left(\begin{smallmatrix}1&0\\1&1\end{smallmatrix}\right), \left(\begin{smallmatrix}1&1\\0&1\end{smallmatrix}\right) \rangle$. Now note that $\left(\begin{smallmatrix}1&0\\1&1\end{smallmatrix}\right) = \left(\begin{smallmatrix}1&0\\3&1\end{smallmatrix}\right)^2$ and that $\left(\begin{smallmatrix}1&0\\3&1\end{smallmatrix}\right)\left(\begin{smallmatrix}4&1\\1&3\end{smallmatrix}\right)\left(\begin{smallmatrix}1&0\\3&1\end{smallmatrix}\right)^{-1} = \left(\begin{smallmatrix}4&1\\3&1\end{smallmatrix}\right)\left(\begin{smallmatrix}1&0\\2&1\end{smallmatrix}\right) = \left(\begin{smallmatrix}1&1\\0&1\end{smallmatrix}\right)$. Hence $\left(\begin{smallmatrix}1&0\\1&1\end{smallmatrix}\right)$ and $\left(\begin{smallmatrix}1&1\\0&1\end{smallmatrix}\right)$ both belong to $\langle \left(\begin{smallmatrix}1&0\\3&1\end{smallmatrix}\right), \left(\begin{smallmatrix}4&1\\1&3\end{smallmatrix}\right) \rangle$. Conclude $SL(2,5) = \langle \left(\begin{smallmatrix}1&0\\3&1\end{smallmatrix}\right), \left(\begin{smallmatrix}4&1\\1&3\end{smallmatrix}\right) \rangle$ from which it follows (Theorem A.10.4) that $PSL(2,5) = \langle j, k \rangle$.

Now define the function $F{:}\,\{X, Y\} \to \{j, k\}$ by $F(X) = j$ and $F(Y) = k$. In order to apply Theorem 2.2.6 it must be shown that the relations of G are "satisfied" in $PSL(2,5)$. This is done as follows: $j^5 = 1$ follows from the Corollary to Theorem 3.1.3; similarly, $[\left(\begin{smallmatrix}1&1\\0&1\end{smallmatrix}\right)]^5 = 1$, but as seen in the previous paragraph, $[\left(\begin{smallmatrix}1&1\\0&1\end{smallmatrix}\right)] = jkj^{-1}$ and so $k^5 = 1$ also; $jk = [\left(\begin{smallmatrix}4&1\\3&1\end{smallmatrix}\right)]$ as seen in the last paragraph. so $(jk)^2 = 1$ by Theorem 3.2.24 since the trace of $\left(\begin{smallmatrix}4&1\\3&1\end{smallmatrix}\right)$ is 0; finally $k^{-1}j = [\left(\begin{smallmatrix}3&4\\4&4\end{smallmatrix}\right)]\cdot[\left(\begin{smallmatrix}1&0\\3&1\end{smallmatrix}\right)] = [\left(\begin{smallmatrix}0&4\\1&4\end{smallmatrix}\right)]$ so $(k^{-1}j)^3 = 1$ by Theorem 3.2.25.

Theorem 2.2.6 may now be applied to yield an isomorphism $D{:}G \to PSL(2,5)$. This proves Part (b). Part (c) follows immediately.

The inequality card(B) $\leqslant 5$ implies card(G)/5 $\leqslant$ card(G)/card(B). This and card(G) = 60 yield $12 \leqslant$ card(G)/card(B), hence $12 \leqslant$ card(H). Consequently 12 = card(H) so $d{:}M \to H$ must be an isomorphism proving Part (d).

RESULT 4.11.2 *G is simple.*

By Result 4.11.1, Part (b), $G \cong PSL(2,5)$ which is simple by Theorem 3.2.22.

RESULT 4.11.3

(a) *G is subdirectly irreducible.*

(b) *G is indecomposable.*

Since G is simple (Result 4.11.2), $\mathrm{H}(G) = G$ proving Part (a). Part (b) follows from Part (a) and Theorem A.10.10.

RESULT 4.11.4

(a) *G has no subgroups of cardinality* 15, 20, *or* 30.

(b) *G does not satisfy extended Sylow property* $\mathrm{E}(\{2,5\})$.

(c) *G does not satisfy extended Sylow property* $\mathrm{E}(\{3,5\})$.

Suppose H is a subgroup of G of cardinality 15, hence index 4, and let $C = \mathrm{Core}_G(H)$. By simplicity (Result 4.11.2), C is trivial. By Theorem A.1.3, $G/\{1\}$, and hence G, can be embedded in S_4. Thus $\mathrm{card}(G) | \mathrm{card}(S_4)$ meaning $60 | 24$ which is false. Consequently G has no subgroups of cardinality 15. Proofs that G has no subgroups of cardinality 20 or 30 are similar.

A $\{2,5\}$-Hall subgroup and a $\{3,5\}$-Hall subgroup would have cardinality 20 and 15 respectively. Thus these subgroups do not exist by Part (a).

RESULT 4.11.5

(a) *G has* 6 *5-Sylow subgroups.*

(b) *If S is a* 5*-Sylow subgroup, then* $\mathrm{card}(\mathrm{N}_G(S)) = 10$.

(c) *If N is any subgroup of G of cardinality* 10, *then there exists a* 5*-Sylow subgroup S of G such that* $N = \mathrm{N}_G(S)$.

By Sylow's theorem (Theorem A.1.6), $n_5(G) \equiv 1 \bmod 5$ and $n_5(G) | 60$. The only possibilities are $n_5(G) = 1$ or 6. Let S be a 5-Sylow subgroup of G. Since $[G:\mathrm{N}_G(S)] = n_5(G)$ (Theorem A.10.15), $\mathrm{N}_G(S)$ has cardinality 60 or 10. In the first case $\mathrm{N}_G(S) = G$ so $S \lhd G$ contradicting simplicity (Result 4.11.2). Thus $\mathrm{card}(\mathrm{N}_G(S)) = 10$, hence $n_5(G) = [G:\mathrm{N}_G(S)] = 6$, proving Parts (a) and (b).

To show Part (c), let $\mathrm{card}(N) = 10$. By Sylow's theorem, $n_5(N) \equiv 1 \bmod 5$ and $n_5(N) | 10$. Therefore $n_5(N) = 1$. Let S be the unique 5-Sylow subgroup of N. Then $S \lhd N$ (Theorem A.10.15, Part (b)) so $N \subseteq \mathrm{N}_G(S)$. But S is also a 5-Sylow subgroup of G so $\mathrm{card}(\mathrm{N}_G(S)) = 10$ by Part (b). A cardinality comparison now shows $N = \mathrm{N}_G(S)$.

RESULT 4.11.6

(a) *G has* 10 *3-Sylow subgroups.*

(b) *If S is a* 3*-Sylow subgroup, then* $\mathrm{card}(\mathrm{N}_G(S)) = 6$.

(c) *If N is any subgroup of G of cardinality* 6, *then there exists a* 3*-Sylow subgroup S of G such that* $N = \mathrm{N}_G(S)$.

By Sylow's theorem (Theorem A.1.6), $n_3(G) \equiv 1 \bmod 3$ and $n_3(G) | 60$. The only possibilities are $n_3(G) = 1$, 4, or 10. Let S be a 3-Sylow subgroup of G. Since $[G:\mathrm{N}_G(S)] = n_3(G)$ (Theorem A.10.15), $\mathrm{N}_G(S)$ has cardinality 60, 15, or 6. The first case implies $\mathrm{N}_G(S) = G$ and hence $S \lhd G$ contradicting simplicity (Result 4.11.2). The second case is eliminated by Result 4.11.4. Conclude that card($N_G(S)$) = 6 and hence $n_3(G) = [G{:}N_G(S)] = 10$, proving Parts (a) and (b).

To show Part (c), let card(N) = 6. By Sylow's theorem, $n_3(N) \equiv 1 \bmod 3$ and $n_3(N) | 6$. Therefore $n_3(N) = 1$. Let S be the unique 3-Sylow subgroup of N. Then $S \lhd N$ (Theorem A.10.15, Part (b)) so $N \subseteq N_G(S)$. But S is also a 3-Sylow subgroup of G so card($N_G(S)$) = 10 by Part (b). A cardinality comparison now shows $N = N_G(S)$.

RESULT 4.11.7

(a) *G has* 5 *2-Sylow subgroups.*

(b) *If S is a 2-Sylow subgroup, then* card($N_G(S)$) = 12.

(c) *If N is any subgroup of G of cardinality* 12, *then there exists a 2-Sylow subgroup S of G such that* $N = N_G(S)$.

By Result 4.11.1, Part (d), G has a subgroup $H \cong PSL(2,3)$. By Result 4.6.1, H has a normal subgroup K of cardinality 4. Then $K \lhd H$ implies $H \subseteq N_G(K)$. But $N_G(K) \neq G$ lest $K \lhd G$ in contradiction to simplicity (Result 4.11.2). Thus by a Lagrange's-Theorem argument (and the fact that $[G{:}H] = 5$), $H = N_G(K)$. Hence (Theorem A.1.4) K has $[G{:}H] = 5$ conjugates. Finally it need only be noted that a subgroup of G is conjugate to K if and only if it is a 2-Sylow subgroup of G. Thus G has 5 2-Sylow subgroups proving Part (a). Part (b) follows from Part (a) and Theorem A.10.15.

To show Part (c), let card(N) = 12. By Sylow's theorem (Theorem A.1.6), $n_3(N) \equiv 1 \bmod 3$ and $n_3(N) | 12$. Thus $n_3(N) = 1$ or 4. Let L be a 3-Sylow subgroup of N. If $n_3(N) = 1$, then $L \lhd N$ (Theorem A.10.15, Part (b)) so $N \subseteq N_G(L)$ and hence card(N) | card($N_G(L)$). Now note that L is also a 3-Sylow subgroup of G. Thus card($N_G(L)$) = 6 by Result 4.11.6, Part (b). This leads to the contradiction $12 | 6$ which implies $n_3(N) \neq 1$ and so $n_3(N) = 4$. Since there are two elements of order 3 in each of the 3-Sylow subgroups of N, and since no element of order 3 can belong to more than one such subgroup (two distinct subgroups of cardinality 3 have trivial intersection by Lagrange's theorem) there must be 8 elements of order 3 in N. Consequently there are $12 - 8 = 4$ elements remaining in N that can belong to the 2-Sylow subgroups of N. Since each 2-Sylow subgroup of N has 4 elements, there can be only 1 such subgroup, call it S. Then $S \lhd N$ (Theorem A.10.15 again) and so $N \subseteq N_G(S)$. Hence $12 |$ card($N_G(S)$). Thus card($N_G(S)$) = 12 or 60. But in the latter case, $N_G(S) = G$ implying $S \lhd G$ in contradiction to simplicity (Result 4.11.2). Thus card($N_G(S)$) = 12 = card(N) so $N = N_G(S)$.

RESULT 4.11.8 *G is minimal simple.*

First note that G is simple by Result 4.11.2.

By Lagrange's theorem and Result 4.11.4, if H is a proper subgroup of G, then card(H) = 1, 2, 3, 4, 5, 6, 10, or 12. Groups of cardinality 1, 2, 3, or 5 are cyclic hence solvable. Groups of cardinality 4 are abelian (Theorem A.1.2) hence solvable. If card(H) = 6, 10, or 12, then $H = N_G(S)$ where S is a 3-, 5- or 2-Sylow subgroup of G respectively by Results 4.11.5, 4.11.6, and 4.11.7. Then $S \lhd H$ in all three cases, S is abelian in all three cases (card(S) = 3, 5, or 4), and H/S is abelian in all three cases (card(H/S) = 2, 2, or 3). Conclude H is solvable in all three cases by Theorem A.6.6.

LEMMA *If G is a (not necessarily finite) group and A and B are conjugate subgroups, then $N_G(A)$ and $N_G(B)$ are also conjugate.*
Suppose $B = x^{-1}Ax$. It is claimed that $N_G(B) = x^{-1}N_G(A)x$. Let $t \in N_G(B)$. It must be shown that $xtx^{-1} \in N_G(A)$ and to do that, $(xtx^{-1})^{-1}a(xtx^{-1}) \in A$ must be shown for all $a \in A$. But for such a, $x^{-1}ax \in B$ and so $t^{-1}(x^{-1}ax)t \in B$ since $t \in N_G(B)$. This in turn implies $x(t^{-1}(x^{-1}ax)t)x^{-1} \in A$ because $xBx^{-1} = A$. Hence $(xtx^{-1})^{-1}a(xtx^{-1}) \in A$ as required. Thus $N_G(B) \subseteq x^{-1}N_G(A)x$. The reverse inclusion is similar.

RESULT 4.11.9 *G satisfies extended Sylow property* C({2,3}).
Since $60 = 2^2 \cdot 3 \cdot 5$, {2,3}-Hall subgroups of G will have cardinality $2^2 \cdot 3 = 12$. Hence {2,3}-Hall subgroups exist by Result 4.11.1, Part (d).

Now let H and K be two {2,3}-Hall subgroups of G. By Result 4.11.7 there are 2-Sylow subgroups of G, A and B, such that $H = N_G(A)$ and $K = N_G(B)$. By Sylow's theorem (Theorem A.1.6) then, A and B are conjugate, hence H and K are conjugate by the lemma.

RESULT 4.11.10 *G does not satisfy extended Sylow property* D({2,3}).
By Result 4.11.6, Part (b), G has subgroups of cardinality 6. Let A denote any such subgroup. If G satisfied D({2,3}), then there would exist a {2,3}-Hall subgroup B such that $A \subseteq B$. Assume this and seek a contradiction. By Result 4.11.1, Part (d), G has a subgroup $H \cong PSL(2,3)$. By Result 4.11.9 then, H and B are conjugate in G and hence $B \cong H \cong PSL(2,3)$. This cannot be true however since B has a subgroup of cardinality 6 (namely A) and $PSL(2,3)$ does not (Result 4.6.2).

RESULT 4.11.11 *G has maximal subgroups of non-prime-power index.*
Let S be a 5-Sylow subgroup of G and let $N = N_G(S)$. By Result 4.11.5 then card(N) = 10. It is claimed that N is a maximal subgroup of G. Suppose $N \subseteq B \subseteq G$. Then by Lagrange's theorem 10 | card(B) and card(B) | 60. Thus card(B) = 10, 20, 30, or 60. But Result 4.11.4 rules out card(B) = 20 or 30. Consequently card(B) = 10 or 60 implying $B = N$ or G. Thus N is a maximal subgroup of G as claimed. Finally note that $[G:N] = 6$.

RESULT 4.11.12 *G has no Carter subgroups.*
By Result 4.11.4 and Lagrange's theorem, the possible cardinalities for subgroups of G are 1, 2, 3, 4, 5, 6, 10, 12, and 60. Suppose C is a Carter subgroup and seek a contradiction. Clearly card(C) $\neq$ 60 for then $C = G$ implying G is nilpotent. If card(C) = 12, then by Results 4.11.1, Part (d) and 4.11.9, C is isomorphic to $PSL(2,3)$, hence $PSL(2,3)$ is nilpotent contrary to Result 4.6.9.

If card(C) = 10, then by Cauchy's theorem there exists $x \in C$ such that $o(x) = 2$. Let $T = \langle x \rangle$. Then $[C:T] = 5$, a prime, so T is a maximal subgroup of C. Suppose $T \lhd C$. Then $C \subseteq N_G(T)$ implying $10 \mid \text{card}(N_G(T))$. But by Sylow's theorem (Theorem A.1.6) there exists a 2-Sylow subgroup S such that $T \subseteq S$. Clearly $T \lhd S$ and so $S \subseteq N_G(T)$ implying $4 \mid \text{card}(N_G(T))$. Since $10 \mid \text{card}(N_G(T))$, $4 \mid \text{card}(N_G(T))$, and since G has no subgroup of cardinality 20 (Result 4.11.4), it is clear that $\text{card}(N_G(T)) = 60$ so $N_G(T) = G$ and hence $T \lhd G$ in contradiction to simplicity (Result 4.11.2). Consequently the assumption $T \lhd C$ must be false so C has nonnormal maximal subgroups. Conclude C is not nilpotent by Theorem A.10.12.

Similarly to the previous case if card(C) = 6, then C has a maximal subgroup T such that card(T) = 2, and $S \subseteq N_G(T)$ for some 2-Sylow subgroup S. Thus $4 \mid \text{card}(N_G(T))$. If $T \lhd C$, then $C \subseteq N_G(T)$ implying $6 \mid \text{card}(N_G(T))$. However $4 \mid \text{card}(N_G(T))$ and $6 \mid \text{card}(N_G(T))$ imply $12 \mid \text{card}(N_G(T))$ so $\text{card}(N_G(T)) = 12$ or 60. Simplicity rules out the second case, consequently $\text{card}(N_G(T))$ must be 12. Thus by Results 4.11.1, Part (d) and 4.11.9 again, $N_G(T) \cong PSL(2,3)$. But $T \lhd N_G(T)$ implying $PSL(2,3)$ has a normal subgroup of cardinality 2 in contradiction to Result 4.6.3. This contradiction shows $T \lhd C$ must be false so as in the previous case C is not nilpotent.

In the cases card(C) = 5, 4, or 3, C is a 5-, 2-, or 3-Sylow subgroup of G hence $\text{card}(N_G(C)) = 10$, 12, or 6 by Results 4.11.5, 4.11.7, and 4.11.6 respectively. In all three cases $C \neq N_G(C)$.

If card(C) = 2, then there exists a 2-Sylow subgroup S such that $C \subseteq S$ and hence $C \lhd S$ by Theorem A.10.2. Thus $S \subseteq N_G(C)$ so $C \subset N_G(C)$.

Finally if card(C) = 1, then $C = \{1\}$ and so $N_G(C) = G$ implying $C \neq N_G(C)$.

EXERCISES

1. Show that if J and H are subgroups of G such that $J \subset H \subset G$, then J is abelian.
2. Show G has a subgroup T such that card(T) = 3 and $T = C_G(T)$.
3. Show any two distinct 2-Sylow subgroups of G have trivial intersection.
4. Show G has one element of order 1, fifteen elements of order 2, twenty elements of order 3, no elements of order 4, and twenty-four elements of order 5. Conclude that all nontrivial elements of G have prime and hence prime-power order.
5. (a) Show there is an automorphism f of G such that $f(x) = y$ and $f(y) = x$.
   (b) Show f is not an inner automorphism.
6. Show G has 59 subgroups (including itself).
7. Describe the decomposition of G into conjugacy classes and write the class equation. In particular show that all elements of order 2 are conjugate and all elements of order 3 are conjugate.
8. Let A be a simple group of cardinality 60. Show $A \cong G$.
9. Let A be a nonsolvable group of cardinality 60. Show $A \cong G$.

10. Show $G \cong \mathrm{G}(R, S \mid R^2 = 1, S^3 = 1, (RS)^5 = 1)$.

11. Show G is isomorphic to the subgroup of S_5 generated by $f = \left(\begin{smallmatrix}1&2&3&4&5\\4&1&2&5&3\end{smallmatrix}\right)$ and $g = \left(\begin{smallmatrix}1&2&3&4&5\\2&4&1&5&3\end{smallmatrix}\right)$. (This subgroup, $\langle f,g\rangle$, is the alternating group A_5.)

NOTES

1. The icosahedral group is one of the most well-known finite groups and surely the most well-known nonabelian simple group. It is isomorphic to, and most popularly known as, the alternating group A_5 (see Exercise 11).
2. As a counterexample G separates the classes of groups satisfying extended Sylow properties $\mathrm{C}(\pi)$ and $\mathrm{D}(\pi)$ (Results 4.11.9 and 4.11.10).
3. A characterization of G is given in Exercise 8 and a slightly more general characterization in Exercise 9. In addition to these, G has a large number of other characterizations, among which are the following.
   (a) If A is a simple group of cardinality $2^2 \cdot p \cdot q^b$ (p and q primes), then $A \cong G$ (Brauer [3]).
   (b) If A is a finite simple group such that every proper subgroup of every maximal subgroup of A is abelian (cf Exercise 1), then $A \cong G$ (Suzuki [34]).

   For further characterizations, see Note 3 to Example 4.12.
4. $\mathrm{Aut}(G) \cong S_5$. This is proven in Scott [33], Theorem 11.4.8.

□□□□□

Example 4.12

Let $G = PSL(2,7)$.

RESULT 4.12.1 $\mathrm{card}(G) = 168 = 2^3 \cdot 3 \cdot 7$.
By Theorem 3.2.23, $\mathrm{card}(G) = 7(7-1)(7+1)/2 = 168$.

RESULT 4.12.2 *G is simple.*
By Theorem 3.2.22.

RESULT 4.12.3
(a) *G is subdirectly irreducible.*
(b) *G is indecomposable.*
Since G is simple (Result 4.12.2), $\mathrm{H}(G) = G$ proving Part (a). Part (b) follows

from Part (a) and Theorem A.10.10.

RESULT 4.12.4

(a) *D_4 can be embedded in G.*

(b) *S_4 can be embedded in G.*

Let $a = [\left(\begin{smallmatrix}0&5\\4&1\end{smallmatrix}\right)]$, $b = [\left(\begin{smallmatrix}3&4\\1&4\end{smallmatrix}\right)]$, and $H = \langle a,b \rangle$. Let $h = (ba)^2$ and let $j = ab$. Since $ba = [\left(\begin{smallmatrix}3&4\\1&4\end{smallmatrix}\right)][\left(\begin{smallmatrix}0&5\\4&1\end{smallmatrix}\right)] = [\left(\begin{smallmatrix}2&5\\2&2\end{smallmatrix}\right)]$, $h = [\left(\begin{smallmatrix}2&5\\2&2\end{smallmatrix}\right)][\left(\begin{smallmatrix}2&5\\2&2\end{smallmatrix}\right)] = [\left(\begin{smallmatrix}0&6\\1&0\end{smallmatrix}\right)]$. Also note that $j = [\left(\begin{smallmatrix}0&5\\4&1\end{smallmatrix}\right)][\left(\begin{smallmatrix}3&4\\1&4\end{smallmatrix}\right)] = [\left(\begin{smallmatrix}5&6\\6&6\end{smallmatrix}\right)]$. Let $K = \langle h,j \rangle$. Thus K is a subgroup of H. It is claimed that $K \cong D_4$. First note that $jh = [\left(\begin{smallmatrix}5&6\\6&6\end{smallmatrix}\right)][\left(\begin{smallmatrix}0&6\\1&0\end{smallmatrix}\right)] = [\left(\begin{smallmatrix}6&2\\6&1\end{smallmatrix}\right)] = \{\left(\begin{smallmatrix}6&2\\6&1\end{smallmatrix}\right), \left(\begin{smallmatrix}1&5\\1&6\end{smallmatrix}\right)\}$ while $hj = [\left(\begin{smallmatrix}0&6\\1&0\end{smallmatrix}\right)][\left(\begin{smallmatrix}5&6\\6&6\end{smallmatrix}\right)] = [\left(\begin{smallmatrix}1&1\\5&6\end{smallmatrix}\right)] = \{\left(\begin{smallmatrix}1&1\\5&6\end{smallmatrix}\right), \left(\begin{smallmatrix}6&6\\2&1\end{smallmatrix}\right)\}$. Thus $jh \neq hj$ so K is nonabelian.

Define $f: \{X, Y\} \to \{j,h\}$ (where X and Y are the generators of D_4) by $f(X) = j$ and $f(Y) = h$. Then the relations of D_4 are "satisfied" in H as seen as follows: $j^2 = [\left(\begin{smallmatrix}5&6\\6&6\end{smallmatrix}\right)][\left(\begin{smallmatrix}5&6\\6&6\end{smallmatrix}\right)] = [\left(\begin{smallmatrix}5&3\\3&2\end{smallmatrix}\right)]$. Since the last matrix has trace zero, $j^4 = 1$ by Theorem 3.2.24; since $h = [\left(\begin{smallmatrix}0&6\\1&0\end{smallmatrix}\right)]$, $h^2 = 1$ again by Theorem 3.2.24; and (as seen previously) $jh = [\left(\begin{smallmatrix}6&2\\6&1\end{smallmatrix}\right)]$ so $(jh)^2 = 1$ by yet a third application of Theorem 3.2.24. Theorem 2.2.5 may now be applied to show that K is a homomorphic image of D_4. Since card$(D_4) = 8$ (Result 4.3.1), it follows that card$(K) = 1, 2, 4$, or 8. But K was seen to be nonabelian above and groups of cardinality 1, 2, and 4 are abelian (cardinality 4 by Theorem A.1.2). Conclude card$(K) = 8$ and hence $K \cong D_4$. Thus $8 \mid$ card(H).

Next note that $a \in H$ and o$(a) = 3$ by Theorem 3.2.25. Therefore $3 \mid$ card(H). This and $8 \mid$ card(H) imply $24 \mid$ card(H). Now let A denote the octohedral group (Example 4.9), $A = G(S, T \mid S^3 = 1, T^2 = 1, (ST)^4 = 1)$. Then card$(A) = 24$ (Result 4.9.1) so card$(A) = 24 \leqslant$ card(H).

Define the function $F: \{S, T\} \to \{a, b\}$ by $F(S) = a$ and $F(T) = b$. In order to apply Theorem 2.2.6 it must be shown that the relations of A are "satisfied" in H. This is done as follows: $a^3 = 1$ as shown in the previous paragraph, $b^2 = 1$ because the matrix defining b has trace zero (Theorem 3.2.24), and $(ab)^4 = j^4 = 1$ as shown previously. Thus Theorem 2.2.6 applies to yield $A \cong H$. Finally $A \cong S_4$ by Result 4.9.1, so $S_4 \cong H$.

RESULT 4.12.5

(a) *G has no subgroups of cardinality* 28, 42, 56, *or* 84.

(b) *G does not satisfy extended Sylow property* E({2,7}).

Suppose H is a subgroup of cardinality 28, hence index 6. Let $C = \text{Core}_G(H)$. By Result 4.12.2 then, C is trivial. Thus by Theorem A.1.3 $G/\{1\}$, and hence G, can be embedded in S_6. Therefore card$(G) \mid$ card(S_6) which implies $7 \mid 6!$ which is false. Conclude G has no subgroups of cardinality 28. The other results in Part (a) are similar. Part (b) follows from Part (a) by noting that a {2,7}-Hall subgroup would have cardinality $2^3 \cdot 7 = 56$.

RESULT 4.12.6

(a) *G has* 8 *7-Sylow subgroups.*

(b) *If S is a 7-Sylow subgroup, then* $\text{card}(N_G(S)) = 21$.
By Sylow's theorem (Theorem A.1.6), $n_7(G) \equiv 1 \bmod 7$ and $n_7(G) \mid 168$. But $n_7(G) \equiv 1 \bmod 7$ implies $n_7(G)$ and 7 are relatively prime. This and the fact that $n_7(G) \mid 2^3 \cdot 3 \cdot 7$ imply $n_7(G) \mid 2^3 \cdot 3$. It is easily seen from examination of the eight divisors of $2^3 \cdot 3$ that the only two congruent to 1 mod 7 are 1 and 8. But if $n_7(G) = 1$, then the 7-Sylow subgroup of G is normal (Theorem A.10.15, Part (b)). This is in contradiction to simplicity (Result 4.12.2) so $n_7(G) = 8$ proving Part (a). Part (b) follows from Part (a) and Theorem A.10.15.

RESULT 4.12.7
(a) *G has* 28 *3-Sylow subgroups.*
(b) *If S is a 3-Sylow subgroup of G, then* $\text{card}(N_G(S)) = 6$.
First note that $2^2 = 4 = -3$ in $Z(7)$ so -3 has a square root. By Theorem 3.2.26 then, G has $7^2 + 7 = 56$ elements of order 3. Note that 3-Sylow subgroups of G have cardinality 3 and thus there are exactly 2 elements of order 3 in each 3-Sylow subgroup. Consequently there are $56/2 = 28$ 3-Sylow subgroups proving Part (a). Part (b) follows from Part (a) and Theorem A.10.15.

RESULT 4.12.8
(a) *G has* 21 *2-Sylow subgroups.*
(b) *If S is a 2-Sylow subgroup of G, then* $S = N_G(S)$ *and hence S is a Carter subgroup of G.*
By Sylow's theorem (Theorem A.1.6), $n_2(G) \equiv 1 \bmod 2$ and $n_2(G) \mid 168 = 2^3 \cdot 3 \cdot 7$. Thus $n_2(G)$ divides $3 \cdot 7$ so it must be equal to 1, 3, 7, or 21. Let S be a 2-Sylow subgroup of G. Since $[G:N_G(S)] = n_2(G)$ (Theorem A.10.15), $N_G(S)$ has cardinality 168, 56, 24, or 8. In the first case, $N_G(S) = G$ so $S \lhd G$ contradicting simplicity (Result 4.12.2). The second case is eliminated by Result 4.12.5. Assume the third case (cardinality 24) and seek a contradiction.

Let H denote $N_G(S)$. Thus $S \lhd H$. By Result 4.12.4, Part (a), G has a 2-Sylow subgroup isomorphic to D_4. Hence (Sylow's theorem) all 2-Sylow subgroups of G are isomorphic to D_4. Thus (Result 4.1.3) $\text{card}(Z(S)) = 2$. It is easily seen that the center of any group is a characteristic subgroup. This, $S \lhd H$, and Theorem A.10.6 yield $Z(S) \lhd H$. Let T be a 3-Sylow subgroup of H. Then $Z(S) \lhd H$ implies $Z(S)T$ is a subgroup of H. Since $\text{card}(Z(S)) = 2$ and $\text{card}(T) = 3$, $Z(S) \cap T = \{1\}$. Thus $Z(S)T/Z(S) \cong T$ and so $\text{card}(Z(S)T) = 6$. Consequently $[Z(S)T:T] = 6/3 = 2$ implying $T \lhd Z(S)T$ by Theorem A.10.2. Hence $Z(S)T = Z(S) \oplus T$. Since $Z(S)$ and T are abelian, so is $Z(S)T$ by the Corollary to Theorem A.2.1. Now $Z(S)T \subseteq N_G(T)$ since $T \lhd Z(S)T$, but note that T is a 3-Sylow subgroup of G so $\text{card}(N_G(T)) = 6$ by Result 4.12.7, Part (b). This, $\text{card}(Z(S)T) = 6$, and $Z(S)T \subseteq N_G(T)$ imply $N_G(T) = Z(S)T$, hence $N_G(T)$ is abelian.

Next note that Result 4.12.4, Part (b) implies S_4 can be embedded in G. Since S_3 can be embedded in S_4 (Theorem A.10.8), this means G has a subgroup $R \cong S_3$. Let P be a 3-Sylow subgroup of R. Then $[R:P] = 6/3 = 2$ so $P \lhd R$ by Theorem A.10.2. Application of Result 4.12.7 again shows $N_G(P) = R$. Finally T

and P are both 3-Sylow subgroups of G so are conjugate in G by Sylow's theorem. This and the Lemma to Result 4.11.9 yield conjugacy (and therefore isomorphism) of $N_G(T)$ and $N_G(P)$. Hence $Z(S)T \cong R \cong S_3$, the required contradiction ($Z(S)T$ is abelian and S_3 is not).

Conclude $n_2(G) = 21$ and $\text{card}(N_G(S)) = 8$. But $\text{card}(S) = 8$ also. Thus S equals its own normalizer and is therefore a Carter subgroup of G (S is nilpotent by Theorem A.5.7).

RESULT 4.12.9 *G has no subgroups of cardinality* 14.
Suppose H is such a subgroup. By Sylow's theorem (Theorem A.1.6), $n_7(H)$ is congruent to 1 mod 7 and also divides 14. Thus $n_7(H) = 1$ and so H has a normal 7-Sylow subgroup, call it S (normality by Theorem A.10.15, Part (b)). Since $S \lhd H$, $H \subseteq N_G(S)$. But S is also a 7-Sylow subgroup of G implying $\text{card}(N_G(S)) = 21$ by Result 4.12.6. Thus $H \subseteq N_G(S)$ implies $14 \mid 21$, the required contradiction.

RESULT 4.12.10
(a) *G satisfies extended Sylow property* $E(\{2,3\})$.
(b) *G does not satisfy extended Sylow property* $C(\{2,3\})$.
By Result 4.12.4, Part (b), G has subgroups of cardinality 24 proving Part (a).

Let A be an arbitrary $\{2,3\}$-Hall subgroup of G. Then by simplicity (Result 4.12.2) A is not a normal subgroup so $N_G(A) \neq G$. This and $A \subseteq N_G(A)$ imply $A = N_G(A)$ since $[G:A] = 7$ is prime. Thus $[G:N_G(A)] = 7$ and so (Theorem A.1.4) A has 7 conjugates.

To prove Part (b), therefore, it suffices to show there are at least 8 $\{2,3\}$-Hall subgroups of G. Assume to the contrary that there are only 7 and seek a contradiction.

By Sylow's theorem (Theorem A.1.6), a $\{2,3\}$-Hall subgroup has either 1 or 3, hence at most 3, 2-Sylow subgroups. There are then at most $7 \times 3 = 21$ 2-Sylow subgroups contained in some $\{2,3\}$-Hall subgroup of G. There may be, however, two distinct $\{2,3\}$-Hall subgroups having a 2-Sylow subgroup in common. Assume that this is true. There are then at most 20 2-Sylow subgroups contained in some $\{2,3\}$-Hall subgroup of G.

Next note that every 2-Sylow subgroup, S, of G is contained in some $\{2,3\}$-Hall subgroup of G. This is seen as follows. Let A be a $\{2,3\}$-Hall subgroup and let T be a 2-Sylow subgroup of A. Then $x^{-1}Tx = S$ for some $x \in G$ by Sylow's theorem, so $x^{-1}Ax$ is a $\{2,3\}$-Hall subgroup of G and $S \subseteq x^{-1}Ax$.

Combining the results of the last two paragraphs shows that G has at most 20 2-Sylow subgroups. This contradicts $n_2(G) = 21$ (Result 4.12.8). The proof then is complete once the assumption made in the fourth paragraph is justified. That is, two distinct $\{2,3\}$-Hall subgroups having a 2-Sylow subgroup in common must be produced.

In the proof of Result 4.12.4 it was shown that if $a = [\begin{pmatrix} 0 & 5 \\ 4 & 1 \end{pmatrix}]$ and $b = [\begin{pmatrix} 3 & 4 \\ 1 & 4 \end{pmatrix}]$, then $H = \langle a,b \rangle$ has cardinality 24 and that $K = \langle j,h \rangle$ is a subgroup of H of cardinality 8 where $j = ab = [\begin{pmatrix} 5 & 6 \\ 6 & 6 \end{pmatrix}]$ and $h = (ba)^2 = [\begin{pmatrix} 0 & 6 \\ 1 & 0 \end{pmatrix}]$. Thus it suffices to produce a subgroup R such that card$(R) = 24, R \neq H$, and $K \subseteq R$.

Let $a^* = [\begin{pmatrix}6&5\\5&2\end{pmatrix}]$, let $b^* = [\begin{pmatrix}1&3\\4&6\end{pmatrix}]$, and let $R = \langle a^*, b^* \rangle$. Then $a^*b^* = [\begin{pmatrix}6&5\\5&2\end{pmatrix}] \cdot [\begin{pmatrix}1&3\\4&6\end{pmatrix}] = [\begin{pmatrix}5&6\\6&6\end{pmatrix}] = j$. Also $b^*a^* = [\begin{pmatrix}1&3\\4&6\end{pmatrix}][\begin{pmatrix}6&5\\5&2\end{pmatrix}] = [\begin{pmatrix}0&4\\5&4\end{pmatrix}]$ so $(b^*a^*)^2a^*b^* = (b^*a^*)^2j = [\begin{pmatrix}0&4\\5&4\end{pmatrix}][\begin{pmatrix}0&4\\5&4\end{pmatrix}][\begin{pmatrix}5&6\\6&6\end{pmatrix}] = [\begin{pmatrix}6&2\\6&1\end{pmatrix}][\begin{pmatrix}5&6\\6&6\end{pmatrix}] = [\begin{pmatrix}0&6\\1&0\end{pmatrix}] = h$. Thus $\langle j,h \rangle \subseteq R$, that is $K \subseteq R$. The demonstration that card$(R) = 24$, via showing $R \cong S_4$ is analogous to the proof that $H \cong S_4$ in Result 4.12.4 and will be omitted.

What remains is to show $R \neq H$. Let $u = ba^2ba$ and let $v = (a^*)^2b^*$. Then $u = ba(ab)a = baja = [\begin{pmatrix}3&4\\1&4\end{pmatrix}][\begin{pmatrix}0&5\\4&1\end{pmatrix}][\begin{pmatrix}5&6\\6&6\end{pmatrix}][\begin{pmatrix}0&5\\4&1\end{pmatrix}] = [\begin{pmatrix}2&5\\2&2\end{pmatrix}][\begin{pmatrix}3&3\\3&1\end{pmatrix}] = [\begin{pmatrix}0&4\\5&1\end{pmatrix}]$ and $v = a^*(a^*b^*) = a^*j = [\begin{pmatrix}6&5\\5&2\end{pmatrix}][\begin{pmatrix}5&6\\6&6\end{pmatrix}] = [\begin{pmatrix}4&3\\2&0\end{pmatrix}]$. Now $u \in H$ and $v \in R$. But $uv = [\begin{pmatrix}0&4\\5&1\end{pmatrix}][\begin{pmatrix}4&3\\2&0\end{pmatrix}] = [\begin{pmatrix}1&0\\1&1\end{pmatrix}]$ which clearly has order 7. Thus $uv \notin R$ (7 does not divide 24) and so $u \notin R$ proving $H \neq R$.

EXERCISES

1. Show G has a subgroup T such that card$(T) = 3$ and $T = C_G(T)$.
2. If B is a subgroup of G of cardinality 6 show there exists a 3-Sylow subgroup P of G such that $B = N_G(P)$.
3. If B is a subgroup of G of cardinality 21 show there exists a 7-Sylow subgroup P of G such that $B = N_G(P)$.
4. Show all $\{2,3\}$-Hall subgroups of G are isomorphic to S_4.
5. Show G is minimal simple.
6. Show G satisfies extended Sylow property D($\{3,7\}$).
7. Show G has one element of order 1, twenty-one elements of order 2, fifty-six elements of order 3, forty-two elements of order 4, and forty-eight elements of order 7. Conclude that every element of G has prime-power order (different primes allowed).
8. If H is a subgroup of G of cardinality 12, show $H \cong PSL(2,3)$.
9. Describe the decomposition of G into conjugacy classes and write the class equation. In particular show all elements of order 2 are conjugate, all elements of order 3 are conjugate, and all elements of order 4 are conjugate.
10. Show maximal subgroups of G have prime-power index (different primes allowed).

NOTES

1. The group G is the second smallest finite nonabelian simple group.
2. As a counterexample G
   (a) separates the classes of groups satisfying extended Sylow properties E(π) and C(π) (Result 4.12.10).
   (b) separates the class of solvable groups and $\{G \mid G$ has Carter subgroups$\}$.
   (c) separates the class of solvable groups and $\{G \mid$ maximal subgroups of G have prime-power index$\}$.

3. The following characterization of G is found in Huppert [21], Theorem 6.15.
   (a) If A is a simple group of cardinality 168, then $A \cong G$.
   The paper Feit and Thompson [10] proves that if A is a finite simple group having a subgroup T of cardinality 3 such that $C_A(T) = T$, then $A \cong G$ or $A \cong PSL(2,5)$. (See Exercise 1 and Exercise 2 of Example 4.11.) This and some of the results of this chapter yield two characterizations for G and one for $PSL(2,5)$, namely
   (b) (combining with Results 4.11.11 and Exercise 10) If A is a finite simple group and A has a self-centralizing subgroup of cardinality 3 and maximal subgroups of A have prime-power index, then $A \cong G$.
   (c) (combining with Results 4.11.12 and 4.12.8) If A is a finite simple group and A has a self-centralizing subgroup of cardinality 3 and A has Carter subgroups, then $A \cong G$.
   (d) (combining with Results 4.11.9 and 4.12.10) If A is a finite simple group and A has a self-centralizing subgroup of cardinality 3, and A satisfies extended Sylow property $C(\{2,3\})$, then $A \cong PSL(2,5)$.
   Three additional characterizations, two for $PSL(2,5)$ and one for G, can of course be derived by negating the last requirements in (b), (c), and (d).
4. The group G has the following presentation: $G \cong G(X, Y \mid X^7 = 1, Y^3 = 1, (XY)^2 = 1, (YX^5)^4 = 1)$. See Miller [27].
5. See Note 1 to Example 4.14.

□□□□□

Example 4.13

Let $G = PSL(2,9)$. (See Note 1.)

RESULT 4.13.1 card$(G) = 360 = 2^3 \cdot 3^2 \cdot 5$.
By Theorem 3.2.23, card$(G) = 9(9 - 1)(9 + 1)/2 = 360$.

RESULT 4.13.2 *G is simple.*
By Theorem 3.2.22.

RESULT 4.13.3
(a) *G is subdirectly irreducible.*
(b) *G is indecomposable.*
Since G is simple (Result 4.13.2), $H(G) = G$ proving Part (a). Part (b) follows from Part (a) and Theorem A.10.10.

RESULT 4.13.4
(a) *G has no subgroups of cardinality* 72, 90, 120, *or* 180.

(b) *G does not satisfy extended Sylow property* E($\{2,3\}$).
Suppose H is a subgroup of G of cardinality 72, hence index 5, and let $C = \mathrm{Core}_G(H)$. By simplicity (Result 4.13.2), C is trivial. By Theorem A.1.3 then, $G/\{1\}$, and hence G, can be embedded in S_5. Thus $\mathrm{card}(G) \mid \mathrm{card}(S_5)$ meaning $360 \mid 120$ which is false. Consequently G has no subgroups of cardinality 72. Proofs that G has no subgroups of cardinality 90, 120, or 180 are similar.

A $\{2,3\}$-Hall subgroup of G would have cardinality $2^3 \cdot 3^2 = 72$. Thus there are no $\{2,3\}$-Hall subgroups of G by Part (a).

RESULT 4.13.5

(a) *PSL*(2,5) *can be embedded in G.*

(b) *G is not minimal simple.*

(c) *G has maximal subgroups of non-prime-power index.*

Let $S = \mathrm{G}(X, Y \mid X^5 = 1,\ Y^5 = 1,\ (XY)^2 = 1,\ (Y^{-1}X)^3 = 1)$ be the icosahedral group. Let

$$a = \left[\begin{pmatrix} 0 & 1 \\ 2 & 1+\sqrt{2} \end{pmatrix}\right] \quad \text{and} \quad b = \left[\begin{pmatrix} 0 & 1 \\ 2 & 1+2\sqrt{2} \end{pmatrix}\right]$$

in G, and let $H = \langle a,b \rangle$. Define $f\colon \{X, Y\} \to \{a,b\}$ by $f(X) = a$ and $f(Y) = b$. In order to apply Theorem 2.2.5, it must be shown that the relations of S are "satisfied" in H. This is done as follows:

$$a^2 = \left[\begin{pmatrix} 0 & 1 \\ 2 & 1+\sqrt{2} \end{pmatrix}\right]\left[\begin{pmatrix} 0 & 1 \\ 2 & 1+\sqrt{2} \end{pmatrix}\right] = \left[\begin{pmatrix} 2 & 1+\sqrt{2} \\ 2+2\sqrt{2} & 2+2\sqrt{2} \end{pmatrix}\right]$$

so $a^4 = a^2a^2 =$

$$\left[\begin{pmatrix} 2 & 1+\sqrt{2} \\ 2+2\sqrt{2} & 2+2\sqrt{2} \end{pmatrix}\right]\left[\begin{pmatrix} 2 & 1+\sqrt{2} \\ 2+2\sqrt{2} & 2+2\sqrt{2} \end{pmatrix}\right] = \left[\begin{pmatrix} 1+\sqrt{2} & 2 \\ 1 & 0 \end{pmatrix}\right]$$

which is equal to a^{-1} implying $a^5 = 1$;

$$b^2 = \left[\begin{pmatrix} 0 & 1 \\ 2 & 1+2\sqrt{2} \end{pmatrix}\right]\left[\begin{pmatrix} 0 & 1 \\ 2 & 1+2\sqrt{2} \end{pmatrix}\right] = \left[\begin{pmatrix} 2 & 1+2\sqrt{2} \\ 2+\sqrt{2} & 2+\sqrt{2} \end{pmatrix}\right]$$

so $b^4 = b^2b^2 =$

$$\left[\begin{pmatrix} 2 & 1+2\sqrt{2} \\ 2+\sqrt{2} & 2+\sqrt{2} \end{pmatrix}\right]\left[\begin{pmatrix} 2 & 1+2\sqrt{2} \\ 2+\sqrt{2} & 2+\sqrt{2} \end{pmatrix}\right] = \left[\begin{pmatrix} 1+2\sqrt{2} & 2 \\ 1 & 0 \end{pmatrix}\right]$$

which is equal to b^{-1} implying $b^5 = 1$;

$$ab = \left[\begin{pmatrix} 0 & 1 \\ 2 & 1+\sqrt{2} \end{pmatrix}\right]\left[\begin{pmatrix} 0 & 1 \\ 2 & 1+2\sqrt{2} \end{pmatrix}\right] = \left[\begin{pmatrix} 2 & 1+2\sqrt{2} \\ 2+2\sqrt{2} & 1 \end{pmatrix}\right]$$

so $(ab)^2 = 1$ since the last matrix in brackets has trace zero (Theorem 3.2.24);

$$b^{-1}a = \left[\begin{pmatrix} 1+2\sqrt{2} & 2 \\ 1 & 0 \end{pmatrix}\right]\left[\begin{pmatrix} 0 & 1 \\ 2 & 1+\sqrt{2} \end{pmatrix}\right] = \left[\begin{pmatrix} 1 & \sqrt{2} \\ 0 & 1 \end{pmatrix}\right]$$

so $(b^{-1}a)^3 = 1$ by the Corollary to Theorem 3.1.3. Theorem 2.2.5 may now be applied to yield a surjective homomorphism $d:S \to H$ such that $d(x) = a$ and $d(y) = b$. Then $\ker d \lhd S$, but S is simple (Result 4.11.2), so $\ker d = \{1\}$ or S. Since the latter case is precluded by $d(x) = a \neq 1$, conclude $\ker d$ is trivial and hence d is an isomorphism. Thus S can be embedded in G. This and $PSL(2,5) \cong S$ (Result 4.11.1) complete the proof of Part (a).

Since $H \subset G$ (card(H) = card(S) = 60 by Result 4.11.1 while card(G) = 360) and since H is nonsolvable ($H \cong S$ so H is simple by Result 4.11.2), G is not a minimal simple group.

Finally G has no subgroups of cardinalities 120 or 180 (Result 4.13.4) consequently application of Lagrange's theorem shows H is a maximal subgroup of G. But H has index 360/60 = 6.

RESULT 4.13.6 *G has 10 3-Sylow subgroups.*
By Sylow's theorem (Theorem A.1.6), $n_3(G) \equiv 1 \bmod 3$ and $n_3(G)|360$. Thus $n_3(G)|40$ so $n_3(G) = 1, 4, 10,$ or 40.

Let H denote the subgroup of $SL(2,9)$ consisting of unitriangular matrices. For convenience let Z denote $Z(SL(2,9))$. Then since $Z \lhd SL(2,9)$, HZ is a subgroup of $SL(2,9)$ and consequently HZ/Z is a subgroup of G. By Theorem 3.2.8, card(H) = 9 so card(HZ/Z) = 9 since H and Z are disjoint. Hence HZ/Z is a 3-Sylow subgroup of G. Let $T = \{\left(\begin{smallmatrix} a & 0 \\ b & 1/a \end{smallmatrix}\right) \mid a \neq 0\}$. It is easily seen that T is a subgroup of $SL(2,9)$ and that $Z \subseteq T$. It is now claimed that $T \subseteq N_{SL(2,9)}(H)$. Let $\left(\begin{smallmatrix} a & 0 \\ b & 1/a \end{smallmatrix}\right) \in T$. Then for any $\left(\begin{smallmatrix} 1 & 0 \\ c & 1 \end{smallmatrix}\right) \in H$, $\left(\begin{smallmatrix} a & 0 \\ b & 1/a \end{smallmatrix}\right)\left(\begin{smallmatrix} 1 & 0 \\ c & 1 \end{smallmatrix}\right)\left(\begin{smallmatrix} a & 0 \\ b & 1/a \end{smallmatrix}\right)^{-1} =$

$$\begin{pmatrix} a & 0 \\ b+(c/a) & 1/a \end{pmatrix}\begin{pmatrix} 1/a & 0 \\ -b & a \end{pmatrix} = \begin{pmatrix} 1 & 0 \\ c/a^2 & 1 \end{pmatrix} \in H,$$

proving the claim. But from $T \subseteq N_{SL(2,9)}(H)$ it follows immediately that $T/Z \subseteq N_G(HZ/Z)$. Since the underlying field has cardinality 9, there are 9 choices for b and $9 - 1 = 8$ choices for a in a matrix $\left(\begin{smallmatrix} a & 0 \\ b & 1/a \end{smallmatrix}\right) \in T$. Thus card$(T) = 9 \cdot 8 = 72$ and consequently card$(T/Z) = 72/2 = 36$. Thus $36 \leqslant$ card$(N_G(HZ/Z))$. Since HZ/Z is a 3-Sylow subgroup of G, $[G:N_G(HZ/Z)] = n_3(G)$ by Theorem A.10.15. Thus from the first paragraph it follows that card$(N_G(HZ/Z))$ = 360, 90, 36, or 9. The requirement $36 \leqslant$ card$(N_G(HZ/Z))$ precludes the last case. The first case would imply $N_G(HZ/Z) = G$ so $HZ/Z \lhd G$ contradicting simplicity (Result 4.13.2), hence the first case is eliminated. Finally the second case is eliminated by Result 4.13.4. Conclude card$(N_G(HZ/Z))$ = 36 and so $n_3(G) = 10$.

EXERCISES

1. Let A be a simple group of cardinality 360 such that A has a subgroup of cardinality 60. Show $A \cong G$.
2. Show G does not satisfy extended Sylow property $E(\{3,5\})$.
3. Show $n_5(G) = 36$.
4. Show G does not satisfy extended Sylow property $E(\{2,5\})$.

5. Show the 3-Sylow subgroups of G are isomorphic to $Z(3) \times Z(3)$.
6. Show S_4 can be embedded in G.
7. If H is a subgroup of G of cardinality 24, show $H \cong S_4$.
8. Show
   (a) $n_2(G) = 45$.
   (b) G has Carter subgroups.
9. Show every element of G has prime-power order (different primes allowed).

NOTES

1. The notation $PSL(2,9)$ can be used without actually specifying the underlying field because all fields of cardinality 9 are isomorphic. On the other hand if one wants to talk about particular elements of $PSL(2,9)$ as we have occasion to do here, a particular field must be specified. The field we have chosen is $Z(3)$ with $\sqrt{2}$ adjoined, usually written $Z(3)(\sqrt{2})$. Formally this field is the set $Z(3) \times Z(3)$ with addition $(a,b) + (a^*,b^*) = (a + a^*, b + b^*)$ and multiplication $(a,b)(a^*,b^*) = (aa^* + 2bb^*, ba^* + ab^*)$. Informally the element (a,b) can be represented as $a + b\sqrt{2}$. Therefore the multiplication rule becomes $(a + b\sqrt{2})(a^* + b^*\sqrt{2}) = aa^* + 2bb^* + (ba^* + ab^*)\sqrt{2}$ as would be the case if $a + b\sqrt{2}$ and $a^* + b^*\sqrt{2}$ were real numbers. For purposes of clarity the informal representation will be used in this example. Thus the field elements are 0, 1, 2, $\sqrt{2}$, $1 + \sqrt{2}$, $2 + \sqrt{2}$, $2\sqrt{2}$, $1 + 2\sqrt{2}$, and $2 + 2\sqrt{2}$.
2. The group G is the third smallest finite nonabelian simple group. It is isomorphic to the alternating group A_6.
3. As a counterexample G shows the class of groups in which Carter subgroups exist is not subgroup closed (G belongs to this class by Exercise 8 but has a subgroup which does not by Results 4.13.5 and 4.11.12).
4. A characterization for G is given in Exercise 1. It can be shown that the 60-element subgroup requirement in this characterization isn't necessary. In other words G is the only (up to isomorphism) simple group of cardinality 360.
5. The group G has the following presentation: $G \cong G(X, Y \mid X^5 = 1, Y^5 = 1, (XY)^2 = 1, (X^{-1}Y)^4 = 1)$. See Todd and Coxeter [35].

□□□□□

Example 4.14

Let $G = PSL(2,11)$.

RESULT 4.14.1 $\text{card}(G) = 660 = 2^2 \cdot 3 \cdot 5 \cdot 11$.
By Theorem 3.2.23, $\text{card}(G) = 11(11 - 1)(11 + 1)/2 = 660$.

RESULT 4.14.2 *G is simple.*
By Theorem 3.2.22.

RESULT 4.14.3
(a) *G is subdirectly irreducible.*
(b) *G is indecomposable.*
Since G is simple (Result 4.14.2) $\mathrm{H}(G) = G$, proving Part (a). Part (b) follows from Part (a) and Theorem A.10.10.

RESULT 4.14.4 *$PSL(2,3)$ can be embedded in G.*
Let A denote the tetrahedral group, $A = \mathrm{G}(X, Y \mid X^3 = 1, Y^3 = 1, (XY)^2 = 1)$. Let $a = [\left(\begin{smallmatrix} 2 & 7 \\ 10 & 8 \end{smallmatrix}\right)]$ and $b = [\left(\begin{smallmatrix} 10 & 6 \\ 9 & 0 \end{smallmatrix}\right)]$ in G, and let $T = \langle a,b \rangle$. Define the function $f: \{X, Y\} \to \{a,b\}$ by $f(X) = a$ and $f(Y) = b$. In order to apply Theorem 2.2.5 it must be shown that the three relations of A are "satisfied" in T. This is done as follows: $a^3 = b^3 = 1$ by Theorem 3.2.25 while $ab = [\left(\begin{smallmatrix} 2 & 7 \\ 10 & 8 \end{smallmatrix}\right)]\,[\left(\begin{smallmatrix} 10 & 6 \\ 9 & 0 \end{smallmatrix}\right)] = [\left(\begin{smallmatrix} 6 & 1 \\ 7 & 5 \end{smallmatrix}\right)]$ so $(ab)^2 = 1$ since the last matrix in brackets has trace zero (Theorem 3.2.24). Theorem 2.2.5 may now be applied to yield existence of a surjective homomorphism $d{:}A \to T$. Thus $A/\ker d \cong T$. Now as shown above $\mathrm{o}(a) = 3$ and $\mathrm{o}(ab) = 2$ so $6 \mid \mathrm{card}(T)$. But $\mathrm{card}(T) \mid \mathrm{card}(A) = 12$ (by Result 4.6.1). Thus $\mathrm{card}(T) = 6$ or 12. The first case implies $\mathrm{card}(\ker d) = 2$ which is false because A has no normal subgroups of cardinality 2 (Results 4.6.3 and 4.6.1, Part (d)). Thus $\mathrm{card}(T) = 12$, hence $\ker d$ is trivial and so $d{:}A \to T$ is an isomorphism showing A can be embedded in G. Finally $PSL(2,3) \cong A$ by Result 4.6.1 again, so $PSL(2,3)$ can be embedded in G.

RESULT 4.14.5
(a) *G has* 55 *3-Sylow subgroups.*
(b) *If S is a 3-Sylow subgroup, then* $\mathrm{card}(\mathrm{N}_G(S)) = 12$.
First note that $Z(11)$ has no square root of -3. This is seen by examining all squares in $Z(11)$ ($0^2 = 0$, $1^2 = 1$, $2^2 = 4$, $3^2 = 9$, $4^2 = 5$, $5^2 = 3$, $6^2 = 3$, $7^2 = 5$, $8^2 = 9$, $9^2 = 4$, $10^2 = 1$) and noting that $8 = -3$ is not a square. This and the fact that $\mathrm{char}(Z(11)) = 11$ means (by Theorem 3.2.26) G has $11^2 - 11 = 110$ elements of order 3. Since 3-Sylow subgroups of G have cardinality 3 there are exactly 2 elements of order 3 in each 3-Sylow subgroup. Consequently there are $110/2 = 55$ 3-Sylow subgroups of G proving Part (a).

By Theorem A.10.15, $55 = [G{:}\mathrm{N}_G(S)]$ where S is any 3-Sylow subgroup of G. Thus $\mathrm{card}(\mathrm{N}_G(S)) = 660/55 = 12$.

RESULT 4.14.6
(a) *G satisfies extended Sylow property* $\mathrm{E}(\{2,3\})$.
(b) *G does not satisfy extended Sylow property* $\mathrm{C}(\{2,3\})$.
Since $\mathrm{card}(G) = 660 = 2^2 \cdot 3 \cdot 5 \cdot 11$, $\{2,3\}$-Hall subgroups have cardinality $2^2 \cdot 3 = 12$. Hence $\{2,3\}$-Hall subgroups exist by Results 4.14.4 and 4.6.1.

By the above, G has a $\{2,3\}$-Hall subgroup A such that $A \cong PSL(2,3)$. Let S

be a 3-Sylow subgroup of G and let $B = N_G(S)$. Then B is also a $\{2,3\}$-Hall subgroup because it has cardinality 12 by Result 4.14.5, Part (b). To show A and B are not conjugate in G it is (more than) sufficient to show they are not isomorphic. By definition of B, B has a normal subgroup of cardinality 3 (namely S). By Result 4.6.3, however, all subgroups of A of cardinality 3 are nonnormal. Thus A and B are nonisomorphic as required, completing the proof of Part (b).

EXERCISES

1. Show
   (a) $PSL(2,5)$ is isomorphic to the subgroup of G generated by $[\begin{pmatrix} 5 & 2 \\ 7 & 3 \end{pmatrix}]$ and $[\begin{pmatrix} 3 & 0 \\ 3 & 4 \end{pmatrix}]$.
   (b) G is not minimal simple.
   (c) G satisfies extended Sylow property $E(\{2,3,5\})$.
2. Show G has no subgroups of cardinality 66, 110, 132, 220, or 330.
3. Show
   (a) $n_{11}(G) = 12$.
   (b) $n_5(G) = 66$.
4. Show G does not satisfy extended Sylow property $E(\pi)$ for any of the following sets of primes π: $\{2,5\}$, $\{2,11\}$, $\{3,5\}$, $\{3,11\}$, $\{2,3,11\}$, $\{2,5,11\}$, $\{3,5,11\}$.
5. Show G satisfies extended Sylow property $D(\{5,11\})$.
6. Find a maximal subgroup of G of non-prime-power index.
7. Show G has no subgroups of cardinality 22.
8. Show $n_2(G) = 55$.
9. If H is a subgroup of G such that $\text{card}(H) = 60$ show $H \cong PSL(2,5)$.
10. If A and B are isomorphic $\{2,3\}$-Hall subgroups of G, show A and B are conjugate in G.
11. Show that $[M] \in G$ has order 6 if and only if $\text{tr}(M) = 5$ or 6.
12. Show G has 55 elements of order 2, 110 elements of order 3, no elements of order 4, 264 elements of order 5, 110 elements of order 6, and 120 elements of order 11. Note in particular that this means G has elements of non-prime-power order.

NOTES

1. It has been seen previously (Example 4.12, $PSL(2,7)$) that a group may satisfy extended Sylow property $E(\pi)$ but not extended Sylow property $C(\pi)$. The group G, however, is a somewhat more dramatic example of this because it has pairs of $\{2,3\}$-Hall subgroups that fail even to be isomorphic (see the proof of Result 4.14.6). Moreover G shows that existence of a supersolvable π-Hall subgroup doesn't guarantee that all π-Hall subgroups are supersolvable.

   Note that the groups $PSL(2,7)$ and G fail to satisfy extended Sylow property $C(\{2,3\})$ for essentially different reasons. In $PSL(2,7)$ all $\{2,3\}$-Hall subgroups are isomorphic (Example 4.12, Exercise 4) but not necessarily conjugate. In G, any two isomorphic $\{2,3\}$-Hall subgroups are conjugate (Exercise 10) but the trouble is that not every two $\{2,3\}$-Hall subgroups are isomorphic.

2. The group G has the following presentation: $G \cong \mathrm{G}(X, Y \mid X^{11} = 1, Y^2 = 1, (XY)^3 = 1, (X^4YX^6Y)^2 = 1)$. See Lewis [23].

□□□□□

MISCELLANEOUS EXERCISES

I Families of Groups

I.1. For $2 \leqslant n$ let

$$SD_n = \mathrm{G}(X, Y \mid X^{2^n} = 1, Y^2 = 1, X^{2^{n-1}+1}Y^{-1}XY = 1).$$

(These are called the **semidihedral groups.**) Show $SD_n \cong Z(2^n) \times_g Z(2)$ where $g: Z(2) \to \mathrm{Aut}(Z(2^n))$ is defined by $g_0 = 1$ and $g_1 = f$ where $f: Z(2^n) \to Z(2^n)$ is defined by $f(b) = (2^{n-1} - 1)b$ for all $b \in Z(2^n)$.

I.2. Let n and m be nonzero even natural numbers. Let $G_{n,m} = \mathrm{G}(R, S \mid R^n = 1, S^m = 1, (RS)^2 = 1, (R^{-1}S)^2 = 1)$. Show
(a) $G_{n,m}$ is isomorphic to the subgroup of $D_n \times D_m$ generated by (x,y) and (y,x).
(b) $\mathrm{card}(G_{n,m}) = nm$.
(c) Every element of $G_{n,m}$ can be expressed as $r^i s^k$ where $0 \leqslant i \leqslant n - 1$ and $0 \leqslant k \leqslant m - 1$.

II Abelian Groups

II.1. Find all subgroups of $Z(2) \times Z(2) \times Z(2)$ and draw it subgroup diagram.

II.2. Give an example of a finite abelian group and two pure subgroups whose intersection is not pure.

III Nilpotent Groups

III.1. Let $G = SD_3$ (see Exercise I.1). Show
(a) $\mathrm{card}(G) = 16$.
(b) G is nilpotent.
(c) G has exactly 5 elements of order 2.

III.2. Let $G = \mathrm{G}(A, B, C \mid A^4 = 1, B^2 = 1, C^2 = 1, CBCA^2B = 1, BABA^{-1} = 1, CACA^{-1} = 1)$. Show
(a) G is isomorphic to the subgroup of $Z(4)$ wr D_4 generated by $(g,1)$, (k,x^2), and (h,y) where $g, k, h \in \mathrm{F}[Z(4), D_4]$ are defined by $g(t) = 1$ for all $t \in D_4$,

$$k(t) = \begin{cases} 0 & \text{if } t = 1, x, x^2, \text{ or } x^3 \\ 2 & \text{if } t = y, xy, x^2y, \text{ or } x^3y \end{cases}$$

and $h(t) = 2$ for all $t \in D_4$.

(b) $\text{card}(G) = 16$.
(c) G is nilpotent.
(d) The elements of G are $1, a, a^2, a^3, b, c, ab, ac, a^2b, a^2c, a^3b, a^3c, bc, abc, a^2bc$, and a^3bc.
(e) G has exactly 7 elements of order 2.
(f) G is subdirectly irreducible.

III.3. Let $G = \text{G}(X, Y \mid X^8 = 1, X^4Y^{-2} = 1, Y^{-1}XYX = 1)$. (See Note 1 to Example 4.4.) Show
(a) G is isomorphic to the subgroup of $SL(2,41)$ generated by $\begin{pmatrix} 3 & 0 \\ 0 & 14 \end{pmatrix}$ and $\begin{pmatrix} 0 & 1 \\ -1 & 0 \end{pmatrix}$.
(b) $\text{card}(G) = 16$.
(c) G is nilpotent.
(d) G has a unique element of order 2.
(e) G is subdirectly irreducible.

III.4. Show no two of the following groups are isomorphic. Example 4.7, Example 4.8, Exercise III.1, Exercise III.2, Exercise III.3, $G_{4,4}$ (Exercise I.2), D_8, $D_4 \times Z(2)$, and $Q \times Z(2)$ where Q is the quaternion group (Example 4.4).

NOTE: The groups of Exercise III.4 are the only (up to isomorphism) nonabelian groups of cardinality 16.

III.5. Let $G = \text{G}(X, Y \mid X^9 = 1, Y^3 = 1, Y^{-1}XYX^5 = 1)$. Show
(a) $G \cong Z(9) \times_g Z(3)$ where $g{:}Z(3) \to \text{Aut}(Z(9))$ is defined by $g_0 = 1$, $g_1 = k$, and $g_2 = k^2$ where $k{:}Z(9) \to Z(9)$ is $k(t) = 7t$.
(b) $\text{card}(G) = 27$.
(c) G is nilpotent.
(d) G is subdirectly irreducible.

III.6. Let $G = \text{G}(A, B, C \mid A^3 = 1, B^3 = 1, C^3 = 1, C^{-1}A^{-1}CA = 1, B^{-1}A^{-1}BA = 1, B^{-1}C^{-1}BCA^{-1} = 1)$. Show
(a) $G \cong UT(3, Z(3))$.
(b) $\text{card}(G) = 27$.
(c) G is nilpotent.
(d) $\exp(G) = 3$.
(e) If A is a group such that $\text{card}(A) = 27$, $\exp(A) = 3$, and A is nonabelian, then $A \cong G$.

NOTE: The groups of Exercises III.5 and III.6 are the only (up to isomorphism) nonabelian groups of cardinality 27.

III.7. Let Q denote the quaternion group (Example 4.4). By the theorem quoted in Note 2 of Example 4.4, $A = Q \times Z(4)$ is nonhamiltonian. Show this by exhibiting a nonnormal subgroup of A.

III.8. Let G be the central pushout of the central pushout system $(Z(2), D_4, D_4, h, h)$ where $h: Z(2) \to Z(D_4)$ is defined by $h(0) = 1$ and $h(1) = x^2$. Show
(a) $\text{card}(G) = 32$.
(b) G is nilpotent.
(c) G is subdirectly irreducible.
(d) G has 19 elements of order 2 and 12 elements of order 4.

III.9. Let G be the central pushout of the central pushout system $(Z(2), D_4, Q, h, k)$ where Q is the quaternion group (Example 4.4), $h: Z(2) \to Z(D_4)$ is defined by $h(0) = 1$ and $h(1) = x^2$, and $k: Z(2) \to Z(Q)$ is defined by $k(0) = 1$ and $k(1) = x^2$. Show
(a) $\text{card}(G) = 32$.
(b) G is nilpotent.
(c) G is subdirectly irreducible.
(d) G has 11 elements of order 2 and 20 elements of order 4.

III.10. Let $G = Z(4) \text{ wr } Z(2)$. Show
(a) $\text{card}(G) = 32$.
(b) G is nilpotent.
(c) G is subdirectly irreducible.

III.11. Let $K = \{A \mid A \in UT(4, Z(2)) \text{ and } A_{21} = 0\}$. (See Section 3.2, Exercise I.5.) Show
(a) $\text{card}(K) = 32$.
(b) K is nilpotent.
(c) $Z(K) \cong Z(2) \times Z(2)$.
(d) K is indecomposable.
(e) K is not subdirectly irreducible.

III.12. Show no two of the groups in Exercises III.7–III.11 are isomorphic.

IV Supersolvable Groups

IV.1. Show $D_6 \cong D_3 \times Z(2)$.

IV.2. (a) Show D_9 has 16 subgroups (including itself) and draw its subgroup diagram.
(b) Show D_9 is subdirectly irreducible.

IV.3. Let $G = \text{Hol}(Z(5))$. Show
(a) $\text{card}(G) = 20$.
(b) G is supersolvable.
(c) G has 5 elements of order 2, 10 elements of order 4, and 4 elements of order 5.
(d) G is complete.

V Solvable Groups

V.1. Give an example of a finite solvable group G such that
(a) G is converse lagrange.
(b) G' is nilpotent.

(c) G has noncyclic minimal normal subgroups.

V.2. Give an example of a finite solvable group G such that

(a) G is converse lagrange.

(b) G' is nilpotent.

(c) Minimal normal subgroups of G are cyclic.

V.3. Show the class of converse lagrange groups is not quotient closed.

V.4. Let $G = GL(2,3)$. (See Chapter 3, Miscellaneous Exercises.) Show

(a) $G' = SL(2,3)$ and hence G' is not nilpotent.

(b) G is converse lagrange.

(c) The normal subgroups of G are G, G', G'', G''', and $\{1\}$.

(d) Minimal normal subgroups of G are cyclic.

V.5. Let $G = \mathrm{G}(X, Y \mid X^3Y^{-3} = 1, (XY)^2 = 1)$. Show

(a) $x^{24} = 1$ and $y^{24} = 1$.

(b) $\mathrm{card}(G) \leqslant 96$.

(c) G is solvable.

V.6. Let

$$G = \left\{ \begin{pmatrix} 1 & 0 & 0 \\ x & a & b \\ y & c & d \end{pmatrix} \middle| \; a,b,c,d,x,y \in Z(3) \text{ and } ad - bc = 1 \right\}.$$

Show

(a) G is a group in which inverses are described by

$$\begin{pmatrix} 1 & 0 & 0 \\ x & a & b \\ y & c & d \end{pmatrix}^{-1} = \begin{pmatrix} 1 & 0 & 0 \\ -dx+by & d & -b \\ cx-ay & -c & a \end{pmatrix}.$$

(b) G is an extension of (a group isomorphic to) $Z(3) \times Z(3)$ by $SL(2,3)$, hence $\mathrm{card}(G) = 216$ and G is solvable.

(c) G has no subgroups of index 2 hence G is not converse lagrange.

(d) $$G' = \left\{ \begin{pmatrix} 1 & 0 & 0 \\ x & a & b \\ y & c & d \end{pmatrix} \middle| \; x,y \in Z(3) \text{ and } \left(\begin{smallmatrix} a & b \\ c & d \end{smallmatrix}\right) \in SL(2,3)' \right\}$$

hence $\mathrm{card}(G') = 72$.

(e) $Z(G') = \{1\}$ hence G' is not nilpotent.

VI Nonsolvable Groups

VI.1. Let $G = PGL(2,7)$ (see Chapter 3, Miscellaneous Exercises). Show

(a) The 2-Sylow subgroups of G are isomorphic to D_8.

(b) G does not satisfy extended Sylow property $\mathrm{E}(\{2,3\})$.

(c) The 2-Sylow subgroups of G are maximal subgroups and also Carter subgroups.

(d) Conclude that extended Sylow property $\mathrm{E}(\pi)$ is not extension closed and that the class of groups in which maximal subgroups have prime-power index is not extension closed.

VI.2. Let $G = PSL(2,13)$. Show
(a) $\text{card}(G) = 1092$.
(b) G is minimal simple.
(c) G has nontrivial elements of non-prime-power order.

VI.3. Show
(a) $S_3 \times S_4$ can be embedded in S_7.
(b) S_7 satisfies extended Sylow property $E(\{2,3\})$.
(c) $PSL(2,9)$ can be embedded in S_7.
(d) Conclude that extended Sylow property $E(\pi)$ is not subgroup closed.

VI.4. Let $G = G(X, Y \mid X^6 = 1, Y^3 = 1, (XY)^4 = 1, (X^2Y)^4 = 1, (X^3Y)^3 = 1, (YX^2Y)^{-2}X^{-2}(YX^2Y)^2X^2 = 1)$. Show G is perfect. (Actually G is simple and has cardinality 5616; but this is much more difficult to show.)

VI.5. Give an example of a nonsolvable group all of whose maximal subgroups have prime-power index (different primes allowed) that has nontrivial elements of non-prime-power order.

VI.6. Give an example of a nonsolvable group that has Carter subgroups and has nontrivial elements of non-prime-power order.

SUGGESTIONS FOR FURTHER READING

For a complete list of the p-groups of cardinalities p^3 and p^4 described by their presentations see Burnside [4], Sections 112–118. Also found in Burnside [4], Section 126, are presentations for all groups of cardinality 24. See Hall and Senior [15] for the groups of cardinalities 2^5 and 2^6. A permutational construction for the family of dicyclic groups (see Example 4.4, Note 1) appears in Scott [33], page 189.

Extensions of the general linear and projective general linear groups, namely the groups $\Gamma L(2,F)$ and $P\Gamma L(2,F)$ are discussed in Scott [33], pages 280–283. Permutational constructions for the Mathieu groups M_{11} and M_{12}, simple groups of cardinalities 7920 and 95040 respectively, are found om Carmichael [5], page 151, Exercise 12. See Greenberg [13] for a thorough treatment of the five Mathieu groups.

Examples of two nonisomorphic simple groups, both of cardinality 20160 are given in Rotman [31], Theorem 8.28.

5

Infinite Abelian Groups

Abelian groups can be neatly categorized into three classes. In the first class there are the groups, called **torsion** groups, all of whose elements have finite order. In the second class there are the groups, called **torsion-free** groups, none of whose elements (except the identity) have finite order. The remaining class consists of groups which have both nontrivial elements of finite order and elements of infinite order. These are called **mixed** groups.

Now note that the first and second classes share a common property. For a group that is either torsion or torsion-free, the torsion subgroup is a direct factor or splits off ("splits off" is another phrase meaning "is a direct factor of"). This suggests dividing the class of mixed abelian groups into two subclasses as follows. A mixed abelian group is termed either **splitting** or **nonsplitting** depending on whether or not its torsion subgroup splits off.

The class of splitting groups turns out to be not too interesting, or rather a splitting group A is usually no more interesting than its torsion subgroup $\mathrm{t}(A)$ and the associated torsion-free factor $A/\mathrm{t}(A)$. If the properties of $\mathrm{t}(A)$ and $A/\mathrm{t}(A)$ are known, then the properties of A itself cannot be too mysterious because $A \cong \mathrm{t}(A) \times (A/\mathrm{t}(A))$.

On the other hand if A is a nonsplitting group, then A is no longer tied to $\mathrm{t}(A)$ and $A/\mathrm{t}(A)$ in so simple a way as direct product and its properties can be very mysterious indeed. There exist many radically different groups all having the same (that is, isomorphic) torsion subgroup and the same associated torsion-free factor.

So the three classes of abelian groups that we will be interested in are *torsion, torsion-free,* and *nonsplitting.*

In comparing examples of nonsplitting groups it will be of interest to examine how close a group comes to splitting. "Closeness to splitting" can be measured in several different ways. The following observations form a basis for defining three classes of mixed abelian groups that are each close to being split.

Every splitting group A has a decomposition $A = t(A) \oplus F$. Thus because of the projection maps (Theorem A.2.3), $t(A)$ is a homomorphic image of A. Also $A = t(A) \oplus F$ implies $A/t(A) \cong F$ and thus $A/t(A)$ can be embedded in A. Thirdly $A/t(A) \cong F$ also shows that F is torsion-free (Theorem A.4.2) and so A has nontrivial torsion-free direct factors.

This motivation leads to the following three definitions (the terminology used here is merely a convenient one for this chapter. The three classes have no standard names).

DEFINITION Let A be a mixed abelian group. Then A is an $\mathbf{M_1}$ **group** if and only if $t(A)$ is a homomorphic image of A.

DEFINITION Let A be a mixed abelian group. Then A is an $\mathbf{M_2}$ **group** if and only if $A/t(A)$ can be embedded in A.

DEFINITION Let A be a mixed abelian group. Then A is an $\mathbf{M_3}$ **group** if and only if A has nontrivial torsion-free direct factors.

In addition to the question of splitting, a major area of investigation in this chapter will be the finiteness properties. Some of these (such as the FC condition) can be immediately eliminated from a discussion of abelian groups due to the fact that all abelian groups satisfy the given property. The number of interesting finiteness properties that are left, however, is still considerable. We will discuss the following in this chapter: *torsion, reduced, hopfian, cohopfian, coatomic, residually finite*, and the (*ascending* and *descending*) *chain conditions*.

As a third major area of concern, the indecomposability criteria play a less important role in the theory of abelian groups than they do for nonabelian groups. The reasons for this are as follows: First of all it is easily seen that a nontrivial abelian simple group has prime cardinality, so there are no infinite abelian simple groups. Secondly, while infinite abelian subdirectly irreducible groups exist, they are far from being prevalent. All of them, in fact, are given in Example 5.4 (see Example 5.4, Note 2). Thirdly, although the concept of indecomposability is quite an important one for infinite abelian groups, it is by no means as crucial as in the finite case. Consider, for example, a group A with a nontrivial decomposition $A = B \oplus C$. In the finite case we are usually justified in considering A as an unfit subject for study. Once the properties of B and C are known, most of the properties of A are deducible, so we turn our attention from A to the groups B and C instead. But in the infinite case it is quite possible for the decomposition $A = B \oplus C$ to be a nontrivial one (that is, B and C are both nontrivial and proper subgroups of A) and yet for A to be isomorphic to one of the factors. If $A \cong B$, we cannot relegate the study of A to the study of B and C

for doing so will gain very little as A and B are essentially the same group. For this reason many examples of infinite groups which are decomposable are worthy of our attention. This is particularly true for abelian groups. Roughly half the groups presented in this chapter are decomposable.

Since all of the examples of this chapter are abelian, additive notation will be predominant. Some of the lemmas, however, are stated for groups that are not necessarily abelian. In these lemmas multiplicative notation will be used.

Example 5.1

Let Q be the group of all rational numbers under the operation of addition. (See Note 1.)

RESULT 5.1.1 *Q is torsion-free.*
Suppose $a/b \in Q$ such that $o(a/b) = n$. Then $na/b = 0$, hence $na = 0$. This and $n \neq 0$ imply $a = 0$. Consequently $a/b = 0$ as required.

RESULT 5.1.2 *Q is divisible.*
Let $a/b \in Q$ and let n be a nonzero natural number. Then $a/bn \in Q$ and $a/b = n(a/bn)$. Consequently Q is divisible.

RESULT 5.1.3 *Q, and each of its subgroups, is indecomposable.*
Let H be a subgroup of Q, suppose $H = A \oplus C$ where both A and C are nontrivial and seek a contradiction. Thus there exist nonzero elements $a/b \in A$ and $j/k \in C$. Since both A and C are subgroups of Q, $n(a/b) \in A$ and $n(j/k) \in C$ for all integers n. In particular when $n = jb$, $jb(a/b) \in A$; and when $n = ak$, $ak(j/k) \in C$. Thus $ja \in A$ and $aj \in C$. That is, $ja \in A \cap C$. But $H = A \oplus C$ implies $A \cap C$ is trivial. Therefore $ja = 0$, hence $j = 0$ or $a = 0$. Consequently, $a/b = 0$ or $j/k = 0$, the desired contradiction in either case.

RESULT 5.1.4 *Every proper subgroup of Q is reduced.*
Suppose H is a nonreduced subgroup of Q, and seek to show $Q = H$. Since H is not reduced, H has a nontrivial divisible subgroup D, so it suffices to show $Q \subseteq D$. Let $x/y \in Q$. Note $y \neq 0$. Since D is nontrivial, D has a nonzero element a/b, which means $a \neq 0$. Consequently $ay \neq 0$, hence there exists $t \in D$ such that $(ay)t = a/b$ by divisibility of D. Thus $t = (a/b)(1/ay) = 1/by$. Then $t \in D$ implies $(xb)t \in D$. That is, $(xb)(1/by) = x/y \in D$.

RESULT 5.1.5 *Q is not residually finite.*
Since Q is not reduced (Q is divisible by Result 5.1.2), it is not residually finite by Theorem 1.3.11.

RESULT 5.1.6 *Q has no maximal subgroups.*
By Result 5.1.2 and the Lemma to Theorem 1.2.13.

RESULT 5.1.7 *If B is a nontrivial subgroup of Q, then Q/B is torsion.*
Let $i/j \in B$, $i/j \neq 0$. Thus $i \neq 0$. Let $B + x \in Q/B$, and seek to show $o(B + x)$ is finite. Since $x \in Q$, $x = m/n$ for some integers m and n. Now $(in)x = im$ which belongs to $\langle i \rangle$. But $i/j \in B$ implies $i = j(i/j) \in B$ so $(in)x \in B$ and hence $(in)(B + x) = 0$. This and $in \neq 0$ (because $i \neq 0$ and $n \neq 0$) show $B + x$ has finite order.

RESULT 5.1.8 *Q, and each of its subgroups, is pure-simple.*
Let H be a subgroup of Q, let B be a nontrivial pure subgroup of H, and seek to show $B = H$. Let $x \in H$. Then $B + x \in Q/B$ hence by Result 5.1.7, $B + x$ has finite order, call it n. Hence $n(B + x) = 0$ implying $nx \in B$. Thus $nx \in nH \cap B$ so purity of B in H yields $nx \in nB$. Thus $nx = nb$ for some $b \in B$ and so $n(x - b) = 0$. This and the fact that Q is torsion-free imply $x = b$ so $x \in B$ as required.

RESULT 5.1.9 *Q is characteristically simple.*
Suppose H is a nontrivial characteristic subgroup of Q and seek to show $Q = H$. It is first claimed that if t is any nonzero element of Q, then the function $d_t : Q \to Q$ defined by $d_t(w) = tw$ is an automorphism. For rational numbers w and z, $d_t(w + z) = t(w + z) = tw + tz = d_t(w) + d_t(z)$. Furthermore d_t is injective because $tw = 0$ implies $0 = tw/t = w$. Finally d_t is surjective because $s = t(s/t) = d_t(s/t)$ for any rational number s. Thus d_t is an automorphism as claimed. Then $d_t[H] = H$ since H is a characteristic subgroup. Now to show $Q = H$, let $y \in Q$. Since H is nontrivial there exists $x \in H$, $x \neq 0$. Therefore $y/x \in Q$ so by the above, $d_{y/x}[H] = H$. Thus $x \in H$ implies $y = (y/x)x \in H$.

RESULT 5.1.10 *Q is locally cyclic.*
Let H be a finitely generated subgroup of Q with generators $a_1/b_1, a_2/b_2, \ldots, a_n/b_n$. Seek to show H is cyclic. Each denominator b_i may be assumed to be positive for if $b_i < 0$, then $-b_i$ is positive and a_i/b_i may be replaced by $-a_i/b_i$. Now from this set $\{b_1, b_2, \ldots, b_n\}$ of positive denominators choose the largest and call it b_k. Thus $b_i \leqslant b_k$ for $1 \leqslant i \leqslant n$. Consequently, $b_i | b_k!$ so there is an integer c_i satisfying $b_i c_i = b_k!$ for all i, $1 \leqslant i \leqslant n$. Consider now the subgroup $S = \langle 1/b_k! \rangle$. Since $1/b_k! \in S$, $a_i c_i (1/b_k!) \in S$ for all i. Therefore $a_i c_i / b_i c_i = a_i/b_i \in S$ for all i. Since all generators of H belong to S, H is a subgroup of S. Finally note that S is cyclic, hence so too is its subgroup H.

LEMMA *If G is a locally cyclic group, then the subgroup lattice* $\mathrm{L}(G)$ *is distributive.*
First note that if x and y are two elements of G, then $x, y \in \langle a \rangle$ for some a so x and y commute. Thus G is abelian.

Hence (see Theorem A.9.2) it must be shown that $A + (B \cap C) = (A + B) \cap (A + C)$ for all subgroups A, B, and C. Since $B \cap C \subseteq B$, it follows that $A +$

$(B \cap C) \subseteq A + B$ and similarly that $A + (B \cap C) \subseteq A + C$. Thus $A + (B \cap C) \subseteq (A + B) \cap (A + C)$. To show the reverse inclusion, let $x \in (A + B) \cap (A + C)$. Then $x = a_1 + b = a_2 + c$ where a_1 and a_2 belong to A, b belongs to B and $c \in C$. Because G is locally cyclic, $\langle b,c \rangle$ is a cyclic group. Let w generate $\langle b,c \rangle$, so $b = mw$ and $c = nw$ for some integers m and n. Let d be the greatest common divisor of m and n. Thus $m = dm^*$ and $n = dn^*$ where m^* and n^* are relatively prime. Consequently there exist integers i and j such that $im^* + jn^* = 1$. Observe that $dm^*n^*w = m^*(dn^*)w = m^*nw = m^*c \in C$. Also $dm^*n^*w = n^*(dm^*)w = n^*mw = n^*b \in B$. Conclude $dm^*n^*w \in B \cap C$. Since $im^* + jn^* = 1$, $x = im^*x + jn^*x = im^*(a_2 + c) + jn^*(a_1 + b) = im^*a_2 + im^*c + jn^*a_1 + jn^*b$. Setting $a_3 = im^*a_2 + jn^*a_1$ yields $a_3 \in A$ and $x = a_3 + im^*c + jn^*b = a_3 + im^*nw + jn^*mw = a_3 + im^*dn^*w + jn^*dm^*w = a_3 + (i + j)dm^*n^*w$. But as noted before, $dm^*n^*w \in B \cap C$. Thus $x \in A + (B \cap C)$ as required.

RESULT 5.1.11 *L(Q) is distributive.*
By Result 5.1.10 and the lemma.

RESULT 5.1.12 *Q, and each of its subgroups, is hopfian.*
Let H be a subgroup of Q. Suppose B is a nontrivial subgroup of H such that $H/B \cong H$ and seek a contradiction. Since B is nontrivial, Q/B is torsion by Result 5.1.7, hence its subgroup H/B is torsion. This and $H/B \cong H$ imply H is torsion. But H is a subgroup of Q and Q is torsion-free. Since H is both torsion and torsion-free it must be trivial. Conclude B is trivial, the desired contradiction.

RESULT 5.1.13 *Q is cohopfian.*
Suppose that B is a subgroup of Q such that $B \cong Q$. Then B is divisible (Result 5.1.2) and nontrivial. Conclude $B = Q$ by Result 5.1.4.

Supplement: Embedding certain subgroups of torsion-free abelian groups into Q

NOTATION If A is a torsion-free abelian group and $x \in A$, $x \neq 0$, then $P(A,x)$ will denote $\{a | a \in A$ and $ma \in \langle x \rangle$ for some $m \in N$, $m \neq 0\}$.

RESULT 5.1.14 *Let A be a torsion-free abelian group and let $x \in A$, $x \neq 0$. Then*

(a) *$P(A,x)$ is a pure subgroup of A.*
(b) *$x \in P(A,x)$.*
(c) *$P(A,x)$ can be embedded in Q. In particular the function $g: P(A,x) \to Q$ defined by $g(a) = j/m$ if and only if $ma = jx$ with $m \neq 0$ is an embedding.*

If a and b belong to $P(A,x)$, then $ma = jx$ and $nb = kx$ where $m,n \neq 0$. Thus $mn \neq 0$ and $mn(a - b) = mna - mnb = njx - mkx = (nj - mk)x$ so $a - b \in P(A,x)$ showing $P(A,x)$ is a subgroup of A. To show purity, let $n \in N$, $n \neq 0$ and let $a \in nA \cap P(A,x)$. Thus $a = ny$ and $ma = jx$ for some $y \in A$, some positive integer m, and some integer j. This implies $mny = jx$. But mn is nonzero because m and n are. Consequently $y \in \mathrm{P}(A,x)$ so $a \in n\mathrm{P}(A,x)$ showing $\mathrm{P}(A,x)$ is a pure subgroup of A and thus completing the proof of Part (a). Part (b) is immediate.

To prove Part (c) it must first be shown that g is a function. Suppose $ma = jx$ and $na = kx$ where $m,n \neq 0$, and seek to show $j/m = k/n$. Two cases arise; either $a = 0$ or $a \neq 0$. In the first case $ma = na = 0$ so $jx = kx = 0$. This and the hypotheses (A torsion-free and $x \neq 0$) imply $j = k = 0$. Hence $j/m = k/n$. In the second case $mka = jkx = jna$. This and A torsion-free and the case hypothesis $a \neq 0$ imply $mk = jn$ so $j/m = k/n$. Thus g is a function. To show g is a homomorphism, suppose $a,b \in \mathrm{P}(A,x)$ with $ma = jx$ and $nb = kx$ where $m,n \neq 0$. Then $mn(a + b) = mna + mnb = (nj + mk)x$. This and $mn \neq 0$ imply $g(a + b) = (nj + mk)/mn = (j/m) + (k/n) = g(a) + g(b)$. Finally to show g is injective suppose $g(a) = 0$ with $ma = jx$, $m \neq 0$. Then $0 = j/m$ implying $0 = j$. Consequently $ma = 0$. This, A torsion-free, and $m \neq 0$ imply $a = 0$ as required.

EXERCISES

1. Show $Q \cong \mathrm{Hom}(Q,Q)$.
2. Show there exists no group G such that $G/Z(G) \cong Q$.
3. Show Q is not subdirectly irreducible.
4. Show $Q/\langle x\rangle \cong Q/Z$ for all nonzero $x \in Q$.
5. Show the following partial converse of Result 5.1.8. If A is a torsion-free, pure-simple abelian group, then A can be embedded in Q.
6. Show that if A is a torsion-free, locally cyclic group, then A can be embedded in Q.
7. Let $A_n = Z$ for all positive integers n. Define $\beta_n : A_n \to A_{n+1}$ by $\beta_n(x) = nx$. For $i \leqslant k$ let $\pi_i^k : A_i \to A_k$ be the identity map if $i = k$ and let it be the composition $\beta_{k-1} \circ \beta_{k-2} \circ \ldots \circ \beta_i$ if $i < k$. Finally let $(D, \leqslant)$ denote the directed set of positive integers with the usual order. Then $(\{A_i\}_{i \in D}, \{\pi_i^k \mid i \leqslant k\})$ is a direct system over $(D, \leqslant)$. Let L be its direct limit. Show $L \cong Q$.
8. Show that if A is a nontrivial, torsion-free, divisible abelian group, then $A \cong Q$ under any one of the additional conditions
   (a) All proper subgroups of A are reduced.
   (b) A is indecomposable.
   (c) A/B is torsion for all nontrivial subgroups B.
9. Show $Q \cong \mathrm{G}(\{X_i \mid i \in N, i \neq 0\} \mid (X_{i+1})^{i+1} X_i^{-1} = 1 \text{ for all } i)$.

NOTES

1. Whenever the phrase "let $a/b \in Q$" or its equivalent "let a/b be a rational number" is used in this chapter we will assume that a/b is reduced to lowest terms. This means that a and b are relatively prime. A rational number can always be reduced to lowest terms by dividing both its numerator and its denominator by their greatest common divisor. We do not, however, impose this

convention in the middle of a calculation. We may, for instance, multiply the rational number a/b by 2 and call the result $2a/b$ even though $2a/b$ may not be reduced to lowest terms. In the middle of a calculation we will allow this. At the outset, however, all fractions are assumed to be reduced to lowest terms.

2. Three characterizations of Q are given in Exercise 8, and a presentation is given in Exercise 9.
3. As a counterexample Q shows that the class of cohopfian groups is not subgroup closed (Q is cohopfian by Result 5.1.13 but its subgroup Z is not cohopfian since $\langle 2 \rangle$ is a proper isomorphic subgroup of Z).
4. Result 5.1.12 (but not its proof) can be generalized to show that every subgroup of $Q \times Q \times \ldots \times Q$ (any finite number of terms) is hopfian. This is best done using the concept of rank, and in fact is an easy corollary of the theorem stating $r_0(A) = r_0(A/B) + r_0(B)$ whenever A, B, and A/B are torsion-free groups.
5. A word of explanation is in order as to why some results were shown for all subgroups of Q while others were shown just for Q itself. The classes *torsion-free* and *locally cyclic* are subgroup closed and easily seen to be so. We will note later that certain subgroups of Q are torsion-free and locally cyclic, and need offer nothing more in the way of proof than the fact that Q has both these properties.

   Result 5.1.8 represents a slightly different situation. Here too the class in question is subgroup closed but this is not so easily seen. It is easier to show every subgroup of Q has the property as we have done. Note, however, that the combination of Result 5.1.8 and Exercise 5 shows that for torsion-free abelian groups at least this class is subgroup closed.

   Results 5.1.3 and 5.1.12 represent an entirely different situation. Both the classes *indecomposable* and *hopfian* are not subgroup closed. (For counterexamples see Note 2 to Example 5.6 and Miscellaneous Exercise II.2.)
6. The converse of the Lemma to Result 5.1.11 is also true.
7. See Miscellaneous Exercises IV.1, IV.5, and IV.6 at the end of this chapter, and Miscellaneous Exercise V.2 at the end of Chapter 6.

□□□□□

Example 5.2

Let p be a prime. Let $Q_p = \{a/b \mid a/b \in Q \text{ and } b \text{ is relatively prime to } p\}$.

RESULT 5.2.1 *Q_p is a subgroup of Q.*

Let a/b and c/d belong to Q_p. Then $(a/b) - (c/d) = (ad - bc)/bd$. But $a/b, c/d \in Q_p$ imply p is relatively prime to both b and d, hence p is relatively prime to bd

since p is a prime. Thus $(a/b) - (c/d) \in Q_p$ establishing the result.

RESULT 5.2.2 *Q_p is torsion-free.*
Since Q is torsion-free (Result 5.1.1) and Q_p is a subgroup of Q.

RESULT 5.2.3 *Q_p is q-divisible for all primes $q \neq p$.*
Seek to show $qQ_p = Q_p$. Let $a/b \in Q_p$, so p is relatively prime to b. But p is also relatively prime to q since p and q are distinct primes. Therefore p must be relatively prime to bq because p is a prime. Thus $a/bq \in Q_p$ and note that $q(a/bq) = a/b$. Thus $a/b \in qQ_p$ as required.

RESULT 5.2.4 $Z \subseteq Q_p$.
Since 1 and p are relatively prime, $1/1 \in Q_p$. Hence $Z \subseteq Q_p$.

RESULT 5.2.5 *Q_p is indecomposable.*
By Result 5.1.3.

RESULT 5.2.6 *Q_p is pure-simple.*
By Result 5.1.8.

RESULT 5.2.7 *Q_p is hopfian.*
By Result 5.1.12.

RESULT 5.2.8 *Q_p is locally cyclic.*
Since Q is locally cyclic (Result 5.1.10) and Q_p is a subgroup of Q.

RESULT 5.2.9 *The subgroup lattice $\mathrm{L}(Q_p)$ is distributive.*
By Result 5.2.8 and the Lemma to Result 5.1.11.

RESULT 5.2.10 $\cap\{p^nQ_p \mid n \in N\} = \{0\}$.
Let $a/b \in \cap\{p^nQ_p \mid n \in N\}$ and seek to show $a = 0$. To do this it suffices to show p^n divides a for all $n \in N$. But for all such n, $a/b \in p^nQ_p$ so $a/b = p^n(c/d)$ where $c/d \in Q_p$. Thus $ad = p^ncb$ so $p^n | ad$. However, $c/d \in Q_p$ implies p and d are relatively prime so p^n and d are also. This and $p^n | ad$ yield $p^n | a$.

RESULT 5.2.11 *For all $n \in N$,*
(a) $Q_p = (p^nQ_p) + Z$.
(b) $\langle p^n \rangle = p^nQ_p \cap Z$.
(c) $Q_p/p^nQ_p \cong Z(p^n)$.
Let $a/b \in Q_p$ and let $n \in N$. Then b and p are relatively prime hence so are b and p^n. Thus there exist integers i and j such that $ip^n + jb = 1$. Multiplication by a/b yields $ip^n(a/b) + ja = a/b$ so $a/b \in (p^nQ_p) + Z$ proving Part (a).

To show Part (b), suppose $k \in p^nQ_p \cap Z$. Then $k = p^n(c/d)$ for some $c/d \in$

Q_p. Hence $kd = p^n c$ so $p^n \mid kd$. But $c/d \in Q_p$ implies p and d are relatively prime, hence so are p^n and d. This and $p^n \mid kd$ imply $p^n \mid k$. Conclude $k \in \langle p^n \rangle$ proving $p^n Q_p \cap Z \subseteq \langle p^n \rangle$. The reverse inclusion is clear.

Finally Parts (a) and (b) yield $Q_p/p^n Q_p \cong Z/\langle p^n \rangle \cong Z(p^n)$ proving Part (c).

RESULT 5.2.12 *Q_p is residually finite.*
Let $\mathcal{R} = \{p^n Q_p \mid n \in N\}$. By Result 5.2.10, $\cap\mathcal{R}$ is trivial, and by Result 5.2.11, Part (c), each $Q_p/p^n Q_p$ is finite. Conclude Q_p is residually finite by Theorem 1.3.9.

RESULT 5.2.13 *Q_p is reduced.*
By Result 5.2.12 and Theorem 1.3.11. Alternately, by Result 5.1.4 (Q_p is proper in Q since $1/p \notin Q_p$).

LEMMA *Let A be an abelian group such that $A/pA \cong Z(p)$ for some prime p and A is q-divisible for all primes $q \neq p$. Then A has a unique maximal subgroup, namely pA.*
Since pA has index p it is a maximal subgroup of A (subgroups of prime index in any group are maximal). Let H now be any maximal subgroup of A. Then since A is abelian, maximality of H implies $A/H \cong Z(q)$ for some prime q. Let $H + a$ generate the cyclic group A/H. If $q \neq p$, then A is q-divisible implying existence of $b \in A$ such that $a = qb$. Hence $H + a = H + qb = q(H + b)$. But $A/H \cong Z(q)$ implies $q(H + b) = 0$, so $H + a = 0$ which is false ($H + a$, being a generator, has order q, not 1). This contradiction shows $q = p$. Thus $A/H \cong Z(p)$ from which it follows that $pA \subseteq H$. As noted above, however, pA is a maximal subgroup of A. Therefore $pA \subseteq H$ implies $H = pA$ or $H = A$. Maximality of H precludes $H = A$, so $H = pA$ as required.

RESULT 5.2.14 *Q_p has a unique maximal subgroup, namely pQ_p.*
By Results 5.2.11, Part (c) and 5.2.3 and the lemma.

EXERCISES

1. Show every element of Q_p/Z has order relatively prime to p.
2. Under the usual multiplication of rational numbers, Q is a ring. Show Q_p is a subring.
3. Let H be an abelian group with the following properties:
   (a) H is torsion-free.
   (b) H is q-divisible for all primes $q \neq p$.
   (c) H is an extension of an infinite cyclic group by a torsion group every element of which has order relatively prime to p.

   Show $H \cong Q_p$.
4. Show that Q_p has no nontrivial p-divisible subgroups.

NOTES

1. A characterization of Q_p is given in Exercise 3.
2. See Miscellaneous Exercises II.3, IV.1, IV.5, and IV.6 at the end of this chapter, and Miscellaneous Exercise V.2 at the end of Chapter 6. See also Exercise 9 of Example 5.5.

□□□□□

Example 5.3

Let p be a prime. Let $Q^p = \{a/p^n \mid a/p^n \in Q\}$.

RESULT 5.3.1 *Q^p is a subgroup of Q.*
Let a/p^n, $c/p^m \in Q^p$. Without loss of generality, $n \leqslant m$. Hence $(a/p^n) - (c/p^m) = (ap^{m-n} - c)/p^m$ which belongs to Q^p establishing the result.

RESULT 5.3.2 *Q^p is torsion-free.*
Since Q is torsion-free (Result 5.1.1) and Q^p is a subgroup of Q.

RESULT 5.3.3 *Q^p is p-divisible.*
For all $a/p^n \in Q^p$, a/p^{n+1} is also an element of Q^p and $a/p^n = p(a/p^{n+1})$.

RESULT 5.3.4 *Let q be a prime such that $q \neq p$. Then Q^p has no nontrivial q-divisible subgroups.*
Suppose F is a q-divisible subgroup and seek to show F is trivial. Let $a/p^n \in F$. To show $a/p^n = 0$ it suffices to show $a = 0$, and to do this it suffices to show $q^k \mid a$ for all natural numbers k. By q-divisibility of F, there exists $b/p^m \in F$ such that $a/p^n = q^k(b/p^m)$. Hence $q^k p^n b = p^m a$ implying $q^k \mid p^m a$. But as q and p are distinct primes, q^k and p^m are relatively prime. So $q^k \mid p^m a$ implies $q^k \mid a$.

RESULT 5.3.5 *$Z \subseteq Q^p$.*
Since $1 = 1/p^0$, $1 \in Q^p$. Hence $Z \subseteq Q^p$.

RESULT 5.3.6 *Q^p is indecomposable.*
By Result 5.1.3.

RESULT 5.3.7 *Q^p is hopfian.*
By Result 5.1.12.

RESULT 5.3.8 *Q^p is locally cyclic.*
By Result 5.1.10, Q is locally cyclic. Hence too, so is any subgroup.

RESULT 5.3.9 *The subgroup lattice* $L(Q^p)$ *is distributive.*
By Result 5.3.8 and the Lemma to Result 5.1.11.

RESULT 5.3.10 *Q^p is pure-simple.*
By Result 5.1.8.

RESULT 5.3.11 $\cap\{qQ^p \mid q \in PR \text{ but } q \neq p\} = \{0\}$.
Let $a/p^n \in \cap\{qQ^p \mid q \in PR \text{ but } q \neq p\}$ and seek to show $a = 0$. To do this it suffices to show that an infinite number of primes divide a. It will be shown that $q \mid a$ for all primes $q \neq p$. For such q, $a/p^n \in qQ^p$ so $a/p^n = q(c/p^m)$ for some $c/p^m \in Q^p$. Then $qcp^n = ap^m$ so $q \mid ap^m$. But q and p^m are relatively prime (since q and p are distinct primes) so $q \mid a$ as required.

RESULT 5.3.12 *For all primes $q \neq p$,*
(a) $Q^p = (qQ^p) + Z$.
(b) $\langle q \rangle = qQ^p \cap Z$.
(c) $Q^p/qQ^p \cong Z(q)$.
Let $a/p^n \in Q^p$. Since p and q are distinct primes, p^n and q must be relatively prime, so there exist integers i and j such that $iq + jp^n = 1$. Multiplication by a/p^n yields $iq(a/p^n) + ja = a/p^n$. Hence $a/p^n \in (qQ^p) + Z$ proving Part (a).

To show Part (b), suppose $k \in qQ^p \cap Z$. Then $k = q(c/p^m)$ for some $c/p^m \in Q^p$. Hence $p^m k = qc$ so $q \mid p^m k$. This and the fact that q and p^m are relatively prime imply $q \mid k$. Conclude $k \in \langle q \rangle$ proving $qQ^p \cap Z \subseteq \langle q \rangle$. The reverse inclusion is clear.

Finally by Parts (a) and (b), $Q^p/qQ^p \cong Z/\langle q \rangle$. But $Z/\langle q \rangle \cong Z(q)$ so $Q^p/qQ^p \cong Z(q)$ proving Part (c).

RESULT 5.3.13 *Q^p is residually finite.*
Let $\mathcal{R} = \{qQ^p \mid q \in PR \text{ but } q \neq p\}$. By Result 5.3.11, $\cap\mathcal{R}$ is trivial, and by Result 5.3.12, Part (c), each Q^p/qQ^p is finite. Conclude Q^p is residually finite by Theorem 1.3.9.

RESULT 5.3.14 *Q^p is reduced.*
By Result 5.3.13 and Theorem 1.3.11. Alternately, by Result 5.1.4 (Q^p is proper in Q since $1/(p+1) \notin Q^p$).

EXERCISES

1. Show
   (a) $Q^p \cap Q_p = Z$.
   (b) $Q^p + Q_p = Q$.
   (c) $Q^p \cap Q^q = Z$ for all primes $q \neq p$.
2. Let q be a prime such that $q \neq p$. Show $\text{Hom}(Q^p, Q^q)$ is trivial.

3. Under the usual multiplication of rational numbers, Q is a ring. Show Q^p is a subring.
4. Let H be an abelian group with the following properties:
   (a) H is torsion-free.
   (b) H is p-divisible.
   (c) H is an extension of an infinite cyclic group by a p-group.
   Show $H \cong Q^p$.
5. Show $Q^p \cong G(\{X_i | i \in N, i \neq 0\} \mid (X_{i+1})^p X_i^{-1} = 1 \text{ for all } i)$.

NOTES

1. A characterization of Q^p is given in Exercise 4. Note, however, that it has not yet been shown that Q^p satisfies Property (c). This follows from Result 5.4.1.
2. A presentation for Q^p is given in Exercise 5.
3. See Miscellaneous Exercises II.3, IV.1, IV.5, and IV.6 at the end of this chapter, and Miscellaneous Exercise V.2 at the end of Chapter 6.

□□□□□

Example 5.4 (The p-Prufer Groups)

Let p be a prime. Let $Z(p^\infty) = Q^p/Z$.

RESULT 5.4.1 *$Z(p^\infty)$ is a p-group. Moreover the element $Z + (a/p^n) \in Z(p^\infty)$ has order p^n.*

Let $x = Z + (a/p^n)$. Then $p^n x = Z + (p^n a/p^n) = Z + a = 0$ since $a \in Z$. Thus $o(x)$ divides p^n, hence it too must be some power of p, say p^m. Thus $p^m | p^n$. Now $p^m(Z + (a/p^n)) = 0$ by definition of order. Consequently $Z + (p^m a/p^n) = 0$ implying $p^m a/p^n \in Z$ and hence $p^n | ap^m$. But since $a/p^n \in Q$, a/p^n is reduced to lowest terms (see Note 1 to Example 5.1) and hence a and p^n are relatively prime. Then $p^n | ap^m$ implies $p^n | p^m$. This and $p^m | p^n$ imply $p^n = p^m = o(x)$.

RESULT 5.4.2 *$Z(p^\infty)$ is divisible.*

By Theorem A.8.7, it suffices to show $Z(p^\infty)$ is q-divisible for every prime q. First suppose $q \neq p$. Let $Z + a \in Z(p^\infty)$. Thus $a \in Q^p$ so by Result 5.3.12, Part (a), $a = qy + z$ where $y \in Q^p$ and $z \in Z$. Therefore $Z + a = Z + (qy + z) = Z + qy = q(Z + y)$ showing $Z(p^\infty)$ is q-divisible when $q \neq p$. When $q = p$, Q^p is q-divisible by Result 5.3.3. Hence, as is easily seen, so is its quotient group $Z(p^\infty)$.

LEMMA *If H is a subgroup of $Z(p^\infty)$ and $Z + (t/p^j) \in H$, then $Z + (1/p^j) \in H$.*

Since $t/p^j \in Q$, t and p^j are relatively prime (see Note 1 to Example 5.1). Therefore there are integers c and d such that $cp^j + dt = 1$. Since H is a subgroup, $Z + (t/p^j) \in H$ implies $d(Z + (t/p^j)) \in H$. It follows then that $Z + (dt/p^j) = Z + ((1 - cp^j)/p^j) = (Z + (1/p^j)) - (Z + c) = Z + (1/p^j) \in H$.

RESULT 5.4.3 *Let $A_m = \langle Z + (1/p^m)\rangle$ for all $m \in N$. Then every proper subgroup of $Z(p^\infty)$ is A_m for some m. Hence every proper subgroup is finite.*
Let H be a proper subgroup. If $Z + (1/p^i) \in H$ for all natural numbers i, then $H = Z(p^\infty)$ which is being assumed false. Thus there is a least natural number n such that $Z + (1/p^n) \notin H$. Note n is nonzero. Let $m = n - 1$. Then $Z + (1/p^m) \in H$ by definition of n, and hence $A_m \subseteq H$. It is claimed that $H = A_m$. Let $Z + (t/p^j) \in H$. By the lemma, then, $Z + (1/p^j) \in H$. If $p^n \leqslant p^j$, then $p^{j-n}(Z + (1/p^j)) = Z + (1/p^n)$ which implies $Z + (1/p^n) \in H$ which is false. Thus $p^j < p^n$ and so $p^j \leqslant p^m$. Consequently $tp^{m-j}(Z + (1/p^m)) = Z + (t/p^j)$ so $Z + (t/p^j) \in A_m$.

RESULT 5.4.4 *The lattice of subgroups $\mathrm{L}(Z(p^\infty))$ is a chain.*
Let H and K be subgroups. If $K = Z(p^\infty)$, then clearly $H \subseteq K$ and likewise if $H = Z(p^\infty)$, then $K \subseteq H$. Therefore it may be assumed both H and K are proper. By Result 5.4.3, then, $H = A_i$ and $K = A_j$ for some i and j. First suppose that $i \leqslant j$. Then $p^{j-i}(Z + (1/p^j)) = Z + (1/p^i)$ so $H = A_i \subseteq A_j = K$. Similarly $K \subseteq H$ if $j \leqslant i$.

RESULT 5.4.5 *$Z(p^\infty)$ is subdirectly irreducible.*
It suffices to show $Z + (1/p)$ belongs to every nontrivial subgroup. Let B be such a subgroup, and let $A_1 = \langle Z + (1/p)\rangle$ (same notation as in Result 5.4.3). If $B = Z(p^\infty)$, then surely $A_1 \subseteq B$ so $Z + (1/p) \in B$. If B is a proper subgroup, then by Result 5.4.3, $B = \langle Z + (1/p^n)\rangle$ for some n. Note that n is nonzero lest B be trivial. Thus $Z + (1/p) = p^{n-1}(Z + (1/p^n)) \in B$.

RESULT 5.4.6 *$Z(p^\infty)$ is indecomposable.*
By Result 5.4.5 and Theorem A.10.10.

RESULT 5.4.7 *$Z(p^\infty)$ is pure-simple.*
It suffices to show that nontrivial proper subgroups are not pure. Let B be such a subgroup. Then by Result 5.4.3, $B = A_n$ for some n. If B were pure, it would follow that $p^n Z(p^\infty) \cap A_n \subseteq p^n A_n$. But $p^n Z(p^\infty) = Z(p^\infty)$ by Result 5.4.2. Thus $A_n \subseteq p^n A_n$ which would further imply $Z + (1/p^n) = p^n(Z + (t/p^n))$ for some t. Hence $Z + (1/p^n) = Z + t = Z$, so $1/p^n$ would belong to Z which is false ($n \neq 0$ since $B = A_n$ is nontrivial). Thus B is not a pure subgroup.

RESULT 5.4.8 *$Z(p^\infty)$ satisfies the descending chain condition.*
It must be shown that an infinite proper descending chain of subgroups such as $\ldots C_3 \subset C_2 \subset C_1 \subset C_0$ cannot exist. If there is such a chain, then C_1 must be a proper subgroup, hence is finite by Result 5.4.3. Thus on one hand C_1 is finite, and on the other it has an infinite proper descending chain of subgroups $\ldots C_4 \subset C_3 \subset C_2 \subset C_1$. This contradiction proves the result.

RESULT 5.4.9

(a) *For all natural numbers n,* $Z(p^\infty)[p^n] = \langle Z + (1/p^n)\rangle$.

(b) *All subgroups of* $Z(p^\infty)$ *are fully invariant.*

Let $x \in \langle Z + (1/p^n)\rangle$. Thus $x = Z + (t/p^n)$ for some t. Consequently $p^n x = 0$ so $x \in Z(p^\infty)[p^n]$. Conversely if $w = Z + (t/p^k) \in Z(p^\infty)[p^n]$, then $p^n w = 0$ so $o(w)$ divides p^n. Applying Result 5.4.1 yields $p^k | p^n$ so $k \leqslant n$. Consequently $w = Z + (t/p^k) = Z + (tp^{n-k}/p^n) \in \langle Z + (1/p^n)\rangle$ completing the proof of Part (a).

For Part (b) let H be a subgroup of $Z(p^\infty)$. If $H = Z(p^\infty)$, then H is surely fully invariant so assume H is proper instead. Therefore $H = \langle Z + (1/p^n)\rangle$ for some n by Result 5.4.3. Conclude H is fully invariant by Part (a) and Theorem A.4.1.

RESULT 5.4.10 $Z(p^\infty)$ *has no maximal subgroups.*

By Result 5.4.2 and the Lemma to Theorem 1.2.13.

RESULT 5.4.11 $Z(p^\infty)$ *is not residually finite.*

By Result 5.4.2, $Z(p^\infty)$ contains a nontrivial divisible subgroup (itself) hence is not reduced. Conclude $Z(p^\infty)$ is not residually finite by Theorem 1.3.11. Alternate proof: By Theorem 1.3.10 and the facts that $Z(p^\infty)$ is subdirectly irreducible (Result 5.4.5) and infinite.

RESULT 5.4.12

(a) *If H is any proper subgroup of* $Z(p^\infty)$, *then* $Z(p^\infty)/H \cong Z(p^\infty)$.

(b) $Z(p^\infty)$ *is nonhopfian.*

Let H be a proper subgroup. Then $H = Z(p^\infty)[p^n]$ for some $n \in N$ by Results 5.4.3 and 5.4.9, Part (a). Application of Theorem A.4.1 then yields $Z(p^\infty)/H \cong p^n Z(p^\infty)$. But $p^n Z(p^\infty) = Z(p^\infty)$ by divisibility (Result 5.4.2). Hence $Z(p^\infty)/H \cong Z(p^\infty)$ proving Part (a). Part (b) follows from Part (a).

RESULT 5.4.13 $Z(p^\infty)$ *is cohopfian.*

By Result 5.4.3, every proper subgroup of $Z(p^\infty)$ is finite. On the other hand $Z(p^\infty)$ itself is infinite so it cannot have proper isomorphic subgroups.

RESULT 5.4.14 $Z(p^\infty)$ *is locally cyclic.*

Let $H = \langle x_1, x_2, \ldots, x_k\rangle$ be a finitely generated subgroup of $Z(p^\infty)$ where $x_i = Z + (a_i/p^{m(i)})$ for $1 \leqslant i \leqslant k$. Let m be the largest $m(i)$. Then

$$x_i = Z + (a_i p^{m-m(i)}/p^m) = a_i p^{m-m(i)}(Z + (1/p^m)),$$

so $x_i \in \langle Z + (1/p^m)\rangle$ for $1 \leqslant i \leqslant k$ showing H is a subgroup of $\langle Z + (1/p^m)\rangle$. Therefore H is cyclic.

EXERCISES

1. Show that if A is a subgroup of $Z(p^\infty)$ and $\mathcal{L}$ is a collection of subgroups of $Z(p^\infty)$, then $A + (\cap\mathcal{L}) = \cap\{A + H \mid H \in \mathcal{L}\}$.

2. Show that every nontrivial homomorphism $g: Z(p^\infty) \to Z(p^\infty)$ is surjective. (A homomorphism g is **nontrivial** if and only if there exists some x such that $g(x) \neq 0$.)
3. Show $Z(p^\infty)$ can be embedded in every nontrivial divisible abelian p-group.
4. Let A be a nontrivial divisible indecomposable abelian p-group. Show $A \cong Z(p^\infty)$.
5. Let A be an unbounded p-group such that the lattice of subgroups $L(A)$ is a chain. Show $A \cong Z(p^\infty)$.
6. Show $Z(p^\infty) \cong Q/Q_p$.
7. Show $Z(p^\infty) \cong G(\{X_i | i \in N, i \neq 0\} \mid (X_{i+1})^p X_i^{-1} = 1 \text{ for all } i \text{ and } X_1{}^p = 1)$.

NOTES

1. As an example of an infinite group all of whose proper subgroups are finite, $Z(p^\infty)$ is indeed a cause for celebration.

   One of the reasons for the importance of $Z(p^\infty)$ is the following theorem which characterizes all divisible abelian groups:

   THEOREM *If A is a divisible abelian group, then A is isomorphic to the restricted direct product $\Sigma\{B_i\}_{i\in I}$ where I is some index set and each B_i is either isomorphic to Q or to $Z(p^\infty)$ for some prime p.*

   For a proof see Rotman [31], Theorem 9.13.
2. Exercises 4 and 5 present two characterizations of $Z(p^\infty)$. The first of these can be improved somewhat: $Z(p^\infty)$ is the only (up to isomorphism) indecomposable infinite abelian p-group. It is also the only infinite abelian p-group all of whose proper subgroups are finite. Finally, every subdirectly irreducible infinite abelian group is isomorphic to $Z(p^\infty)$ for some prime p (Fuchs [11], vol I, Theorem 3.1, p 16).
3. As a counterexample, $Z(p^\infty)$ shows that none of the classes *hopfian, residually finite,* or *reduced abelian* are quotient closed. This is because Q^p belongs to these classes (Results 5.3.7, 5.3.13 and 5.3.14 respectively) while $Z(p^\infty) = Q^p/Z$ does not (Results 5.4.12, 5.4.11 and 5.4.2).
4. Expressed in lattice-theoretic terminology, Exercise 1 says that the subgroup lattice $L(Z(p^\infty))$ is **infinitely join distributive.** The group of integers Z furnishes an example of an abelian group whose subgroup lattice is distributive but not infinitely join distributive (see Miscellaneous Exercise II.1).
5. A presentation for $Z(p^\infty)$ is given in Exercise 7.
6. See Miscellaneous Exercises IV.4 and IV.6 at the end of this chapter, and Miscellaneous Exercise V.2 at the end of Chapter 6. See also Exercise 10 of Example 5.5; note that this exercise shows Aut($Z(p^\infty)$) is abelian.

□□□□□

Example 5.5 (The p-adic Integers)

Let p be a prime. Let $J_p = \mathrm{Hom}(Z(p^\infty), Z(p^\infty))$.

RESULT 5.5.1 *J_p is torsion-free.*
By Result 5.4.2, $Z(p^\infty)$ is divisible. Hence J_p is torsion-free by Theorem 1.8.2.

RESULT 5.5.2 *J_p is q-divisible for all primes $q \neq p$.*
Since $Z(p^\infty)$ is a p-group (Result 5.4.1), the result follows from Theorem 1.8.3.

THEOREM 5.5.3 *For all natural numbers n, $J_p/p^nJ_p \cong Z(p^n)$.*
Define $d{:}J_p \to Z(p^n)$ by $d(f) = t$ if and only if $f(Z + (1/p^n)) = Z + (t/p^n)$. First it must be shown that d is a function. Suppose $d(f)$ equals both t and w. Then $Z + (t/p^n) = Z + (w/p^n)$ so $(t - w)/p^n \in Z$ implying $p^n \mid (t - w)$. This means that $t \equiv w \bmod p^n$ so that t and w represent the same element of $Z(p^n)$ proving d is a function. If $d(f) = t$ and $d(g) = s$, then $(f + g)(Z + (1/p^n)) = f(Z + (1/p^n)) + g(Z + (1/p^n)) = [Z + (t/p^n)] + [Z + (s/p^n)] = Z + ((t + s)/p^n)$. Therefore $d(f + g) = t + s = d(f) + d(g)$ showing that d is a homomorphism. It is claimed that $\ker d = p^nJ_p$. Let $g = p^nf \in p^nJ_p$ and observe that $g(Z + (1/p^n)) = p^nf(Z + (1/p^n)) = f(Z + (p^n/p^n)) = f(Z) = 0 = Z + (0/p^n)$. Hence $d(g) = 0$ so $g \in \ker d$. Conversely suppose $g \in \ker d$. Therefore $d(g) = 0$ so $g(Z + (1/p^n)) = Z + (0/p^n) = 0$. Define $h{:}Z(p^\infty) \to Z(p^\infty)$ by $h(Z + (a/p^i)) = g(Z + (a/p^{n+i}))$. To see that h is a function suppose $Z + (a/p^i) = Z + (c/p^k)$. First note $p^i = o(Z + (a/p^i)) = o(Z + (c/p^k)) = p^k$ by Result 5.4.1. Thus $Z + (a/p^i) = Z + (c/p^i)$ so $(a - c)/p^i \in Z$ and hence $a - c = p^ie$ for some integer e. Consequently $[Z + (a/p^{n+i})] - [Z + (c/p^{n+i})] = Z + ((a - c)/p^{n+i}) = Z + (p^ie/p^{n+i}) = Z + (e/p^n) = e(Z + (1/p^n))$. However, $g(e(Z + (1/p^n))) = eg(Z + (1/p^n)) = e0 = 0$. Consequently $g(Z + (a/p^{n+i})) = g(Z + (c/p^{n+i}))$ completing the proof that h is a function. To see that h is a homomorphism, let $x = Z + (a/p^i)$ and $y = Z + (c/p^k)$ belong to $Z(p^\infty)$. Without loss of generality, $i \leqslant k$. Therefore letting $e = ap^{k-i}$ yields $x = Z + (e/p^k)$. Thus $x + y = Z + ((e + c)/p^k)$, and so $h(x + y) = g(Z + ((e + c)/p^{n+k}))$ while $h(x) + h(y) = g(Z + (e/p^{n+k})) + g(Z + (c/p^{n+k})) = g(Z + (e + c)/p^{n+k}))$ also. Thus $h \in J_p$. For any $Z + (a/p^i) \in Z(p^\infty)$, $p^nh(Z + (a/p^i)) = p^ng(Z + (a/p^{n+i}))$ which equals $g(Z + (ap^n/p^{n+i})) = g(Z + (a/p^i))$. Conclude that $g = p^nh \in p^nJ_p$ proving $p^nJ_p = \ker d$ as claimed.

Applying the first isomorphism theorem yields $J_p/p^nJ_p \cong \mathrm{Im}\, d$, so it suffices to show that d is onto $Z(p^n)$. Let $t \in Z(p^n)$. Define $k{:}Z(p^\infty) \to Z(p^\infty)$ by $k(x) = tx$. It is clear that k is a homomorphism, that is, $k \in J_p$. And furthermore $k(Z + (1/p^n)) = Z + (t/p^n)$. Therefore $d(k) = t$, so d is onto $Z(p^n)$ as required.

RESULT 5.5.4 $\cap\{p^nJ_p \mid n \in N\} = \{0\}$.
Let $g \in \cap\{p^nJ_p \mid n \in N\}$. To show $g = 0$, it must be shown that $g(x) = 0$ for all $x \in Z(p^\infty)$. For such an x, $o(x) = p^n$ for some n. But $g \in p^nJ_p$ implies $g = p^nf$ for some f. Therefore $g(x) = p^nf(x) = f(p^nx) = f(0) = 0$.

RESULT 5.5.5

(a) *J_p is residually finite.*

(b) *J_p is reduced.*

Let $\mathcal{R} = \{p^n J_p \mid n \in N\}$. By Result 5.5.4, $\cap \mathcal{R}$ is trivial and by Result 5.5.3, each $J_p/p^n J_p$ is finite. Conclude J_p is residually finite by Theorem 1.3.9. This proves Part (a). Part (b) follows from Part (a) and Theorem 1.3.11.

LEMMA *Direct factors of abelian groups are pure subgroups.*

Let A be an abelian group with a direct factor B. That is, $A = B \oplus C$ for some subgroup C. Let n be a positive integer and seek to show $nA \cap B \subseteq nB$. For $x \in nA \cap B$, $x \in B$ and $x = na$ for some a. But $a \in A$ implies $a = b + c$ where $b \in B$ and $c \in C$. Consequently $x = nb + nc$ implying $nc = x - nb \in B \cap C$. This further implies $nc = 0$, hence $x = nb \in nB$.

RESULT 5.5.6 *J_p is indecomposable.*

Suppose $J_p = B \oplus C$ and seek to show either B or C is trivial. Let $M = pJ_p$. By Theorem A.4.1, M is a fully invariant subgroup so

$$J_p/M = [(M + B)/M] \oplus [(M + C)/M] \tag{1}$$

by Theorem A.2.8. But by Result 5.5.3, $J_p/M \cong Z(p)$. Now $Z(p)$ is indecomposable (it has, in fact, only two subgroups, itself and the trivial subgroup), hence so too is J_p/M. Thus one of the direct factors in the decomposition in Line (1) must be trivial. Without loss of generality assume the trivial factor is $(M + B)/M$. Thus $B \subseteq M = pJ_p$ and so $pJ_p \cap B = B$. Apply the lemma and conclude $pB = B$. Consequently $B = p^n B$ for all natural numbers n and so $B \subseteq p^n J_p$. Thus $B \subseteq \cap\{p^n J_p \mid n \in N\}$ and so B is trivial by Result 5.5.4.

LEMMA *If G is a (not necessarily abelian) group having a collection of fully invariant subgroups $\mathcal{L}$ such that G/H is hopfian for all $H \in \mathcal{L}$ and $\cap \mathcal{L} = \{1\}$, then G is hopfian.*

Suppose that $g{:}G/B \to G$ is an isomorphism, but that B is nontrivial and seek a contradiction. Let $e{:}G \to G/B$ be the natural map and let $s{:}G \to G$ be $s = g \circ e$. Note that s is surjective. Since B is nontrivial, there exists $b \in B$, $b \neq 1$. Therefore $b \notin \cap \mathcal{L}$. Thus there exists $H \in \mathcal{L}$ such that $b \notin H$. By hypothesis, H is fully invariant, so $s[H] \subseteq H$. Thus by Theorem A.10.1, there is an induced surjective homomorphism $s'{:}G/H \to G/H$ defined by $s'(Hx) = Hs(x)$.

Applying the first isomorphism theorem yields $(G/H)/(\ker s') \cong G/H$. This and the fact that G/H is hopfian imply $\ker s'$ is trivial. Now $s(b) = g \circ e(b) = g(Bb) = 1$ since $Bb = 1$. Thus $s'(Hb) = Hs(b) = H$. Conclude $Hb \in \ker s'$. This and triviality of $\ker s'$ imply $Hb = H$ hence $b \in H$, the required contradiction.

RESULT 5.5.7 *J_p is hopfian.*

By Theorem A.4.1, $\{p^n J_p \mid n \in N\}$ is a collection of fully invariant subgroups of J_p. By Result 5.5.4, this collection has trivial intersection, and by Result 5.5.3,

J_p/p^nJ_p is finite, hence hopfian for all n. Conclude J_p is hopfian by the lemma.

RESULT 5.5.8 *J_p has a unique maximal subgroup, namely pJ_p.*
By Result 5.5.3, $J_p/pJ_p \cong Z(p)$. By Result 5.5.2, J_p is q-divisible for all primes $q \neq p$. Conclude the result from the Lemma to Result 5.2.14.

LEMMA *Let p be a prime, let A be an abelian p-group, and let S be the set $S = \{0,1,\ldots,p-1\}$. For any sequence $f:N \to S$, define $f^*:A \to A$ by*

$$f^*(x) = \Sigma_{i \in N} f_i p^i x,$$

noting that this sum can be considered finite since $p^i x = 0$ for all $p^i \geqslant o(x)$. Then
(a) *f^* is a homomorphism.*
(b) *If A is unbounded and f and g are sequences, then $f^* = g^*$ implies $f = g$.*
Let $x,y \in A$. Then $f^*(x+y) = \Sigma f_i p^i(x+y) = (\Sigma f_i p^i x) + (\Sigma f_i p^i y) = f^*(x) + f^*(y)$, proving Part (a).

To show Part (b), suppose $f \neq g$ and seek a contradiction. Let n be the least natural number such that $f_n \neq g_n$. Since A is unbounded it has an element x of order p^{n+1}. Then $f^*(x) = \Sigma_{i=0}^n f_i p^i x$ and $g^*(x) = \Sigma_{i=0}^n g_i p^i x$. But $f^*(x) = g^*(x)$ and $f_i = g_i$ for $i < n$ (definition of n). Therefore $f_n p^n x = g_n p^n x$ which implies $(f_n - g_n)p^n x = 0$. Consequently $o(x) | (f_n - g_n)p^n$ and hence $p | f_n - g_n$. This and $f_n, g_n \in S$ imply $f_n = g_n$, the required contradiction.

RESULT 5.5.9 *J_p is uncountable.*
Let $S = \{0, 1, \ldots, p-1\}$ as in the lemma. Let S^N denote the set of all sequences with values in S, and let $P(N)$ be the set of all subsets of N (the power set of N). By Cantor's theorem, $\text{card}(N) < \text{card}(P(N))$ and so $P(N)$ is uncountable. However, the sets $P(N)$ and $\{0,1\}^N$ (the set of all sequences with values in $\{0,1\}$) are equivalent so $\{0,1\}^N$ is uncountable. Now note that $\{0,1\}^N$ is a subset of S^N so S^N is uncountable.

Define $\lambda: S^N \to J_p$ by $\lambda(f) = f^*$ where f^* is defined as in the lemma for the group $A = Z(p^\infty)$. Since $Z(p^\infty)$ is unbounded, the lemma shows $f^* = g^*$ implies $f = g$. That is, λ is injective. This and S^N uncountable imply J_p is uncountable.

EXERCISES

1. Show J_p is not locally cyclic.
2. Show J_p is not pure-simple.
3. Show that pure subgroups of J_p are hopfian.
4. Show that pure subgroups of J_p are indecomposable.
5. Show $p^nJ_p = \{f \mid f(Z + (1/p^n)) = 0\}$ for all natural numbers n.
6. Show that if H is any subgroup of J_p having finite index, then $H = p^mJ_p$ for some m.
7. Let H be the cyclic subgroup of J_p generated by the identity homomorphism $i:Z(p^\infty) \to Z(p^\infty)$. Show for all primes $q \neq p$ that J_p/H has an element of

order q. Show J_p/H has no elements of order p.

8. Show
   (a) If B is a subgroup of J_p such that $B \not\subseteq pJ_p$, then J_p/B is divisible.
   (b) If B is a nontrivial pure subgroup of J_p, then J_p/B is divisible.
   (c) J_p/H is divisible (H as in Exercise 7).
9. Show Q_p can be embedded in J_p.
10. Show that J_p is a ring when multiplication is defined by composition of functions. In addition, show that
    (a) J_p is an integral domain.
    (b) $p^m J_p$ is an ideal of J_p for all natural numbers m.
    (c) $Z(p^\infty)$ is a J_p-module.
11. (a) Let i be defined as in Exercise 7. Suppose $\alpha, \beta: J_p \to J_p$ are homomorphisms such that $\alpha(i) = \beta(i)$. Show $\alpha = \beta$.
    (b) Suppose $\alpha: J_p \to J_p$ is any homomorphism. Show there exists $f \in J_p$ such that $\alpha(g) = f \circ g$ for all $g \in J_p$.

NOTES

1. The group J_p is an extremely important example not only in group theory but in the theory of topological groups and in the theory of rings as well.
2. One of the most important properties of J_p has, unfortunately, been omitted as being somewhat too deep for inclusion. This property is that J_p is algebraically compact. (An abelian group A is **algebraically compact** if and only if whenever A can be embedded in an abelian group G such that the image of the embedding is a pure subgroup, then this image is a direct factor of G.)

   To prove algebraic compactness, J_p can be represented as an inverse limit of finite cyclic p-groups (Result 5.5.3 is helpful here) and then Proposition 39.4 of Fuchs [11], vol I can be applied.

   A characterization of J_p can now be stated. The groups J_p are the only (up to isomorphism) indecomposable reduced algebraically compact infinite abelian groups. (See Fuchs [11], vol I, Corollary 40.4 and succeeding paragraph.)
3. Result 5.5.9 can be sharpened. Not only is $\text{card}(S^N) \leqslant \text{card}(J_p)$, but these two cardinalities are equal, and so J_p has the same cardinality as the continuum. This can be shown by proving every element of J_p is f^* for some sequence $f: N \to S$.
4. See Miscellaneous Exercises II.3, IV.5, and IV.6 at the end of this chapter, and Miscellaneous Exercise V.2 at the end of Chapter 6.

□□□□□

Example 5.6

Let p, q, and r be distinct primes. Consider the following subgroups of $Q \times Q$: $B = Q^p \times \{0\}$, $C = \{0\} \times Q^q$, and $D = \langle(1/r, 1/r)\rangle$. Let $A = B + C + D$.

LEMMA
(a) *If* $(x,0) \in A$, *then* $x \in Q^p$.
(b) *If* $(0,y) \in A$, *then* $y \in Q^q$.
Any $(x,0) \in A$ can be written as $(x,0) = (c_1/p^i,0) + (0,c_2/q^j) + c_3(1/r,1/r)$. Therefore $0 = (c_2/q^j) + (c_3/r)$ implying $c_3q^j = -c_2r$ and hence $r|c_3q^j$. But since r and q are distinct primes, this implies $r|c_3$ so $c_3 = rt$ for some integer t. Thus $x = (c_1/p^i) + t$ so $x \in Q^p$ because $Z \subseteq Q^p$ (Result 5.3.5). This proves Part (a). Part (b) is similar.

RESULT 5.6.1
(a) *A is torsion-free.*
(b) *A/B is torsion-free.*
(c) *A/C is torsion-free.*
(d) $\exp(A/(B+C)) = r$, *hence $A/(B+C)$ is torsion.*
Let (x,y) be an element of A of order n. Thus $nx = ny = 0$. But $x,y \in Q$ which is torsion-free (Result 5.1.1) hence $x = y = 0$ so $(x,y) = (0,0)$ proving Part (a).

Next suppose $B + (x,y)$ is an element of A/B of order n. Thus $(nx,ny) \in B$. This implies $ny = 0$ so $y = 0$ since Q is torsion-free. Thus $(x,0) \in A$, so $x \in Q^p$ by the lemma. Consequently $(x,0) \in B$ and hence $B + (x,y) = 0$ proving Part (b). Part (c) is similar.

To prove Part (d) it suffices to show $r(x,y) \in B + C$ for all $(x,y) \in A$. Since $(x,y) \in A$, $(x,y) = (c_1/p^i,0) + (0,c_2/q^j) + c_3(1/r,1/r)$ for some c_1, c_2, c_3, i, and j. Thus $rx = (rc_1/p^i) + c_3$ which belongs to Q^p because $Z \subseteq Q^p$ (Result 5.3.5). Similarly $ry \in Q^q$. Thus $(rx,ry) = (rx,0) + (0,ry) \in B + C$.

LEMMA *Let p be a prime, G be an abelian p-divisible group, and H be a subgroup of G. Then G/H is also p-divisible.*
It must be shown that $p(G/H) = G/H$. Let $H + x \in G/H$. Then $x \in G$ implying $x = py$ for some y by p-divisibility. Therefore $H + x = H + py = p(H + y) \in p(G/H)$ as required.

RESULT 5.6.2
(a) $B = \cap\{p^nA \mid n \in N\}$ *and* $C = \cap\{q^nA \mid n \in N\}$.
(b) *B and C are fully invariant subgroups of A.*
Denote $\cap\{p^nA \mid n \in N\}$ by W. If $(x,0) \in B$, then for any natural number n, there exists $y \in Q^p$ such that $x = p^ny$ (Result 5.3.3). Therefore $(x,0) = (p^ny,0) = p^n(y,0) \in p^nA$. Consequently $(x,0) \in W$, so $B \subseteq W$.

Before showing the reverse inclusion, some auxiliary notation will be introduced. Let $H = B + \langle(1/r,0)\rangle$, $K = C + \langle(0,1/r)\rangle$, and $F = H + K$. Note H, K, and

F are not subgroups of A. They are, however, subgroups of $Q \times Q$.

It is claimed that Hom(W,K) is trivial. Suppose $f \in$ Hom(W,K). Let $d:K \to K$ be the function $d(x,y) = r(x,y) = (rx,ry)$. It is easily seen that d is a homomorphism. If $(x,y) \in K$, then $(x,y) = (0,a/q^m) + i(0,1/r)$ for some a, m, and i. So $d(x,y) = r(x,y) = (0,ar/q^m) + (0,i)$. But $Z \subseteq Q^q$ (Result 5.3.5) so $(0,i) \in C$ and hence $d(x,y) \in C$. This shows Im d is a subgroup of C, and hence $d \circ f: W \to C$. Now W is p-divisible because if $t \in W$ (t here is an ordered pair), then $t \in pA$ so $t = ps$ for some $s \in A$. But s must belong to W because for all natural numbers n, $t \in p^{n+1}A$ implies $t = p^{n+1}a$ for some $a \in A$, and so $ps = p^{n+1}a$ hence $s = p^n a$ because A is torsion-free (Result 5.6.1). Thus $s \in W$ and so $t = ps$ completes the demonstration that W is p-divisible. Application of the lemma then shows that Im $d \circ f$ is p-divisible. But $C \cong Q^q$ so conclude Im $d \circ f$ is trivial by Result 5.3.4. To show $f = 0$, let $(x,y) \in W$. Then $d \circ f(x,y) = 0$, which is to say $rf(x,y) = 0$. Consequently $f(x,y) = 0$ because C is torsion-free. Thus $f = 0$ so Hom(W,K) is trivial as claimed.

Next note that all second coordinates of elements of H are zero, as are all first coordinates of elements of K. Thus $H \cap K = \{(0,0)\}$, which means $F = H \oplus K$. Let $\pi_K:F \to K$ be the projection homomorphism defined in Appendix A.2. Since $(1/r,1/r) = (1/r,0) + (0,1/r)$, $D \subseteq F$. Also $B,C \subseteq F$ since $B \subseteq H$ and $C \subseteq K$. Consequently $A = B + C + D \subseteq F$.

Let λ be the restriction of π_K to W ($W \subseteq F$ because $W \subseteq A$). Then λ belongs to Hom(W,K) so $\lambda = 0$ by the third paragraph. To show now that $W \subseteq B$, let $(x,y) \in W$. Then $\lambda(x,y) = 0$. Thus $(x,y) \in A$ and $y = 0$ so $(x,y) \in B$ by the Lemma to Result 5.6.1. This completes the proof of Part (a) for the subgroup B. The proof for C is similar.

Part (b) follows immediately from Part (a), Theorem A.4.1, and the easily-seen fact that the intersection of fully invariant subgroups is fully invariant.

RESULT 5.6.3 *A is indecomposable.*

Suppose $A = M \oplus L$ and seek to show either M or L is trivial. By Result 5.6.2, Part (b) and Theorem A.2.7, $B = (M \cap B) \oplus (L \cap B)$ and $C = (M \cap C) \oplus (L \cap C)$. However, B and C are both indecomposable by Result 5.3.6 ($B \cong Q^p$ and $C \cong Q^q$). Thus either $B \subseteq M$ or $B \subseteq L$ by indecomposability of B, and either $C \subseteq M$ or $C \subseteq L$ by indecomposability of C. Hence four cases arise. It will be shown, however, that two of these cases are impossible.

First suppose $B \subseteq M$ and $C \subseteq L$. Then $B + C \subseteq B + L$. This and $A/(B + C)$ torsion (Result 5.6.1, Part (d)) imply $A/(B + L)$ is torsion. Now it is easily seen that $B \subseteq M \cap (B + L)$. Conversely if $t \in M \cap (B + L)$ (t here is an ordered pair), then $t = b + e$ for some $b \in B$ and $e \in L$. Therefore $e = t - b \in M \cap L$ and hence $e = 0$. This implies $t = b \in B$ showing $M \cap (B + L) = B$. It follows, therefore, that $A/(B + L) = (M + L)/(B + L) = (M + (B + L))/(B + L) \cong M/(M \cap (B + L)) = M/B$. Consequently M/B is torsion. However, M/B, being a subgroup of A/B, is torsion-free by Result 5.6.1, Part (b). Conclude $M = B$. Similarly $L = C$. Thus $A \subseteq B + C$. This, however, is clearly false so the first case has been eliminated. The second case, $B \subseteq L$ and $C \subseteq M$, is similarly dealt with.

Assume now $B,C \subseteq M$. Then $B + C \subseteq M$. This and $A/(B + C)$ torsion imply A/M is torsion. But $A/M \cong L$ so L is torsion. However, L is also torsion-free, being a subgroup of A. Thus L is trivial.

Similarly the last case $(B,C \subseteq L)$ yields M is trivial.

RESULT 5.6.4 *A cannot be embedded in Q.*
Since $B + C$ is a subgroup of A, if A could be embedded in Q, then so could $B + C$. But note that B and C intersect trivially and so $B + C = B \oplus C$. Therefore $B + C$ is decomposable, and hence cannot be embedded in Q by Result 5.1.3.

RESULT 5.6.5 *A is reduced.*
Suppose H is a divisible subgroup of A and seek to show H is trivial. For all natural numbers n, $H = p^nH$ by divisibility so $H \subseteq p^nA$. Thus $H \subseteq B$ by Result 5.6.2, Part (a). Similarly $H \subseteq C$ so $H \subseteq B \cap C = \{0\}$.

EXERCISES

1. Show
   (a) $C \cap (B + D) = \langle(0,1)\rangle$.
   (b) $B \cap (C + D) = \langle(1,0)\rangle$.
   (c) $D \cap (B + C) = \langle(1,1)\rangle$.
2. Show A is not pure-simple.
3. Show
   (a) The subgroup lattice L(A) is not distributive.
   (b) A is not locally cyclic.
4. Show $A/(C + D) \cong Z(p^\infty)$ and $A/(B + D) \cong Z(q^\infty)$.
5. Show A is hopfian.
6. Let H, K, and F be defined as in Result 5.6.2. Show
   (a) H and K are residually finite.
   (b) A is residually finite.

NOTES

1. The group A is of interest as an example of an indecomposable torsion-free abelian group that cannot be embedded in Q. A torsion-free abelian group is said to have **rank** n if it can be embedded in the direct product of n copies of Q, but cannot be embedded in the direct product of $n - 1$ copies. Thus A has rank 2. Examples of indecomposable rank n torsion-free groups where $2 \leqslant n$ are abundant, but showing indecomposability is usually as hard as or harder than the proof of Result 5.6.3.
2. As a counterexample, A shows that subgroups of abelian indecomposable groups need not be indecomposable (as shown in Result 5.6.4, $B + C$ is a decomposable subgroup of A).

3. See Miscellaneous Exercises II.3, IV.5, and IV.6 at the end of this chapter and Miscellaneous Exercise V.2 at the end of Chapter 6.

□□□□□

Example 5.7

Let $G = \Sigma\{Z(p)\}_{p \in PR}$.

RESULT 5.7.1 *If $f \in G$ and $S(f) = \{p_1, p_2, \ldots, p_n\}$, then $o(f) = p_1 p_2 \ldots p_n$. Consequently G is torsion and moreover every element has square-free order.*
Since $f(p_i) \in Z(p_i)$ for each i, $p_i f(p_i) = 0$. Let $m = p_1 p_2 \ldots p_n$. Then $mf(p_i) = 0$ for all $p_i \in S(f)$. Also, of course, $mf(q) = 0$ for all $q \notin S(f)$. Conclude $mf = 0$ and hence $o(f) | m$. Thus $o(f) = p_1{}^{e(1)} p_2{}^{e(2)} \ldots p_n{}^{e(n)}$ where each $e(i) = 0$ or 1. But if some $e(i) = 0$, then since p_i does not divide $o(f)$, p_i and $o(f)$ are relatively prime so there exist integers c and d such that $cp_i + do(f) = 1$. Then $f = cp_i f + do(f)f = cp_i f$. But then $f(p_i) = cp_i f(p_i) = 0$ (because $p_i f(p_i) = 0$). This contradiction (to $p_i \in S(f)$) shows that $e(i) = 1$ for all i, and hence that $o(f) = m$.

RESULT 5.7.2 *For all primes p,*
(a) $pG = D_G(PR \setminus \{p\})$.
(b) $G/pG \cong Z(p)$.
Let J denote the set $PR \setminus \{p\}$. Let $f \in pG$. Then $f = ph$ for some h and so $f(p) = ph(p) = 0$. Therefore $f \in \{k | k(p) = 0\} = D_G(J)$. Conversely let $f \in D_G(J)$. By Theorem 1.4.9, $D_G(J) \cong \Sigma\{Z(q)\}_{q \in J}$. By Theorem A.8.6, each $Z(q)$, where $q \in J$, is p-divisible. Consequently $\Sigma\{Z(q)\}_{q \in J}$, and hence $D_G(J)$, is p-divisible by Theorem 1.4.16. Therefore $f = ph$ for some $h \in D_G(J)$ and hence $f \in pG$. This proves Part (a).

By Part (a) and Theorem 1.4.9, $G = pG \oplus D_G(\{p\})$, so $G/pG \cong D_G(\{p\}) \cong Z(p)$ by the Corollary to Theorem 1.4.9.

RESULT 5.7.3
(a) *Let H be any subgroup of G. Let $S = \{p | H[p] \neq \{0\}\}$. Then $H = D_G(S)$.*
(b) *Let T be any subset of PR. Let $K = D_G(T)$. Then $T = \{p | K[p] \neq \{0\}\}$.*
Let $f \in D_G(S)$ and let $S(f) = \{p_1, p_2, \ldots, p_n\}$. For $1 \leq i \leq n$ define d_i by

$$d_i(q) = \begin{cases} f(p_i) & \text{if } q = p_i, \\ 0 & \text{otherwise.} \end{cases}$$

Then

$$(d_1 + d_2 + \ldots + d_n)(q) = \begin{cases} f(p_i) & \text{if } q = p_i \text{ for some i,} \\ 0 & \text{if } q \notin S(f). \end{cases}$$

Hence $d_1 + d_2 + \ldots + d_n = f$. Thus to show $f \in H$, it suffices to show $d_i \in H$ for all relevant i. For such i, $p_i \in S$ because if $p_i \notin S$, then $f(p_i) = 0$ since $f \in \mathrm{D}_G(S)$ contradicting $p_i \in \mathrm{S}(f)$. But $p_i \in S$ implies $H[p_i]$ is nontrivial so there exists a nonzero $h_i \in H[p_i]$. For all primes $q \neq p_i$, q and p_i are relatively prime so there exist integers c and k such that $cp_i + kq = 1$. Hence $cp_ih_i + kqh_i = h_i$. But $p_ih_i = 0$ so $h_i = kqh_i$. Consequently $h_i(q) = kqh_i(q) = 0$ because $h_i(q) \in Z(q)$. Thus $h_i(q) = 0$ for all primes $q \neq p_i$. But recall h_i is nonzero so $h_i(p_i) \neq 0$, and hence $h_i(p_i)$ generates $Z(p_i)$ (as does any nontrivial element). Thus $d_i(p_i) \in Z(p_i)$ implies $d_i(p_i) = mh_i(p_i) = (mh_i)(p_i)$ for some integer m. For all primes $q \neq p_i$, $d_i(q) = 0$ by definition of d_i and $(mh_i)(q) = 0$ because $h_i(q) = 0$ as shown before. Thus $d_i(q) = (mh_i)(q)$ for all primes q. Conclude $d_i = mh_i \in H$ as required. This proves $\mathrm{D}_G(S) \subseteq H$.

Conversely, let $h \in H$. By Result 5.7.1, the order of h is $p_1p_2 \ldots p_n$ where $\mathrm{S}(h) = \{p_1, p_2, \ldots, p_n\}$. It is claimed that $p_i \in S$ for $1 \leqslant i \leqslant n$. For such i let w_i denote the product of all primes in $\mathrm{S}(h)$, except p_i. Let $g_i = w_ih$. Then $g_i \in H$ and $p_ig_i = 0$ so $g_i \in H[p_i]$. But $g_i \neq 0$ because if $g_i = 0$, then $o(h) | w_i$ which implies $p_i | w_i$ which is false. Conclude $g_i \neq 0$. Consequently $p_i \in S$ by definition of S, establishing the claim. Thus for all primes $q \notin S$, $q \notin \mathrm{S}(h)$ and hence $h(q) = 0$. Thus $q \in \mathrm{D}_G(S)$ proving $H \subseteq \mathrm{D}_G(S)$.

To show Part (b) let $p \in \{p \mid K[p]$ is nontrivial$\}$. Then there exists $k \in K$ such that $k \neq 0$ and $pk = 0$. Suppose $p \notin T$. Then $k \in \mathrm{D}_G(T)$ implies $k(p) = 0$. Now $k \neq 0$ implies $k(q) \neq 0$ for some prime q. Therefore $p \neq q$ so p and q are relatively prime and hence $cp + dq = 1$ for some integers c and d. Then $k = cpk + dqk = dqk$ because $pk = 0$, so $k(q) = dqk(q) = 0$. This contradiction shows $p \notin T$ was false hence $\{p \mid K[p]$ is nontrivial$\} \subseteq T$.

Conversely let $p \in T$. Let $k = \lambda_p(1)$. Then $k \in \mathrm{D}_G(T) = K$ and $k \neq 0$. Also clearly $pk = 0$. Thus $K[p]$ is nontrivial showing $T \subseteq \{p \mid K[p]$ is nontrivial$\}$.

RESULT 5.7.4 *Every subgroup of G is a direct factor.*
Let H be any subgroup. Then by Result 5.7.3, $H = \mathrm{D}_G(S)$ for some S. Hence by Theorem 1.4.9, $G = H \oplus \mathrm{D}_G(PR \setminus S)$.

RESULT 5.7.5
(a) *G is residually finite.*
(b) *G is reduced.*
Part (a) is immediate because G is a subgroup of $\mathsf{X}\{Z(p)\}_{p \in PR}$ and each $Z(p)$ is finite. Part (b) follows from Part (a) and Theorem 1.3.11.

RESULT 5.7.6
(a) *M is a maximal subgroup of G if and only if* $M = \mathrm{D}_G(PR \setminus \{p\})$ *for some prime p.*
(b) *G is coatomic.*
Suppose $M = \mathrm{D}_G(PR \setminus \{p\})$. Then $G/M \cong Z(p)$ by Result 5.7.2. Therefore M is a maximal subgroup.

Conversely let M be a maximal subgroup. Let $S = \{p \mid M[p]$ is nontrivial$\}$. If $S = PR$, then $D_G(S) = D_G(PR) = G$. Thus by Result 5.7.3, Part (a), $G = M$ which is false. Hence there is some prime $q \notin S$. Let F denote the set $PR \setminus \{q\}$ and let K denote $D_G(F)$. Thus $S \subseteq F$ which implies $D_G(S) \subseteq K$ and hence, again by Result 5.7.3, Part (a), $M \subseteq K$. If $K = G$, then $\{p \mid K[p]$ is nontrivial$\} = \{p \mid G[p]$ is nontrivial$\} = PR$. But $F = \{p \mid K[p]$ is nontrivial$\}$ by Result 5.7.3, Part (b). Thus F would equal PR. This contradiction shows $M \subseteq K \neq G$. Hence $M = K$ by maximality of M.

To show Part (b), let H be a proper subgroup of G. Let $S = \{p \mid H[p]$ is nontrivial$\}$. If $S = PR$, then, as above, $H = G$. Thus there is some $q \notin S$. Letting K denote $D_G(PR \setminus \{q\})$ again shows $D_G(S) \subseteq K$ which is a maximal subgroup of G by Part (a). Finally (Result 5.7.3, Part (a)) $H \subseteq K$.

RESULT 5.7.7 *G is hopfian.*
Let $\mathcal{L} = \{pG \mid p$ is a prime$\}$. Then by Theorem A.4.1, $\mathcal{L}$ is a collection of fully invariant subgroups of G and by Result 5.7.2, each G/pG is finite and hence hopfian. Now suppose $f \in \cap\mathcal{L}$. Then for all primes p, $f \in pG$ implies $f = pg$ for some g. Hence $f(p) = pg(p) = 0$ since $g(p) \in Z(p)$. Thus $f = 0$. Conclude $\cap\mathcal{L}$ is trivial so G is hopfian by the Lemma to Result 5.5.7.

RESULT 5.7.8 *G is cohopfian.*
Suppose H is a subgroup of G such that $H \cong G$, and seek to show $H = G$. For all primes p, $H[p]$ is nontrivial because $G[p]$ is nontrivial as is easily seen. Therefore $H = D_G(PR)$ (Result 5.7.3) $= G$.

Alternate proof: By Result 5.7.4, $G = H \oplus K$ for some K. Then $G \cong H$ implies $G \cong G/K$ so K is trivial because G is hopfian (Result 5.7.7). Thus $G = H$.

EXERCISES

1. Show G is locally cyclic.
2. Show that the subgroup lattice $L(G)$ is distributive.
3. Show every subgroup of G is fully invariant.
4. Show G does not satisfy the descending chain condition.
5. Characterize the natural numbers n such that $Z(n)$ can be embedded in G.
6. Let A be an abelian group such that for all primes p, A has an element of order p. Show G can be embedded in A.
7. Let A be an abelian group with the following properties;
   (a) For all primes p, A has an element of order p.
   (b) Every subgroup of A is a direct factor.
   (c) A is locally cyclic.
   Show $A \cong G$.
8. Let $P(PR)$ be the set of all subsets of PR (the power set of PR). Show that the lattices $(L(G),\subseteq)$ and $(P(PR),\subseteq)$ are isomorphic.
9. Show that the subgroup lattice $L(G)$ is self-dual.

10. Show G is countable.

NOTES

1. As a counterexample G separates the classes of coatomic and finitely generated groups (Miscellaneous Exercise IV.6 shows G is not finitely generated), and also separates the class of torsion groups and the class of groups satisfying the descending chain condition (Exercise 4).
2. A characterization of G is given in Exercise 7.
3. Expressed in lattice-theoretic terminology, Result 5.7.8 says that the subgroup lattice of G is complemented. By Exercise 2, it is also distributive and hence is a boolean algebra (boolean algebras are defined as complemented distributive lattices).
4. See Note 1 to Example 5.8. See also Miscellaneous Exercises IV.1, IV.4, and IV.6 at the end of this chapter.

□□□□□

Example 5.8

Let p be a prime. Let $G = \Sigma\{Z(p^n)\}_{n\in N}$.

RESULT 5.8.1 *G is a p-group.*
By Theorem 1.4.6.

LEMMA *Let A be a finite cyclic group of cardinality p^n for some $n \neq 0$. Then $pA \cong Z(p^{n-1})$.*
Since subgroups of cyclic groups are cyclic, pA is a cyclic group. Then because any two finite cyclic groups of the same cardinality are isomorphic, it suffices to show $\mathrm{card}(pA) = p^{n-1}$. Let a generate A. Then clearly $pA = \langle pa\rangle$. However, $p^{n-1}(pa) = 0$, and if $k(pa) = 0$, then $(kp)a = 0$ so $p^n \mid kp$ which implies $p^{n-1} \mid k$. This shows $o(pa) = p^{n-1}$, hence $\mathrm{card}(pA) = p^{n-1}$.

RESULT 5.8.2 *$pG \cong G$.*
Let N' denote the set $N\setminus\{0\}$. By Theorem 1.4.9, $G = \mathrm{D}_G(\{0\}) \oplus \mathrm{D}_G(N')$. But $\mathrm{D}_G(\{0\}) \cong Z(p^0) = \{0\}$ by the Corollary to Theorem 1.4.9. Hence $G = \mathrm{D}_G(N')$. Thus (again by Theorem 1.4.9) $G \cong \Sigma\{Z(p^n)\}_{n\in N'}$. Applying Theorem 1.4.16 yields $pG \cong \Sigma\{pZ(p^n)\}_{n\in N'}$ consequently the lemma shows that $pG \cong B$ where B denotes the group $\Sigma\{Z(p^{n-1})\}_{n\in N'}$. By Theorem 1.4.9, $B = \mathrm{D}_B(\{1\}) \oplus \mathrm{D}_B(N'')$ where $N'' = N \setminus \{0,1\}$. But $\mathrm{D}_B(\{1\}) \cong Z(p^{1-1}) = \{0\}$ (Corollary to Theorem 1.4.9). Thus $B = \mathrm{D}_B(N'') \cong \Sigma\{Z(p^{n-1})\}_{n\in N''}$. Let-

ting $m = n - 1$ yields $B \cong \Sigma\{Z(p^m)\}_{m\in N'} \cong G$. Conclude $pG \cong G$.

RESULT 5.8.3 *G is nonhopfian.*
By Theorem A.4.1, $G/G[p] \cong pG$. Hence by Result 5.8.2, $G/G[p] \cong G$. Thus it suffices to show $G[p]$ is nontrivial. Define f by

$$f(n) = \begin{cases} 1 & \text{if } n = 1, \\ 0 & \text{if } n = 0 \text{ or } 2 \leqslant n. \end{cases} \tag{1}$$

Then $pf(n) = 0$ for all n hence $f \in G[p]$ and $f \neq 0$ as required.

RESULT 5.8.4 *G is noncohopfian.*
By Result 5.8.2, $G \cong pG$ so it suffices to show pG is a proper subgroup. Let f be defined as in Equation (1) (Result 5.8.3). Then $f \notin pG$ because if $f = pg$, $1 = f(1) = pg(1)$ implying $p|1$. Conclude pG is proper.

RESULT 5.8.5 *G is noncoatomic.*
Define $e_n : Z(p^n) \to Z(p^\infty)$ by $e_n(t) = Z + (t/p^n)$. Clearly e_n is a homomorphism for all n. Applying Theorem 1.4.13, there is a unique $g: G \to Z(p^\infty)$ such that $g \circ \lambda_n = e_n$ for all n. Then g is onto $Z(p^\infty)$ because if $Z + (t/p^n) \in Z(p^\infty)$, then $Z + (t/p^n) = e_n(t) = g(\lambda_n(t))$. Conclude $G/\ker g \cong Z(p^\infty)$. If M is a maximal subgroup of G such that $\ker g \subseteq M$, then $M/\ker g$ would be a maximal subgroup of $G/\ker g$. Hence maximal subgroups of $Z(p^\infty)$ would exist in contradiction to Result 5.4.10. Thus there is no maximal subgroup of G containing $\ker g$.

RESULT 5.8.6
(a) *G is residually finite.*
(b) *G is reduced.*
Part (a) is immediate since each $Z(p^n)$ is finite. Part (b) follows from Part (a) and Theorem 1.3.11.

EXERCISES

1. Show that the subgroup lattice L(G) is not distributive and that G is not locally cyclic.
2. Show G does not satisfy the descending chain condition.
3. Show G has subgroups that are not direct factors.

NOTES

1. Compare the tremendous amount of difference between Examples 5.7 and 5.8. Both are restricted direct products of finite cyclic groups, but while Example 5.7 is coatomic, hopfian, cohopfian, locally cyclic, has only direct factors for subgroups, and has a distributive subgroup lattice, Example 5.8 has none of these properties.

2. See Miscellaneous Exercise IV.6 at the end of this chapter.

□□□□□

Example 5.9

Let p be a prime. Let $X = \times\{Z(p^n)\}_{n\in N}$, $T = \mathrm{t}(X)$, and $S = \Sigma\{Z(p^n)\}_{n\in N}$. Note that S is a subgroup of X and is torsion so $S \subseteq T$. Let $G = T/S[p]$.

RESULT 5.9.1 *Both T and G are p-groups.*
It suffices to show T is a p-group. For $f \in T$, f has finite order, call it n. Let p^a be the largest power of p that divides n. Then $n = p^a d$ where d and p are relatively prime. It suffices to show $p^a f = 0$, and to do this $(p^a f)(i) = 0$ must be shown for all natural numbers i. Since d and p are relatively prime, so are d and p^i. Consequently there exist integers c and k such that $cd + kp^i = 1$. Then $f(i) = cdf(i) + kp^i f(i)$. But $f(i) \in Z(p^i)$ implies $p^i f(i) = 0$. Hence $f(i) = cdf(i)$ which implies that $p^a f(i) = cp^a df(i) = cnf(i) = 0$ as required.

LEMMA *If A is an abelian p-group and $x \in p^k A$ for all natural numbers k, then $x \in \mathrm{U}(A)$.*
Let n be a positive integer and seek to show $x \in nA$. Let p^a be the largest power of p dividing n. Thus $n = p^a d$ where d and p are relatively prime. Let $\mathrm{o}(x) = p^t$. Then d and p^t are relatively prime so there exist integers i and k which satisfy $id + kp^t = 1$. Hence $x = idx + kp^t x = idx$. But by hypothesis, $x \in p^a A$ so $x = p^a y$ for some y. Hence $x = idp^a y = niy \in nA$.

RESULT 5.9.2
(a) $T[p]/S[p] \subseteq \mathrm{U}(G)$.
(b) $\mathrm{U}(G)$ *is nontrivial.*
Let $S[p] + f \in T[p]/S[p]$. It is claimed that $S[p] + f \in p^k G$ for all natural numbers k. First note that $f \in T[p]$ so $pf = 0$. Suppose $n \in N$ such that $k < n$. Then $pf = 0$ implies $pf(n) = 0$ in $Z(p^n)$ which means $pf(n) \equiv 0 \bmod p^n$ and hence p^n divides $pf(n)$. This implies $p^{n-1} | f(n)$ and so $f(n) = p^{n-1}c_n$ for some c_n. Now $k \leqslant n - 1$ so $0 \leqslant n - 1 - k$. Define g by

$$g(n) = \begin{cases} p^{n-1-k}c_n & \text{if } k < n, \\ 0 & \text{if } n \leqslant k. \end{cases}$$

Now for $k < n$, $p^{k+1}g(n) = p^n c_n = pf(n) = 0$. Also, clearly, $p^{k+1}g(n) = 0$ for $n \leqslant k$. Thus $p^{k+1}g = 0$ and so $g \in T$. Next note that

$$(p^k g - f)(n) = \begin{cases} f(n) - f(n) = 0 & \text{if } k < n, \\ -f(n) & \text{if } n \leqslant k. \end{cases}$$

Since $\{n \mid n \leqslant k\}$ is finite, $p^k g - f$ has finite support so belongs to S, and furthermore belongs to $S[p]$ since $p(p^k g - f) = p^{k+1} g - pf = 0 - 0 = 0$. Therefore $S[p] + f = p^k(S[p] + g) \in p^k G$ for all k as claimed. This, Result 5.9.1, and the lemma imply $T[p]/S[p] \subseteq \mathrm{U}(G)$ proving Part (a).

The function f defined by $f(n) = p^{n-1}$ clearly has order p but infinite support. Thus $S[p] + f$ is a nonzero element of $T[p]/S[p]$. Conclude from Part (a) that $\mathrm{U}(G)$ is nontrivial.

RESULT 5.9.3 *G is not residually finite.*
By Result 5.9.2 and Theorem 1.3.11.

LEMMA 1 *If A is a bounded abelian group, then A is reduced.*
Suppose $\exp(A) = n$. Seek to show a divisible subgroup D of A is trivial. Let $x \in D$. Then there exists $d \in D$ such that $x = nd$ by divisibility. Then $0 = nd = x$.

LEMMA 2 *If A is an abelian group and B is a subgroup such that B and A/B are reduced, then A is reduced.*
Seek to show a divisible subgroup D of A is trivial. Since D is divisible, so too then, by Theorem A.8.1, is $D/(D \cap B) \cong (B + D)/B$. Thus $(B + D)/B$ is a divisible subgroup of A/B, hence is trivial as A/B is reduced. Consequently $D \subseteq B$. This and B reduced imply D is trivial as required.

RESULT 5.9.4 *G is reduced.*
By Theorem A.4.1, $T/T[p] \cong pT$. Now pT is a subgroup of X which is a direct product of finite (hence reduced) groups. Thus X is reduced by Theorem 1.3.3 and hence so too is the subgroup pT. Thus $T/T[p]$ is reduced. Next note that $T[p]/S[p]$ is a group of exponent p, hence is reduced by Lemma 1. Finally note that $(T/S[p])/(T[p]/S[p]) \cong T/T[p]$. Conclude $T/S[p]$ is reduced by Lemma 2.

EXERCISES

1. (a) Show T is uncountable.
   (b) Show G is uncountable.
2. Show S can be embedded in G.
3. Show that the subgroup lattice $\mathrm{L}(G)$ is not distributive and that G is not locally cyclic.
4. Show G does not satisfy the descending chain condition.
5. Show T/S is divisible.

NOTES

1. The group G is of interest as an example of a reduced abelian group that is not residually finite or equivalently a reduced abelian group with nontrivial Ulm subgroup. (It can be shown that an abelian group A is residually finite

if and only if U(A) is trivial.)

2. As a counterexample G shows the class of residually finite groups is not extension closed. As was seen in the proof of Result 5.9.4, G is an extension of $T[p]/S[p]$ by $T/T[p]$. But $T[p]/S[p]$ is bounded, hence isomorphic to a restricted direct product of finite cyclic groups by Prufer's theorem (Rotman [31], Corollary 9.32), hence residually finite, and $T/T[p]$ can be embedded in X so it is residually finite. On the other hand G is not residually finite.
3. See Miscellaneous Exercise IV.6 at the end of this chapter and Miscellaneous Exercise V.2 at the end of Chapter 6.

□□□□□

Example 5.10

Let $SF = \{a/b \mid a/b \in Q$ and b is square-free$\}$.

RESULT 5.10.1 *SF is a subgroup of Q.*
Suppose a/b, $a^*/b^* \in SF$. Let c be the product of all the primes dividing b^* but not dividing b, let d be the product of all the primes dividing b but not dividing b^*, and let e be the product of all the primes that divide both b and b^*. Thus $b^* = ce$ and $b = de$. Therefore $bc = dec = db^*$, so $(a/b) - (a^*/b^*) = (ac/bc) - (a^*d/b^*d) = (ac/bc) - (a^*d/bc) = (ac - a^*d)/bc$. Consequently, to show $(a/b) - (a^*/b^*) \in SF$ it suffices to show that bc is square-free. Suppose to the contrary that $p^2 \mid bc$ for some prime p. Then $p \mid bc$ so $p \mid c$ or $p \mid b$. In the first case, $pw = c$ for some integer w. This and $p^2 \mid bc$ implies that $p^2 \mid bpw$ so $p \mid bw$ which implies $p \mid b$ or $p \mid w$. But $p \mid c$ implies p does not divide b by definition of c, so $p \mid w$. Thus $pk = w$ for some integer k and hence $c = pw = p^2k$ implying $p^2 \mid c$ which further implies $p^2 \mid b^*$ in contradiction to the fact that b^* is square-free. In the second case, if $p \mid b$, then $ip = b$ for some integer i, so $p^2 \mid bc$ implies $p^2 \mid ipc$ which implies $p \mid ic$. Hence either $p \mid i$ or $p \mid c$. If $p \mid c$, then the first case holds so a contradiction is reached as has been shown. If $p \mid i$, then $pt = i$ for some integer t so $b = ip = p^2t$ in contradiction to the fact that b is square-free.

RESULT 5.10.2 *SF is torsion-free.*
Since Q is torsion-free (Result 5.1.1) and SF is a subgroup of Q.

RESULT 5.10.3 *SF is indecomposable.*
By Result 5.1.3.

RESULT 5.10.4 *SF is pure-simple.*
By Result 5.1.8.

RESULT 5.10.5 *SF is hopfian.*
By Result 5.1.12.

RESULT 5.10.6 *SF is locally cyclic.*
Since Q is locally cyclic (Result 5.1.10) and SF is a subgroup of Q.

RESULT 5.10.7 *The subgroup lattice* L(*SF*) *is distributive.*
By Result 5.10.6 and the Lemma to Result 5.1.11.

RESULT 5.10.8 $Z \subseteq SF$.
Since 1 is square-free, $1/1 \in SF$. Thus $Z \subseteq SF$.

RESULT 5.10.9 $\cap\{pSF \mid p \text{ is a prime}\} = Z$.
For all primes p, $1/p \in SF$ and $1 = p(1/p)$. Thus $1 \in pSF$ for all primes p, and hence $Z \subseteq \cap\{pSF \mid p \text{ is a prime}\}$. Conversely suppose a/b belongs to the intersection. It may be assumed that a and b are relatively prime (see Note 1 to Example 5.1). It is claimed that $b = 1$. If not, then there is some prime p that divides b so $b = pk$ for some integer k. Then $a/b \in pSF$ implies $a/b = p(c/d)$ for some $c/d \in SF$. Hence $ad = pbc = p(pk)c$ so $p^2 \mid ad$. However, p does not divide a because $p \mid b$ and a and b are relatively prime. Thus a and p^2 are also relatively prime. From this and $p^2 \mid ad$ conclude $p^2 \mid d$ which is false since $c/d \in SF$. Thus $b = 1$ as claimed. Hence $a/b = a \in Z$.

RESULT 5.10.10 $\cap\{p^2SF \mid p \text{ is a prime}\} = \{0\}$.
Let $a/b \in \cap\{p^2SF \mid p \text{ is a prime}\}$. Then $a/b \in \cap\{pSF \mid p \text{ is a prime}\} = Z$ by Result 5.10.9, hence $a/b = k$ for some integer k. Seek to show $k = 0$. It suffices to show that k is divisible by every prime. For all primes p, $k \in p^2SF$ so $k = p^2(r/s)$ for some $r/s \in SF$, and so

$$p^2 \mid ks. \tag{1}$$

Either $p \mid s$ or p and s are relatively prime. In the second case, p^2 and s are also relatively prime. This and Line (1) imply $p^2 \mid k$, hence $p \mid k$. In the first case, $pt = s$ for some integer t. This and Line (1) imply $p^2 \mid kpt$ so $p \mid kt$, implying $p \mid k$ or $p \mid t$. But $p \mid t$ can be eliminated since it would imply $p^2 \mid s$ in contradiction to the fact that s is square-free. Thus $p \mid k$.

RESULT 5.10.11 $SF/Z \cong \Sigma\{Z(p)\}_{p \in PR}$.
Define $d_p{:}Z(p) \to SF/Z$ by $d_p(a) = Z + (a/p)$. First it has to be checked that d_p is a function. If a and a^* represent the same element of $Z(p)$, then $a \equiv a^* \bmod p$ so $p \mid a - a^*$. Thus $(a - a^*)/p \in Z$ and so $Z + (a/p) = Z + (a^*/p)$ hence d_p is a function. Moreover d_p is a homomorphism since $d_p(a + a^*) = Z + ((a + a^*)/p) = (Z + (a/p)) + (Z + (a^*/p)) = d_p(a) + d_p(a^*)$. Apply Theorem 1.4.13 to get a homomorphism $h{:}\Sigma\{Z(p)\}_{p \in PR} \to SF/Z$ such that $h \circ \lambda_p = d_p$. It will be shown that this is the desired isomorphism.

To show h is surjective, let $Z + (a/b) \in SF/Z$. That $Z + (a/b) \in \operatorname{Im} h$ will be shown by induction on the number of primes dividing b. First suppose this number is one. Thus b is a prime so $Z + (a/b) = d_b(a) = h \circ \lambda_b(a) \in \operatorname{Im} h$. Hence induction begins. Assume now the induction hypothesis that $Z + (d/e) \in \operatorname{Im} h$ whenever there are $n - 1$ or fewer primes dividing e and suppose n primes divide b. Thus $b = p_1p_2 \ldots p_{n-1}p$. Let b' denote $p_1p_2 \ldots p_{n-1}$ so $b = b'p$. Then b' and p are relatively prime so there exist integers i and k such that $ib' + kp = 1$. Hence $a = ib'a + kpa$. Consequently $(ia/p) + (ka/b') = (iab' + kap)/pb' = a/pb' = a/b$. Hence $Z + (a/b) = [Z + (ia/p)] + [Z + (ka/b')]$. Since both of the summands in brackets belong to $\operatorname{Im} h$ (the first is $d_p(ia) = h \circ \lambda_p(ia)$ while the second belongs to $\operatorname{Im} h$ by the induction hypothesis), so does $Z + (a/b)$.

Finally it must be shown that h is injective. Suppose $h(f) = 0$ but $f \neq 0$, and seek a contradiction. Since f is nonzero, it has a nonempty support, say $S(f) = \{p_1, p_2, \ldots, p_k\}$. Denote $f(p_i)$ by x_i for relevant i. Then (Theorem 1.4.3)

$$f = \lambda_{p_1}(x_1) + \lambda_{p_2}(x_2) + \ldots + \lambda_{p_k}(x_k),$$

so

$$0 = h(f) = h \circ \lambda_{p_1}(x_1) + h \circ \lambda_{p_2}(x_2) + \ldots + h \circ \lambda_{p_k}(x_k)$$

thus

$$0 = d_{p_1}(x_1) + d_{p_2}(x_2) + \ldots + d_{p_k}(x_k). \tag{1}$$

Now either $k = 1$ or $2 \leqslant k$. In the first case,

$$0 = d_{p_1}(x_1).$$

Thus $Z + (x_1/p_1) = Z$ so $x_1 = cp_1$ for some integer c. This, however, implies $x_1 = 0$ in the group $Z(p_1)$ which is false ($p_1 \in S(f)$ so $f(p_1) \neq 0$). Assume then the second case, $2 \leqslant k$. Therefore p_1 and m are relatively prime where m denotes $p_1p_2 \ldots p_k$. Consequently there exist integers α and β such that $\alpha p_1 + \beta m = 1$. Now note that for $2 \leqslant i$, $p_ix_i = 0$ so $\beta m x_i = 0$. Also, $\beta m x_1 = (1 - \alpha p_1)x_1 = x_1$. Multiply Equation (1) by βm to get

$$0 = d_{p_1}(\beta m x_1) + d_{p_2}(\beta m x_2) + \ldots + d_{p_k}(\beta m x_k) = d_{p_1}(x_1).$$

This implies the contradiction $0 = f(p_1)$ as in the first case.

RESULT 5.10.12 *For all primes p,*
(a) *$p(SF/Z) = pSF/Z$.*
(b) *$SF/pSF \cong Z(p)$.*
(c) *SF/p^2SF is finite* (see Exercise 1).
Let $Z + x \in pSF/Z$. Thus $x = py$ for some $y \in SF$. Hence $Z + x = Z + py = p(Z + y) \in p(SF/Z)$. Conversely let $Z + x \in p(SF/Z)$. Thus $Z + x = p(Z + y)$ for some y and hence $x - py \in Z$. But $Z \subseteq pSF$ by Result 5.10.9. Hence $x - py = pw$ for some w and so $Z + x = Z + p(y + w) \in pSF/Z$. This proves Part (a).

By Result 5.10.11, $SF/Z \cong G$ where G denotes the group $\Sigma\{Z(p)\}_{p \in PR}$.

This further implies $pG \cong p(SF/Z) = pSF/Z$ by Part (a). Consequently $G/pG \cong (SF/Z)/(pSF/Z) \cong SF/pSF$. But $G/pG \cong Z(p)$ by Result 5.7.2. Combining these isomorphisms yields Part (b).

By Theorem A.4.1, $SF/SF[p] \cong pSF$. But SF is torsion-free so $SF[p]$ is trivial. Thus $SF \cong pSF$. Hence $pSF/p^2SF \cong SF/pSF \cong Z(p)$, so SF/pSF and pSF/p^2SF are both finite groups. But $(SF/p^2SF)/(pSF/p^2SF) \cong SF/pSF$ so SF/p^2SF is an extension of a finite group by a finite group, hence is finite.

RESULT 5.10.13 *SF is residually finite.*
Let $\mathcal{R} = \{p^2SF \mid p \text{ is a prime}\}$. By Result 5.10.10, $\cap\mathcal{R}$ is trivial, and by Result 5.10.12, Part (c), each SF/p^2SF is finite. Conclude SF is residually finite by Theorem 1.3.9.

RESULT 5.10.14 *SF is reduced.*
By Result 5.10.13 and Theorem 1.3.11. Alternate proof: Since $1/4 \notin SF$, SF is a proper subgroup of Q, hence is reduced by Result 5.1.4.

LEMMA *Central extensions of coatomic groups by (not necessarily abelian) coatomic groups are coatomic.*
Let G be a group with a subgroup H such that $H \subseteq Z(G)$ and both H and G/H are coatomic. Suppose A is a proper subgroup of G. Then either $HA \neq G$ or $HA = G$. In the first case, $HA/H \neq G/H$ so there exists M/H such that M/H is a maximal subgroup of G/H containing HA/H. Then M is a maximal subgroup of G and $A \subseteq HA \subseteq M$.

Assume now the second case. Then $H \cap A \neq H$ because equality would imply $H \subseteq A$, hence $HA = A$ and so $G = A$ which is false. Thus since H is coatomic, there exists a maximal subgroup M of H containing $H \cap A$. Then M is a subgroup of $Z(G)$ so M is normal in G, hence MA is a subgroup. Now $HA \subseteq HMA$ and so $G = H(MA)$. This and normality of H imply $[G:MA] = [H:H \cap MA]$ by Theorem A.10.7.

Since M is contained in both H and MA, $M \subseteq H \cap MA$. Conversely if $x \in H \cap MA$, then $x = ma$ where $m \in M$ and $a \in A$. Then $a = m^{-1}x$ so $a \in H \cap A$ and therefore $a \in M$. This further implies $x = ma \in M$ showing $H \cap MA \subseteq M$. Thus $H \cap MA = M$ so that $[G:MA] = [H:M]$ by the above. Now since H is abelian ($H \subseteq Z(G)$), maximal subgroups have prime index. Conclude $[G:MA] = p$ for some prime p so MA must be a maximal subgroup of G with $A \subseteq MA$.

RESULT 5.10.15 *SF is coatomic.*
Since Z is finitely generated, it is coatomic (Theorem A.10.11). By Results 5.10.11 and 5.7.6, SF/Z is also coatomic. Then SF is coatomic by the lemma.

Supplement: Embedding certain subgroups of torsion-free abelian groups into SF

NOTATION If A is a torsion-free abelian group and $x \in A$, $x \neq 0$, then $S(A,x)$ will denote $\{a \mid a \in A$ and $ma \in \langle x \rangle$ for some $m \in N$, $m \neq 0$ such that m is square-free$\}$.

RESULT 5.10.16 *Let A be a torsion-free abelian group and let $x \in A$, $x \neq 0$. Then*

(a) *$S(A,x)$ is a subgroup of A.*
(b) *$x \in S(A,x)$.*
(c) *$S(A,x) \subseteq P(A,x)$.*
(d) *$S(A,x)$ can be embedded in SF. Moreover $S(A,x) \cong SF$ if $P(A,x)$ is divisible.* (Note: $P(A,x)$ is divisible if and only if $P(A,x) \cong Q$.)

Let $a,b \in S(A,x)$. Thus there exist integers m,n,j, and k such that $ma = jx$ and $nb = kx$ where both m and n are nonzero and square-free. Let d be the greatest common divisor of m and n. Thus $m = de$ and $n = df$ for some integers e and f and furthermore e and f are relatively prime. Then $def(a - b) = defa - defb = fma - enb = fjx - ekx$ which belongs to $\langle x \rangle$. Clearly def is nonzero, so $a - b \in S(A,x)$ will follow once it is shown that def is square-free. Suppose to the contrary that $p^2 \mid def$ for some prime p. Then p^2 and f are not relatively prime because if they were, $p^2 \mid def$ would imply $p^2 \mid de$, that is, $p^2 \mid m$ which is false. Thus $p \mid f$. Similarly $p \mid e$. This contradiction (recall e and f are relatively prime) completes the proof of Part (a). Parts (b) and (c) are immediate.

To prove Part (d) define $g{:}P(A,x) \to Q$ by $g(a) = n/m$ if and only if $ma = nx$ with $m \neq 0$. By Result 5.1.14, Part (c), g is an injective homomorphism. Let h be the restriction of g to $S(A,x)$. It is claimed that Im $h \subseteq SF$. For $a \in S(A,x)$, $ma = nx$ for some nonzero square-free m. Hence $h(a) = g(a) = n/m \in SF$. At this stage of the proof $h{:}S(A,x) \to SF$ is an injective homomorphism (injective because g is). To complete the proof it suffices to show that h is onto SF when $P(A,x)$ is divisible. So assume divisibility and let $j/k \in SF$. Then there exists $w \in P(A,x)$ such that $kw = x$. Let $t = jw$. Then $kw = x$ and k square-free (since $j/k \in SF$) imply $w \in S(A,x)$. This and $t = jw$ imply $t \in S(A,x)$. Also $t = jw$ implies $kt = kjw = jx$ so $h(t) = g(t) = j/k$ as required.

EXERCISES

1. Show $SF/p^2SF \cong Z(p^2)$ for all primes p.
2. Under the usual multiplication of rational numbers, Q is a ring. Show SF is not a subring.
3. Let H be an abelian group with the following properties:
   (a) H is torsion-free.
   (b) H has a subgroup B such that B is infinite cyclic, every element of H/B

has square-free order, and $B = \cap\{pH \mid p \text{ is a prime}\}$.

Show $H \cong SF$.

4. Find a nonzero element $x \in SF$ such that $SF/\langle x\rangle \ncong SF/Z$. (cf. Section 5.1, Exercise 4.)
5. Show Z is a fully invariant subgroup of SF.
6. Show $\mathrm{Aut}(SF) \cong Z(2)$.
7. Show every subgroup of SF is characteristic.
8. Show $\Phi(SF) = Z$.

NOTES

1. The group SF has an interesting application. It can be used to show that every abelian group A is isomorphic to the Frattini subgroup of some abelian group B. Sketch of proof: Represent A as a quotient group of a free abelian group, $A \cong \Sigma\{Z_i\}_{i\in I}/K$. By Exercise 8, $Z = \Phi(SF)$. It follows that A is isomorphic to $\Phi(\Sigma\{SF_i\}_{i\in I}/K)$.
2. A characterization of SF is given in Exercise 3.
3. See Miscellaneous Exercises II.3, IV.1, and IV.4–IV.6 at the end of this chapter. For a generalization of SF see Miscellaneous Exercise II.11.

□□□□□

Example 5.11

Let $G = \mathsf{X}\{Z(p)\}_{p\in PR}$.

RESULT 5.11.1 $\mathrm{t}(G) = \Sigma\{Z(p)\}_{p\in PR}$.
By Theorems 1.4.1 and 1.4.6, $\Sigma\{Z(p)\}_{p\in PR}$ is a subgroup of G and is torsion. Thus $\Sigma\{Z(p)\}_{p\in PR} \subseteq \mathrm{t}(G)$. To show the reverse inclusion, let $f \in \mathrm{t}(G)$. Thus $nf = 0$ for some nonzero natural number n. It must be shown that $\mathrm{S}(f)$ is finite, and to do that it suffices to show $p \mid n$ for all $p \in \mathrm{S}(f)$. For such p, $f(p) \neq 0$. This and $f(p) \in Z(p)$ imply $f(p)$ has order p. But $nf = 0$ yields $nf(p) = 0$ so $f(p)$ divides n. That is, $p \mid n$.

RESULT 5.11.2
(a) *G is residually finite.*
(b) *G is reduced.*
Since each $Z(p)$ is finite, Part (a) is immediate. Part (b) follows from Part (a) and Theorem 1.3.11.

RESULT 5.11.3 *$G/\mathrm{t}(G)$ is divisible.*
By Theorem 1.4.17.

RESULT 5.11.4
(a) *G is not an* M_2 *group.*
(b) *G is nonsplitting.*
Making use of the fact that G is reduced (Result 5.11.2) it need only be shown that $G/t(G)$ is a nontrivial divisible group. It is divisible by Result 5.11.3, and it is nontrivial because the function f defined by $f(p) = 1$ for all p belongs to G but does not have finite support (its support is PR), hence does not belong to $t(G)$ by Result 5.11.1. This proves Part (a). Part (b) follows immediately.

LEMMA *If A is a nonsplitting abelian group such that $A/t(A)$ is divisible and $t(A)$ is both hopfian and coatomic, then A is not an* M_1 *group.*
Suppose to the contrary that $d{:}A \to t(A)$ is a surjective homomorphism and seek a contradiction. Let $K = \ker d$. Then $A/K \cong t(A)$. It is claimed that

$$t(A) + K = A. \tag{1}$$

Let $B = t(A) + K$. If B is a proper subgroup of A, then B/K is a proper subgroup of A/K. But $A/K \cong t(A)$ implies A/K is coatomic hence there exists a maximal subgroup M/K of A/K that contains B/K. First this implies M is a maximal subgroup of A. Secondly, M contains B hence M also contains $t(A)$. This means $M/t(A)$ is a subgroup of $A/t(A)$ and is furthermore a maximal subgroup since M is a maximal subgroup of A. However $A/t(A)$, being divisible abelian, has no maximal subgroups by the Lemma to Theorem 1.2.13. Thus the claim in Line (1) has been established. This in turn implies $t(A)/(t(A) \cap K) \cong A/K \cong t(A)$. Applying the hopfian hypothesis shows that $t(A) \cap K$ is trivial. This and Line (1) contradict the hypothesis that A is nonsplitting.

RESULT 5.11.5 *G is not an* M_1 *group.*
By Result 5.11.3, $G/t(G)$ is divisible. By Result 5.11.1, $t(G) = \Sigma\{Z(p)\}_{p \in PR}$ so $t(G)$ is coatomic (Result 5.7.6) and hopfian (Result 5.7.7). Finally G is nonsplitting by Result 5.11.4. Application of the lemma yields the result.

RESULT 5.11.6 *G is not an* M_3 *group.*
Suppose $G = A \oplus B$ where B is torsion-free and seek to show B is trivial. Since $t(G)$ is fully invariant (Theorem A.4.5), $t(G) = (t(G) \cap A) \oplus (t(G) \cap B)$ by Theorem A.2.7. But $t(G) \cap B$ is trivial so $t(G) = t(G) \cap A$ implying $t(G) \subseteq A$. Thus $(G/t(G))/(A/t(G)) \cong G/A$. Then since $G/t(G)$ is divisible (Result 5.11.3), so too is G/A (Theorem A.8.1). But $G = A \oplus B$ implies $G/A \cong B$ so B is divisible. This and Result 5.11.2 show B is trivial.

RESULT 5.11.7 *For all primes p,*
(a) $pG = \bar{D}_G(PR \setminus \{p\})$.
(b) $G/pG \cong Z(p)$.
By The Corollary to Theorem 1.3.7, $\bar{D}_G(PR \setminus \{p\}) = \ker \pi_p$ so to prove Part (a) it suffices to show $\ker \pi_p = pG$. Let $f = ph \in pG$. Then $\pi_p(f) = f(p) = ph(p) = 0$

so $pG \subseteq \ker \pi_p$. Conversely suppose $h \in \ker \pi_p$, so $h(p) = 0$. For all primes $q \neq p$ Theorem A.8.6 shows there exists a unique $x_q \in Z(q)$ satisfying $h(q) = px_q$. Define f by

$$f(r) = \begin{cases} x_r & \text{if } r \neq p, \\ 0 & \text{if } r = p. \end{cases}$$

Then clearly $pf = h$ so $h \in pG$ completing the proof of Part (a).

Since $\pi_p : G \to Z(p)$ is surjective, $G/\ker \pi_p \cong Z(p)$. This and Part (a) imply Part (b).

RESULT 5.11.8 *G is hopfian.*
Let $\mathcal{L} = \{pG \mid p \text{ is a prime}\}$. By Theorem A.4.1, $\mathcal{L}$ is a collection of fully invariant subgroups and by Result 5.11.7, each G/pG is finite hence hopfian. Now suppose $f \in \cap\mathcal{L}$. Then for all primes p, $f \in pG$ implies $f = pg$ for some g and hence $f(p) = pg(p) = 0$ since $g(p) \in Z(p)$. Thus $f(p) = 0$ for all primes p, so $f = 0$. Conclude $\cap\mathcal{L}$ is trivial so the Lemma to Result 5.5.6 shows G is hopfian.

RESULT 5.11.9 *G can be embedded in* $\times\{Z(p^\infty)\}_{p \in PR}$ *and* $\times\{Z(p^\infty)\}_{p \in PR}$ *is a splitting group.*
Denote $\times\{Z(p^\infty)\}_{p \in PR}$ by A. For all primes p, define $e_p : Z(p) \to Z(p^\infty)$ by $e_p(b) = Z + (b/p)$. It is easily seen that e_p is an injective homomorphism. Define $d : G \to A$ by $d(f)(p) = e_p(f(p))$ for all primes p. Then d is a homomorphism because for $f,g \in G$, $d(f+g)(p) = e_p(f(p) + g(p)) = e_p(f(p)) + e_p(g(p)) = d(f)(p) + d(g)(p) = (d(f) + d(g))(p)$, so $d(f+g) = d(f) + d(g)$. Moreover d is injective because if $d(f) = 0$, then $d(f)(p) = 0$ for all primes p, hence $e_p(f(p)) = 0$ which implies $f(p) = 0$ for all primes p by injectivity of e_p. Hence $f = 0$ proving d is injective. Conclude G can be embedded in A.

It remains to show that A is splitting. By Theorem A.4.4, $t(A)$ is a pure subgroup of A. Now A is divisible by Result 5.4.2 and Theorem 1.3.4. Thus $t(A)$ is a pure subgroup of a divisible group hence is divisible by Theorem A.8.5. Then A is splitting by Theorem A.8.4.

EXERCISES

1. Show G is uncountable.
2. Let $Z_p = Z$ for all primes p. Show there exists a surjective homomorphism $h : \times\{Z_p\}_{p \in PR} \to G$.
3. Show how the result quoted in Note 4 implies G is cohopfian.

NOTES

1. The group G shows that a nonsplitting group need not belong to any of the classes M_1, M_2, or M_3.
2. As a counterexample G shows:

(a) Neither the class of splitting groups nor any of the classes M_1, M_2, or M_3 are subgroup closed. This is by Result 5.11.9.
(b) Neither the class of splitting groups nor any of the classes M_1, M_2, or M_3 are quotient closed. This follows from Exercise 2.
(c) Neither the class of splitting groups nor any of the classes M_1, M_2, or M_3 are preserved by direct products.
(d) The class of cohopfian groups is not quotient closed. A sketch of this result is as follows: G is cohopfian by Exercise 3; $G/\mathrm{t}(G)$, however, is divisible and torsion-free hence $G/\mathrm{t}(G) \cong \Sigma\{Q_i\}_{i\in I}$ for some index set I by Scott [33], Theorem 5.2.7. Now $\mathrm{t}(G)$ is countable (Section 5.7, Exercise 10) while G itself is uncountable (Exercise 4). Thus $G/\mathrm{t}(G)$ must also be uncountable, hence the index set I cannot be finite. It is easily seen then that $\Sigma\{Q_i\}_{i\in I}$, and hence $G/\mathrm{t}(G)$, is noncohopfian (cf. Section 1.4, Exercise 17).

3. Result 5.11.9 can be vastly generalized. In fact, every nonsplitting group can be embedded in a splitting group. This is because every abelian group A can be embedded in a divisible group B (Rotman [31], Theorem 9.22), then the argument in the second paragraph of Result 5.11.9 can be used to show B is splitting.
4. Algebraic compactness of G follows from Fuchs [11], vol I, Corollary 38.3.
5. See Miscellaneous Exercises III.4, III.5, IV.5, and IV.6 at the end of this chapter, and Miscellaneous Exercise V.2 at the end of Chapter 6.

□□□□□

Example 5.12

Let $X = \mathsf{X}\{Z(p)\}_{p\in PR}$ (the group of Example 5.11). Define $f \in X$ by $f(p) = 1$ for all p. Let $G/\mathrm{t}(X) = \mathrm{P}(X/\mathrm{t}(X), \mathrm{t}(X) + f)$.

RESULT 5.12.1 $\mathrm{t}(G) = \mathrm{t}(X) = \Sigma\{Z(p)\}_{p\in PR}$.
Since $\mathrm{t}(X) \subseteq G$, it follows that $\mathrm{t}(G) = \mathrm{t}(X)$ from Theorem A.4.3. The second equality follows from Result 5.11.1.

RESULT 5.12.2 $G/\mathrm{t}(G) \cong Q$.
By Result 5.1.14, $G/\mathrm{t}(X)$ is a pure subgroup of $X/\mathrm{t}(X)$. But $X/\mathrm{t}(X)$ is divisible by Result 5.11.3, hence $G/\mathrm{t}(X)$ is also by Theorem A.8.5. Now $G/\mathrm{t}(X)$ can be embedded in Q by Result 5.1.14 again, and $G/\mathrm{t}(X)$ is nontrivial since $f \in G$ but $f \notin \mathrm{t}(X)$. However, the only nontrivial divisible subgroup of Q is Q itself (Result 5.1.4), so $G/\mathrm{t}(X) \cong Q$. This and Result 5.12.1 imply $G/\mathrm{t}(G) \cong Q$.

RESULT 5.12.3

(a) *G is residually finite.*

(b) *G is reduced.*

By Result 5.11.2 X is residually finite. Hence G, being a subgroup of X, is also. This proves Part (a). Part (b) follows from Part (a) and Theorem 1.3.11.

RESULT 5.12.4

(a) *G is not an* M_2 *group.*

(b) *G is nonsplitting.*

By Results 5.12.2 and 5.1.2, $G/t(G)$ is divisible. But G is reduced by Result 5.12.3. Hence $G/t(G)$ cannot be embedded in G proving Part (a). Part (b) follows immediately.

RESULT 5.12.5 *G is not an* M_1 *group.*

By Results 5.12.2 and 5.1.2, $G/t(G)$ is divisible. Since $t(G) = \Sigma\{Z(p)\}_{p \in PR}$ by Result 5.12.1, $t(G)$ is coatomic (Result 5.7.6) and hopfian (Result 5.7.7). Finally G is nonsplitting by Result 5.12.4. Conclude $t(G)$ is not a homomorphic image of G be the Lemma to Result 5.11.5.

LEMMA *If G is a (not necessarily abelian) group, H is a fully invariant subgroup of G, and H and G/H are hopfian, then G is hopfian.*

Let $s{:}G \to G$ be a surjective homomorphism and seek to show s is injective as well. By full invariance of H, $s[H] \subseteq H$. Thus by Theorem A.10.1, $s'{:}G/H \to G/H$ is also a surjective homomorphism. Then s' must also be injective because G/H is hopfian. Next let $r{:}H \to H$ be the restriction of s to H. It is claimed that r is surjective. If $c \in H$, then $c = s(y)$ for some y by surjectivity of s. Thus $H = Hs(y) = s'(Hy)$. But then $Hy = H$ by injectivity of s' so $y \in H$ and hence $r(y) = c$ proving the claim. By this claim and the fact that H is hopfian conclude r is injective. Finally it can be shown that s is injective. Suppose $s(x) = 1$. Then $s'(Hx) = H$ so $Hx = H$ by injectivity of s'. Thus $x \in H$ hence $1 = s(x) = r(x)$. Then $x = 1$ by injectivity of r, proving injectivity of s.

RESULT 5.12.6 *G is hopfian.*

By Results 5.12.1 and 5.77, $t(G)$ is hopfian. Also $G/t(G)$ is hopfian by Results 5.12.2 and 5.1.12. Finally $t(G)$ is a fully invariant subgroup (Theorem A.4.5). Therefore G is hopfian by the lemma.

LEMMA *If G is a (not necessarily abelian) group, H is a fully invariant subgroup of G, and H and G/H are cohopfian, then G is cohopfian.*

Let $s{:}G \to G$ be an injective homomorphism and seek to show s is surjective as well. Let $x \in G$ and seek an element u such that $s(u) = x$. By full invariance of H, $s[H] \subseteq H$. Let $r{:}H \to H$ be the restriction of s to H. Then r is an injective homomorphism. But H is cohopfian so r must also be surjective. Now $s[H] \subseteq H$ also implies $s'{:}G/H \to G/H$ is a homomorphism by Theorem A.10.1. It is claimed that s' is injective. Suppose $s'(Hw) = H$. Then $Hs(w) = H$ so $s(w) \in H$. Then by surjectivity of r, there exists $h \in H$ such that $r(h) = s(w)$. Hence $s(h) = s(w)$ so $h = w$ by injectivity of s. Thus $w \in H$ so $Hw = H$ proving s' is injective as claimed. Next note G/H is cohopfian so s' must also be surjective. Thus there exists $Ha \in G/H$ such that $s'(Ha) = Hx$. Hence $Hs(a) = Hx$ and so $xs(a)^{-1} \in H$. Now by surjectivity of r there exists $y \in H$ such that $r(y) = xs(a)^{-1}$. Thus $s(y) = xs(a)^{-1}$, hence $x = s(y)s(a) = s(ya)$.

RESULT 5.12.7 *G is cohopfian.*
By Results 5.12.1 and 5.7.8, $\mathrm{t}(G)$ is cohopfian. Also $G/\mathrm{t}(G)$ is cohopfian by Results 5.12.2 and 5.1.13. Finally $\mathrm{t}(G)$ is a fully invariant subgroup (Theorem A.4.5). Therefore G is cohopfian by the lemma.

EXERCISES

1. Show G is not an M_3 group.
2. Show G is a pure subgroup of X.
3. For all primes p show $G = \bar{D}_X(\{p\}) \oplus pG$. Conclude G is decomposable.
4. If H is any subgroup of G, show that either H or G/H is torsion.
5. Show G is countable.

NOTES

1. Some justification must be given for inclusion of this example since its properties are almost identical with those of Example 5.11. First of all it was included to illustrate the method of construction which is a most useful one for obtaining mixed groups. Secondly it will be used to help construct another group, Example 5.14.
2. In Note 1 to Example 5.6 we defined what it means for a torsion-free abelian group to have finite rank. A mixed group H is said to have **torsion-free rank** n if and only if $H/\mathrm{t}(H)$ has rank n. Thus G has torsion-free rank 1.
3. See Miscellaneous Exercises III.4, III.5, IV.5, and IV.6 at the end of this chapter, and Miscellaneous Exercise V.2 at the end of Chapter 6.

□□□□□

Example 5.13

Let $X = \mathsf{X}\{Z(p)\}_{p \in PR}$ (the group of Example 5.11). Define $f \in X$ by $f(p) = 1$ for all p. Let $G/\mathrm{t}(X) = \mathrm{S}(X/\mathrm{t}(X), \mathrm{t}(X) + f)$.

RESULT 5.13.1 $\mathrm{t}(G) = \mathrm{t}(X) = \Sigma\{Z(p)\}_{p \in PR}$.

Since $\mathrm{t}(X) \subseteq G$, it follows that $\mathrm{t}(G) = \mathrm{t}(X)$ from Theorem A.4.3. The second equality follows from Result 5.11.1.

RESULT 5.13.2 $G/\mathrm{t}(G) \cong SF$.

Let $P/\mathrm{t}(X) = \mathrm{P}(X/\mathrm{t}(X), \mathrm{t}(X)+f)$ (thus P is the group of Example 5.12). Then by Result 5.10.16, $G/\mathrm{t}(X) \cong SF$ because $P/\mathrm{t}(X)$ is divisible (Result 5.12.2). Finally, $\mathrm{t}(G) = \mathrm{t}(X)$ by Result 5.13.1, so $G/\mathrm{t}(G) \cong SF$.

RESULT 5.13.3

(a) *G is residually finite.*

(b) *G is reduced.*

Since G is a subgroup of X and X is residually finite and reduced (Result 5.11.2).

RESULT 5.13.4

(a) *G is not an* M_2 *group.*

(b) *G is nonsplitting.*

Because of Result 5.13.2, it suffices to show SF cannot be embedded in G. It will, in fact, be shown that SF cannot be embedded in X. Suppose to the contrary that $d{:}SF \to X$ is an injective homomorphism. Then for all primes p, both 1 and $1/p$ belong to SF and $1 = p(1/p)$ so $d(1) = pd(1/p)$. Denote $d(1)$ by h and $d(1/p)$ by g_p. Thus $h = pg_p$ for all primes p. But $g_p(p)$ belongs to $Z(p)$ and so $pg_p(p) = 0$. Thus $h(p) = 0$ for all primes p. Conclude $h = 0$ and hence $d(1) = 0$. Applying injectivity yields the desired contradiction $(1 = 0)$ which proves Part (a). Part (b) follows immediately.

RESULT 5.13.5 *G is an* M_1 *group.*

By Results 5.13.1 and 5.10.11, there exists an isomorphism $h{:}SF/Z \to \mathrm{t}(G)$. By Result 5.13.2, there exists an isomorphism $g{:}G/\mathrm{t}(G) \to SF$. Let e and e^* be the natural maps $e{:}G \to G/\mathrm{t}(G)$ and $e^*{:}SF \to SF/Z$. Finally take the composition of these four surjective homomorphisms $k = h \circ e^* \circ g \circ e$. Then $k{:}G \to \mathrm{t}(G)$ is a surjective homomorphism as required.

RESULT 5.13.6 *G is not an* M_3 *group.*

Suppose $G = B \oplus C$ where C is nontrivial and torsion-free, and seek a contradiction. For convenience denote $\mathrm{t}(G)$ by T. Since (Theorem A.4.5) T is a fully invariant subgroup, $G/T = [(B + T)/T] \oplus [(C + T)/T]$ by Theorem A.2.8. However, G/T is indecomposable by Results 5.13.2 and 5.10.3. Thus either $(B + T)/T$ or $(C + T)/T$ is trivial. But the second case implies $C \subseteq T$ so C is torsion, which contradicts the assumptions made on C. Thus $(B + T)/T$ is trivial and hence $B \subseteq T$. Therefore $G = B + C = T + C$. Also $T \cap C$ is trivial because C is torsion-free. Conclude $G = T \oplus C$, the desired contradiction (to Result 5.13.4).

RESULT 5.13.7 *G is coatomic.*

By Results 5.13.1 and 5.7.6, $\mathrm{t}(G)$ is coatomic. By Results 5.13.2 and 5.10.15,

$G/t(G)$ is coatomic. Conclude G is coatomic by the Lemma to Result 5.10.15.

EXERCISES

1. Show G is hopfian.
2. Show G is decomposable.
3. If H is any subgroup of G, show that either H or G/H is torsion.

NOTES

1. The group G is of interest as a nonsplitting group which belongs to class M_1 but not to classes M_2 or M_3.
2. The torsion-free rank of G is 1 (see Note 2 to Example 5.12).
3. See Miscellaneous Exercises III.4, III.5, IV.5, and IV.6 at the end of this chapter.

□□□□□

Example 5.14

Let G be the group of Example 5.12. Let $A = G \times [\Sigma\{Q_i\}_{i\in N}]$.

RESULT 5.14.1 $t(A) \cong t(G) = \Sigma\{Z(p)\}_{p\in PR}$.
By Theorem A.4.6, $t(A) = t(G) \times t(\Sigma\{Q_i\}_{i\in N})$. But $\Sigma\{Q_i\}_{i\in N}$ is torsion-free by Result 5.1.1 and Theorem 1.4.7 so $t(\Sigma\{Q_i\}_{i\in N})$ is trivial. Thus $t(A) \cong t(G)$ proving the isomorphism. The equality follows from Result 5.11.1.

RESULT 5.14.2 $A/t(A) \cong \Sigma\{Q_i\}_{i\in N}$.
Let F denote $\Sigma\{Q_i\}_{i\in N}$. By Theorem A.4.7, $A/t(A) \cong (G/t(G)) \times (F/t(F))$. But $G/t(G) \cong Q$ by Result 5.12.2 and $t(F)$ is trivial since F is torsion-free. Thus $A/t(A) \cong Q \times F \cong F$ by Theorem 1.4.12.

LEMMA *Direct factors of splitting abelian groups are splitting.*
Suppose $A = B \oplus C = t(A) \oplus K$ and seek to show B is splitting. It will suffice to show $B = t(B) \oplus [(t(C) + K) \cap B]$.

To show $B = t(B) + [(t(C) + K) \cap B]$ let $b \in B$. Then $A = t(A) \oplus K$ implies $b = x + k$ where $x \in t(A)$ and $k \in K$. Then $A = B \oplus C$ implies $x = b^* + c$ where $b^* \in B$ and $c \in C$. Let x have order n. Then $0 = nx = n(b^* + c) = nb^* + nc$ so $nb^* = -nc \in B \cap C$ and hence $nb^* = 0$ so $b^* \in t(B)$. Also $nc = -nb^* = -0 = 0$ so $c \in t(C)$. Now $b - b^* = x + k - b^* = c + k \in (t(C) + K) \cap B$. Combining this with $b^* \in t(B)$ yields $b \in t(B) + [(t(C) + K) \cap B]$ as required.

To show t(B) and [(t(C) + K) $\cap$ B] intersect trivially let x be an element of their intersection. Then $x = c + k$ where $c \in$ t(C) and $k \in K$. Also $x \in$ t(B) so x has finite order. Let m = o(c) and n = o(x). Therefore $mnx = mn(c + k) = mnc + mnk = 0 + mnk$. Also $nx = 0$ implies $mnx = 0$ and hence $mnk = 0$. Thus $k \in$ t(A) $\cap$ K so $k = 0$. Hence $x = c \in B \cap C$ so $x = 0$.

RESULT 5.14.3 *A is nonsplitting.*
By Theorem A.2.4, G is isomorphic to a direct factor of A. But G is nonsplitting by Result 5.12.4. Hence A is nonsplitting by the lemma.

RESULT 5.14.4 *A is not an* M_1 *group.*
By Result 5.14.2, A/t(A) $\cong \Sigma\{Q_i\}_{i \in N}$ and so A/t(A) is divisible (Result 5.1.2 and Theorem 1.4.5). Also t(A) $\cong \Sigma\{Z(p)\}_{p \in PR}$ (Result 5.14.1) and so t(A) is coatomic (Result 5.7.6) and hopfian (Result 5.7.7). Finally A is nonsplitting by Result 5.14.3, so conclude t(A) is not a homomorphic image of A by the Lemma to Result 5.11.5.

RESULT 5.14.5 *A is an* M_2 *group.*
By Result 5.14.2.

RESULT 5.14.6 *A is an* M_3 *group.*
By Theorem A.2.4, $\Sigma\{Q_i\}_{i \in N}$ is isomorphic to a direct factor of A. This direct factor then is clearly nontrivial, and is torsion-free by Result 5.1.1 and Theorem 1.4.7.

EXERCISES

1. Show A is not residually finite.
2. Show for all primes p that $A/pA \cong Z(p)$.
3. Show A is nonhopfian.
4. Show A is noncohopfian.

NOTES

1. The group A is of interest as a nonsplitting group which belongs to classes M_2 and M_3 but not to class M_1.
2. See Miscellaneous Exercises III.4, III.5, IV.5, and IV.6 at the end of this chapter and Miscellaneous Exercise V.2 at the end of Chapter 6.

□□□□□

Example 5.15

Let G be the group of Example 5.13. Let $A = (\Sigma\{G_i\}_{i\in N}) \times (\Sigma\{SF_i\}_{i\in N})$.

RESULT 5.15.1 $t(A) \cong \Sigma\{t(G_i)\}_{i\in N}$.
By Theorem A.4.6, $t(A) = t(\Sigma\{G_i\}_{i\in N}) \times t(\Sigma\{SF_i\}_{i\in N})$. But $t(\Sigma\{SF_i\}_{i\in N})$ is trivial since $\Sigma\{SF_i\}_{i\in N}$ is torsion-free (Result 5.10.2 and Theorem 1.4.7). Thus $t(A) \cong t(\Sigma\{G_i\}_{i\in N})$. Application of Theorem 1.4.14 to the right-hand side of this last isomorphism yields the result.

RESULT 5.15.2 $A/t(A) \cong \Sigma\{SF_i\}_{i\in N}$.
Let B denote $\Sigma\{G_i\}_{i\in N}$ and let C denote $\Sigma\{SF_i\}_{i\in N}$. By Theorem A.4.7, $A/t(A) \cong (B/t(B)) \times (C/t(C))$. But $t(B) = \Sigma\{t(G_i)\}_{i\in N}$ by Theorem 1.4.14 and $t(C)$ is trivial since C is torsion-free. So $A/t(A) \cong [\Sigma\{G_i\}_{i\in N}/\Sigma\{t(G_i)\}_{i\in N}] \times C \cong [\Sigma\{G_i/t(G_i)\}_{i\in N}] \times C$ by Theorem 1.4.15. But $G/t(G) \cong SF$ by Result 5.13.2 so $A/t(A) \cong \Sigma\{SF_i\}_{i\in N} \times C = C \times C \cong C$ by Result 1.4.12.

RESULT 5.15.3 *A is nonsplitting.*
By the Corollary to Theorem 1.4.9, G is isomorphic to a direct factor of the group $\Sigma\{G_i\}_{i\in N}$ which, in turn, is isomorphic to a direct factor of A by Theorem A.2.4. Thus, since G is nonsplitting (Result 5.11.4), two applications of the Lemma to Result 5.14.3 show A is nonsplitting.

RESULT 5.15.4 *A is an* M_1 *group.*
By Result 5.13.5, there exists a surjective homomorphism $g_i{:}G_i \to t(G_i)$ for all natural numbers i. Applying Theorem 1.4.10 yields a surjection $g{:}\Sigma\{G_i\}_{i\in N} \to \Sigma\{t(G_i)\}_{i\in N}$. But this last group is isomorphic to $t(A)$ by Result 5.15.1 hence there is a surjection $h{:}\Sigma\{G_i\}_{i\in N} \to t(A)$. By Theorem A.2.4, there exist J, K, and e such that $A = J \oplus K$ and $e{:}J \to \Sigma\{G_i\}_{i\in N}$ is an isomorphism. Finally let $\pi{:}A \to J$ be the projection map (see Appendix A.2). Then π is surjective, and hence $h\circ e\circ\pi{:}A \to t(A)$ is surjective.

RESULT 5.15.5 *A is an* M_2 *group.*
By Result 5.15.2.

RESULT 5.15.6 *A is an* M_3 *group.*
The group $\Sigma\{SF_i\}_{i\in N}$ is torsion-free (Result 5.10.2 and Theorem 1.4.7) and is isomorphic to a direct factor of A by Theorem A.2.4.

EXERCISES

1. Show A is nonhopfian.
2. Show A is noncohopfian.
3. Show A is both reduced and residually finite.
4. Show A has a quotient group isomorphic to Q.

NOTES

1. The group A is of interest as a nonsplitting group which belongs to classes M_1, M_2, and M_3.
2. See Miscellaneous Exercises III.4, III.5, IV.5, and IV.6 at the end of this chapter and Miscellaneous Exercise V.2 at the end of Chapter 6.

□□□□□

MISCELLANEOUS EXERCISES

I Torsion Groups

I.1. Show $Q/Z \cong \Sigma\{Z(p^\infty)\}_{p\in PR}$.

I.2. Give examples of torsion abelian groups A, B, and C such that $A \times B \cong A \times C$ but $B \ncong C$.

I.3. Let $A = \Sigma\{P_i\}_{i\in N}$ where $P_i = Z(p^\infty)$ for all $i \in N$. Let $B = \Sigma\{S_i\}_{i\in N}$ where $S_0 = Z(p)$ and $S_i = Z(p^\infty)$ for $1 \leqslant i$. Show A can be embedded in B, and B can be embedded in A, but $A \ncong B$.

I.4. Give examples of torsion abelian groups A and B such that A is isomorphic to a quotient group of B, and B is isomorphic to a quotient group of A, but $A \ncong B$.

I.5. Show the group $Z(p^\infty) \times Z(p)$ has a unique maximal subgroup.

I.6. Let $B = \Sigma\{Z(p)_i\}_{i\in N}$. Show
(a) B is bounded.
(b) B is coatomic.

I.7. (a) If A is an abelian group and $B \subseteq \mathrm{U}(A)$, show $\mathrm{U}(A/B) = \mathrm{U}(A)/B$.
(b) Use Part (a) and Example 5.9 to show existence of an abelian p-group C such that $\mathrm{U}(C) \cong Z(p)$.

II Torsion-free Groups

II.1. Find $\mathcal{L}$ and A such that A is a subgroup of Z, $\mathcal{L}$ is a collection of subgroups of Z, and $A + (\cap\mathcal{L}) \neq \cap\{A + C \mid C \in \mathcal{L}\}$.

II.2. Let $A = \times\{Q_p\}_{p\in PR}$. Show
(a) $A/p^kA \cong Z(p^k)$ for all primes p, and all natural numbers k.
(b) A is hopfian.
(c) If $Z_i = Z$ for all $i \in N$, show $\times\{Z_i\}_{i\in N}$ can be embedded in A.
Conclude that the class of hopfian groups is not subgroup closed.

II.3. Show that if A is a torsion-free nondivisible abelian group, then A is non-cohopfian. Conclude that Examples 5.2, 5.3, 5.5, 5.6, and 5.10 are noncohopfian.

II.4. Give examples of torsion-free abelian groups A, B, and C such that $A \times B \cong A \times C$ but $B \ncong C$.

II.5. Give examples of torsion-free abelian groups A and B such that A can be embedded in B, and B can be embedded in A, but $A \ncong B$.

II.6. Give examples of torsion-free abelian groups A and B such that A is isomorphic to a quotient group of B, and B is isomorphic to a quotient group of A, but $A \ncong B$.

II.7. Show $Q \times Q$ is hopfian.

II.8. Show $Q \times Q$ is cohopfian.

II.9. Show $Q \times Q$ is characteristically simple.

II.10. Let p and q be distinct primes. Let $Q_{p,q} = \{a/b \mid a/b \in Q$ and b is relatively prime to both p and $q\}$. Show $Q_{p,q}$ is a subgroup of Q and investigate its properties.

II.11. For all natural numbers $n \geqslant 2$, let $G_n = \{a/b \mid a/b \in Q$ and if p is any prime, then p^n does not divide $b\}$. (Note $G_2 = SF$.) Show G_n is a subgroup of Q, and investigate its properties.

III Mixed Groups

III.1. Let G be the group of Example 5.11. For a prime p, let $A = pG$. Show A is p-divisible, and, in addition, shares the following properties with G:
(a) A is reduced.
(b) A is hopfian.
(c) A is not an M_1 group.
(d) A is not an M_2 group.
(e) A is not an M_3 group.

III.2. Let $A = (\mathsf{X}\{Z(p)\}_{p \in PR}) \times SF$. Show:
(a) $t(A) \cong \Sigma\{Z(p)\}_{p \in PR}$.
(b) A is an M_1 group.
(c) A is reduced.
(d) A is an M_3 group.
(e) A is not an M_2 group.

III.3. Give an example of an M_3 group that is neither an M_1 nor an M_2 group.

III.4. If G is a mixed abelian group, show the subgroup lattice $L(G)$ is not distributive. Conclude that Examples 5.11–5.15 have nondistributive subgroup lattices. Further conclude that these groups are not locally cyclic.

III.5. Give an example of a nonsplitting p-divisible group that is both hopfian and cohopfian.

III.6. Give an example of a reduced nonsplitting group with nontrivial Ulm subgroup.

IV General

IV.1. (a) Show that an arbitrary lattice L is distributive if and only if for all $x, y, z \in L$, $x \wedge (y \vee z) = (x \wedge y) \vee (x \wedge z)$.
(b) Conclude that if G is any of the groups of Examples 5.1–5.3, 5.7,

or 5.10, then $A \cap (B+C) = (A \cap B) + (A \cap C)$ for all subgroups A, B, and C of G.

IV.2. Give an example to show that the class of pure-simple abelian groups is not quotient closed.

IV.3. Give an example to show that the class of indecomposable abelian groups is not quotient closed.

IV.4. Let H be an abelian group all of whose subgroups are characteristic. Let G be a (not necessarily abelian) extension of H by a hypercyclic group K. Show G is hypercyclic. Conclude extensions of Examples 5.4, 5.7, and 5.10 by hypercyclic groups are hypercyclic.

IV.5. Show all (not necessarily abelian) groups satisfying the descending chain condition are torsion. Conclude Examples 5.1–5.3, 5.5, 5.6, and 5.10–5.15 do not satisfy the descending chain condition.

IV.6. Show that

(a) If A is a (not necessarily abelian) group that satisfies the ascending chain condition, then every subgroup of A, including A itself, is finitely generated.

(b) If A is a finitely generated abelian group, then A satisfies the ascending chain condition.

(c) If A is a finitely generated abelian group, then $t(A)$ is finite.

(d) None of the examples in this chapter are finitely generated, and none satisfy the ascending chain condition.

IV.7. Give an example of an abelian group A and two classes of groups $\mathcal{G}$ and $\mathcal{H}$ such that A is residually $\mathcal{G}$ and A is residually $\mathcal{H}$ but A is not residually $\mathcal{G} \cap \mathcal{H}$.

SUGGESTIONS FOR FURTHER READING

Many examples of torsion-free abelian groups are given in vol II of Fuchs [11]. Most of these, however, are rather difficult to construct. Among the more notable examples are:

(a) Groups A and B such that $A \times A \cong B \times B$ but $A \not\cong B$ (Theorem 90.3).

(b) A group with no nontrivial indecomposable direct factors (Theorem 91.5).

(c) A group A such that $A \cong A \times A \times A$ but $A \not\cong A \times A$ (Theorem 91.6).

(d) Groups whose automorphism groups are respectively $Z(4)$, $Z(6)$, the quaternion group, Example 4.5, and $SL(2,3)$ (pages 271–273).

6

Infinite Nonabelian Groups

In the opening remarks to Chapter 4 it was suggested that it is reasonable to classify finite groups in stages depending on how far a group falls from the concept of commutativity. The stages considered in Chapter 4 were

(1) abelian
(2) nilpotent
(3) hypercyclic
(4) solvable
(5) arbitrary

A classification procedure such as this makes equally good sense for infinite groups, and indeed the arrangement of groups into such a classification scheme is the primary theme of this chapter.

Infinite group theory, however, offers somewhat more of a selection than classes 1–5 above. The classification scheme used in this chapter is as follows:

(1) abelian
(2) nilpotent
(3) generalized nilpotent
  (3a) hypercentral
  (3b) locally nilpotent
  (3c) MN
  (3d) $\{G \mid$ elements of relatively prime order in G commute$\}$
  (3e) residually nilpotent
(4) hypercyclic
(5) solvable

(6) generalized solvable
- (6a) locally solvable
- (6b) residually solvable
- (6c) hyperabelian

(7) arbitrary

The word *generalized* as used in 3 and 6 above has a special meaning. A class of groups $\mathcal{H}$ is a **generalized nilpotent class** if and only if (i) all nilpotent groups belong to $\mathcal{H}$, and (ii) if G is a finite group and $G \in \mathcal{H}$, then G is nilpotent. The phrase *generalized solvable* is defined similarly.

Relations among the classes 1, 2, 4, 5, and 7 are shown below.

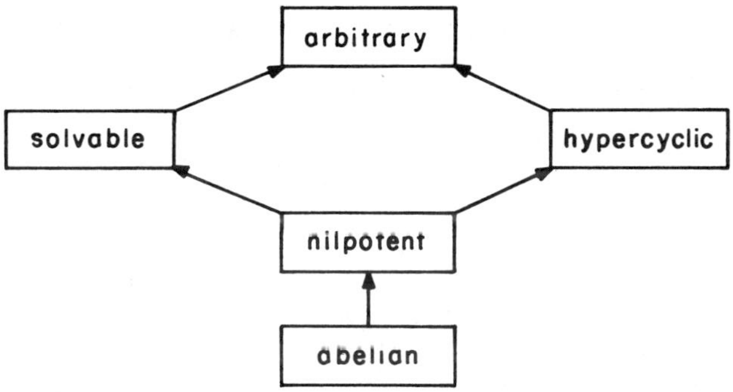

An arrow here from one class to another means all groups in the first class also belong to the second. Relations among the generalized nilpotent and generalized solvable classes to be considered in this chapter are shown in the following two diagrams.

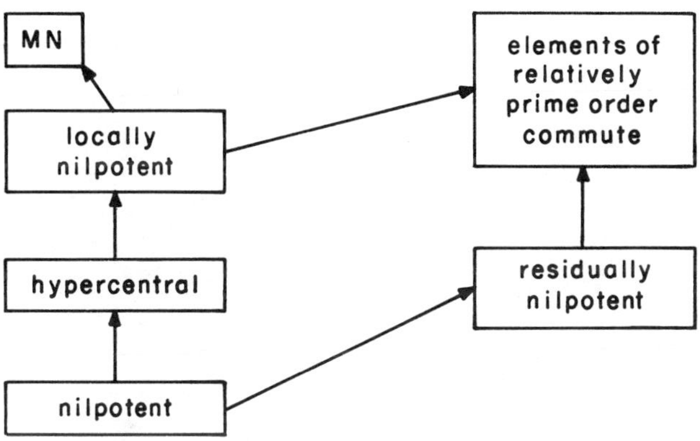

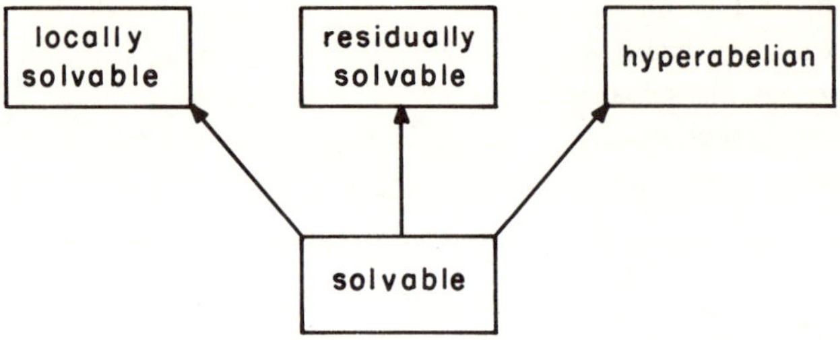

It should also be noted that hypercentral groups are hypercyclic (Theorem A.7.2).

With regards to nilpotent groups, one area of investigation will be the question: If G is nilpotent and $Z(G)$ belongs to some class $\mathcal{H}$, does $G \in \mathcal{H}$? Two positive results of this nature are proved in the appendixes namely for the classes of indecomposable and subdirectly irreducible groups (Theorems A.5.9 and A.5.10) and the property can also be shown without difficulty for the class of bounded groups. Negative results for several classes are presented in this chapter.

With regards to hypercyclic and solvable groups the major area of concern will be the determination of what generalized nilpotent properties the groups have, and with regards to arbitrary groups, the determination of generalized solvable properties will be of primary concern.

The chapter closes with a rather spectacular example – an infinite simple group that contains an isomorphic copy of every finite group. Although only two examples of infinite simple groups are given, it is hoped that the differences between them (the aforementioned group is torsion and divisible, the other is neither) are sufficient to demonstrate the diversity of infinite simple groups.

As in Chapter 5, a second major area of investigation in this chapter will be the various finiteness conditions. The most important of these is the class of finitely generated groups, four examples of which will be given. In addition to finitely generated groups and the finiteness conditions examined in Chapter 5, the FC property and the classes of groups satisfying the ascending or descending chain conditions on normal subgroups will be of interest here.

Example 6.1

Let $G = UT(3,Q)$.

RESULT 6.1.1 *G is nilpotent.*
By Theorem 3.2.3.

RESULT 6.1.2 $Z(G) \cong Q$.
By Theorem 3.2.6.

RESULT 6.1.3 *G is indecomposable.*
By Result 6.1.1, G is nilpotent. By Results 6.1.2 and 5.1.3, $Z(G)$ is indecomposable. Hence G is indecomposable by Theorem A.5.9.

RESULT 6.1.4 $G/Z(G) \cong Q \times Q$.
By Theorem 3.2.7.

RESULT 6.1.5 *G is torsion-free.*
First note $Q \times Q$ is torsion-free because $n(x,y) = (0,0)$ implies $nx = 0$ and $ny = 0$ which in turn imply $x = 0$ and $y = 0$ by Result 5.1.1. Thus $G/Z(G)$ is torsion-free by Result 6.1.4. Results 6.1.2 and 5.1.1 show $Z(G)$ is torsion-free, so G is torsion-free by the Lemma to Theorem 1.2.19.

LEMMA *Central extensions of divisible groups by divisible groups are divisible.*
Suppose A/B and B are divisible where $B \subseteq Z(A)$. Let $x \in A$ and let n be a positive integer. Then $Bx \in A/B$ so there exists By such that $Bx = (By)^n$. Hence $xy^{-n} \in B$ so by divisibility of B, $xy^{-n} = b^n$ for some $b \in B$ and so $x = b^n y^n$. But $b \in Z(A)$ implies $b^n y^n = (by)^n$. Conclude $x = (by)^n$ completing the proof that A is divisible.

RESULT 6.1.6 *G is divisible.*
First note $Q \times Q$ is divisible because if $(x,y) \in Q \times Q$ and n is a positive integer, then by Result 5.1.2 there exist a and b such that $na = x$ and $nb = y$. Hence $n(a,b) = (x,y)$ so $Q \times Q$ is divisible. Thus $G/Z(G)$ is divisible by Result 6.1.4. Results 6.1.2 and 5.1.2 show $Z(G)$ is divisible, so G is divisible by the lemma.

RESULT 6.1.7 *G has no maximal subgroups.*
Suppose to the contrary that M is a maximal subgroup of G. Then $M \triangleleft G$ by Theorem A.10.12. Hence G/M has no proper nontrivial subgroups so must be isomorphic to $Z(p)$ for some prime p. But G is divisible (Result 6.1.6) hence has no nontrivial finite quotient groups by Theorem A.8.2.

RESULT 6.1.8
(a) *G is solvable.*
(b) *G is hypercyclic.*
Part (a) by nilpotency (Result 6.1.1) and Theorem A.6.7. Part (b) by nilpotency (Result 6.1.1) and the Corollary to Theorem A.7.2.

EXERCISES

1. Show G is not finitely generated.

2. Show G
   (a) does not satisfy the ascending chain condition on normal subgroups.
   (b) does not satisfy the descending chain condition on normal subgroups.
3. Show G is not residually finite.
4. Show G is hopfian.
5. Show G is cohopfian.
6. Let n be a positive integer. Define $f_n:G \to G$ by

$$f_n(A) = \begin{pmatrix} 1 & 0 & 0 \\ A_{21}/n & 1 & 0 \\ A_{31}/n & A_{32} & 1 \end{pmatrix}.$$

   Show $f_n \in \mathrm{Aut}(G)$.

NOTE

1. See Miscellaneous Exercises I.2, and V.1–V.3 at the end of this chapter.

□□□□□

Example 6.2

Let p be a prime. Let $M = M(Z(p^\infty))$.

RESULT 6.2.1 *M is nilpotent.*
By Theorem 1.2.17.

RESULT 6.2.2 $Z(M) \cong Z(p^\infty)$.
By Theorem 1.2.16.

RESULT 6.2.3
(a) *M is subdirectly irreducible.*
(b) *M is indecomposable.*
Part (a) by nilpotency (Result 6.2.1), Results 6.2.2 and 5.4.5, and Theorem A.5.10. Part (b) follows from Part (a) and Theorem A.10.10.

RESULT 6.2.4
(a) *M is solvable.*
(b) *M is hypercyclic.*
Part (a) by nilpotency (Result 6.2.1) and Theorem A.6.7. Part (b) by nilpotency and the Corollary to Theorem A.7.2.

RESULT 6.2.5 *M is not residually finite.*
By Result 6.2.3, Theorem 1.3.10, and the fact that M is infinite.

RESULT 6.2.6 *M is neither torsion nor torsion-free.*
By Result 1.2.18, $M/Z(M) \cong Z(p^\infty) \times Z$ so is nontorsion, hence the same is true for M. On the other hand M cannot be torsion-free since it has a subgroup isomorphic to $Z(p^\infty)$.

RESULT 6.2.7 *M is nonhopfian.*
Define $e{:}Z(p^\infty) \to Z(p^\infty)$ by $e(x) = px$. By Theorem 1.2.20 then, there is a homomorphism $e^*{:}M \to M$ defined by $e^*(x,y,i) = (e(x),e(y),i)$. Clearly e^* is onto M because e is onto $Z(p^\infty)$ ($Z(p^\infty)$ is divisible by Result 5.4.2). Thus $M/\ker e^* \cong M$. Since $(Z + (1/p), Z + (1/p), 0) \in \ker e^*$, $\ker e^*$ is nontrivial showing M is nonhopfian.

EXERCISES

1. Show
   (a) M does not satisfy the ascending chain condition on normal subgroups.
   (b) M does not satisfy the descending chain condition on normal subgroups.
2. Show all subgroups of M of finite index are normal.
3. Show M is not finitely generated.
4. Show M is nondivisible.
5. Define $h{:}Z(p^\infty) \to Z(p^\infty)$ by $h(x) = (1+p)x$. First show h is an automorphism. Define $e{:}M \to M$ by $e(a,b,i) = (a, h^{-1}(b), (1+p)i)$. Show
   (a) e is a homomorphism.
   (b) e is injective but not surjective. Thus M is noncohopfian.

NOTES

1. The group M shows that nilpotent groups with torsion, divisible, or cohopfian centers need not be torsion, divisible, or cohopfian.
2. Secondly, M shows subdirectly irreducible nilpotent groups need not be torsion. This is in contrast to the abelian case since an abelian subdirectly irreducible group can be embedded in $Z(p^\infty)$ for some prime p (Fuchs [11], vol I, pages 15 and 16), hence must be torsion.
3. Finally M shows that there are indecomposable nilpotent groups that are neither torsion nor torsion-free. This too is in contrast to the abelian case, see Rotman [31], Corollary 9.36.
4. See Miscellaneous Exercises V.1 and V.2 at the end of this chapter.

□□□□□

Example 6.3

Let $M = M(SF)$.

RESULT 6.3.1 *M is nilpotent.*
By Theorem 1.2.17.

RESULT 6.3.2 $Z(M) \cong SF$.
By Theorem 1.2.16.

RESULT 6.3.3 *M is indecomposable.*
By Results 6.3.2 and 5.10.3, $Z(M)$ is indecomposable. This and nilpotency (Result 6.3.1) imply M is indecomposable by Theorem A.5.9.

RESULT 6.3.4 *M is coatomic.*
By Results 6.3.2 and 5.10.15, $Z(M)$ is coatomic. Also $M/Z(M) \cong SF \times Z$ (Theorem 1.2.18). Now Z, being finitely generated, is coatomic by Theorem A.10.11, and $SF \times Z$ is a central extension of SF by Z so is coatomic by the Lemma to Result 5.10.15. Consequently $Z(M)$ and $M/Z(M)$ are both coatomic. This and a second application of the Lemma to Result 5.10.15 prove the result.

RESULT 6.3.5 *M is torsion-free.*
By Theorem 1.2.19 and Result 5.10.2.

RESULT 6.3.6
(a) *M is solvable.*
(b) *M is hypercyclic.*
Part (a) by nilpotency (Result 6.3.1) and Theorem A.6.7. Part (b) by nilpotency and the Corollary to Theorem A.7.2.

RESULT 6.3.7 *M is not an* FC *group.*
By Theorem 1.2.15, $(a,0,1)$ is conjugate to $(1,0,1)$ for all $a \in SF$. Thus $(1,0,1)$ has an infinite number of conjugates.

EXERCISES

1. Show M is residually finite.
2. Show M is not finitely generated.
3. Show
   (a) M does not satisfy the ascending chain condition on normal subgroups.
   (b) M does not satisfy the descending chain condition on normal subgroups.
4. Show M is noncohopfian.
5. Show M is hopfian.

NOTE

1. See Miscellaneous Exercises I.2 and V.3 at the end of this chapter.

□□□□□

Example 6.4

Let L be the central pushout of the countable central pushout system $(Z(2), \{H_i\}_{i \in N}, \{f_i\}_{i \in N})$ where $H_i = D_4$ for all $i \in N$ and $f_i{:}Z(2) \to Z(H_i)$ is defined by $f_i(0) = 1$ and $f_i(1) = x^2$ for all $i \in N$.

RESULT 6.4.1 *L is infinite.*
By Theorem 1.6.1, each G_n can be embedded in L where, recall, G_n is the central pushout of $(Z(2), G_{n-1}, H_n, k_{n-1}, f_n)$. Note that each $f_i{:}Z(2) \to Z(H_i)$ is an isomorphism ($\{1, x^2\}$ is all of $Z(D_4)$ by Result 4.1.3). Hence each $k_i{:}Z(2) \to Z(G_i)$ is an isomorphism by Theorem 1.6.5. Thus Theorem 1.1.10 may be applied, so $\text{card}(G_n) = \text{card}(G_{n-1}) \cdot \text{card}(H_n)/\text{card}(Z(2)) = \text{card}(G_{n-1}) \cdot 8/2 = 2^2 \cdot \text{card}(G_{n-1})$. Now $G_0 = H_0$ so $\text{card}(G_0) = 8$. Thus $\text{card}(G_0) = 2^3$, $\text{card}(G_1) = 2^5$, $\text{card}(G_2) = 2^7$, etcetera. Since each G_n can be embedded in L, this shows L is infinite.

RESULT 6.4.2 $\exp(L) = 4$.
It is easily seen that $\exp(D_4) = 4$. The result then follows from Theorem 1.6.3.

RESULT 6.4.3 *L is a class-2 nilpotent group.*
Each H_i is class-2 nilpotent by Result 4.1.5. Note that each $f_i{:}Z(2) \to Z(H_i)$ is an isomorphism because $\{1, x^2\} = Z(D_4)$ by Result 4.1.3. Conclude L is class-2 nilpotent from Theorem 1.6.9.

RESULT 6.4.4 $Z(L) \cong Z(2)$.
By Theorem 1.6.8, since each $f_i{:}Z(2) \to Z(H_i)$ is an isomorphism.

RESULT 6.4.5
(a) *L is subdirectly irreducible.*
(b) *L is indecomposable.*
Part (a) by Results 6.4.3, 6.4.4, and Theorem A.5.10. Part (b) follows from Part (a) and Theorem A.10.10.

RESULT 6.4.6 $L' = Z(L)$ *and* $L' \cong Z(2)$.
It follows from Result 6.4.3 that $L' \subseteq Z(L)$. But $Z(L) \cong Z(2)$ by Result 6.4.4.

Thus L' is either trivial or is equal to $Z(L)$. The first case contradicts Result 6.4.3 (L is class-2 nilpotent, not class-1 nilpotent) so $L' = Z(L)$. This and $Z(L) \cong Z(2)$ further imply $L' \cong Z(2)$.

LEMMA *Let G be a group whose commutator subgroup is finite. Then G is an* FC *group.*

Let $\text{card}(G') = n$ and let $a \in G$. It is claimed that a has at most n conjugates. Suppose to the contrary that a has $n + 1$ distinct conjugates, $b_1, b_2, \ldots, b_{n+1}$, and seek a contradiction. For $1 \leqslant i \leqslant n + 1$, there exists $y_i \in G$ such that $b_i = y_i^{-1}ay_i$ so $b_ia^{-1} = y_i^{-1}ay_ia^{-1} \in G'$. Hence the set $\{b_ia^{-1} \mid 1 \leqslant i \leqslant n + 1\}$ is a subset of G'. However it is easily seen that this set has cardinality $n + 1$. Thus $n + 1 \leqslant \text{card}(G') = n$, the desired contradiction.

RESULT 6.4.7 *L is an* FC *group.*

By Result 6.4.6 and the lemma. (See Exercise 2 for an alternate proof.)

RESULT 6.4.8 *L is not residually finite.*

By Result 6.4.5, Part (a), Result 6.4.1, and Theorem 1.3.10.

RESULT 6.4.9

(a) *L is solvable.*

(b) *L is hypercyclic.*

Part (a) by Result 6.4.3 and Theorem A.6.7. Part (b) by Result 6.4.3 and the Corollary to Theorem A.7.2.

RESULT 6.4.10 *L is hopfian.*

Suppose $L/B \cong L$, where B is nontrivial, and seek a contradiction. By Result 6.4.4, $Z(L)$ is nontrivial so $H(L) \subseteq Z(L)$. But $Z(L) \cong Z(2)$ (Result 6.4.4 again) so either $H(L) = \{1\}$ or $Z(L)$. Result 6.4.5 precludes the first case, hence $H(L) = Z(L)$. Then since B is a nontrivial normal subgroup of L, $H(L) \subseteq B$, so $Z(L) \subseteq B$. Consequently $(L/Z(L))/(B/Z(L)) \cong L/B \cong L$. But $L/Z(L)$ is abelian by Result 6.4.3 and Theorem A.5.5 so L is abelian, the desired contradiction.

RESULT 6.4.11 $\exp(L/L') = 2$.

First note that $Z(D_4) = \{1, x^2\}$ (Result 4.1.3) hence the group $D_4/Z(D_4)$ has cardinality $8/2 = 4$ so it must be abelian (Theorem A.1.2) and consequently $D_4' \subseteq Z(D_4)$ (Theorem A.6.4). However D_4 is nonabelian so D_4' is nontrivial hence must equal $Z(D_4)$ since $\text{card}(Z(D_4)) = 2$.

It is similarly claimed that $G_n' = Z(G_n)$ for each group G_n is the direct system whose direct limit is L. This will be shown by induction. When $n = 0, G_n = D_4$ so induction begins. Assume then the induction hypothesis, $G_{n-1}' = Z(G_{n-1})$. Since G_n is the central pushout of $(Z(2), G_{n-1}, H_n, k_{n-1}, f_n)$, Theorems 1.1.9, 1.1.12 and A.2.5 combine to show

$$G_n/Z(G_n) \cong [G_{n-1}/Z(G_{n-1})] \times [D_4/Z(D_4)]. \quad (1)$$

But $G_{n-1}/Z(G_{n-1})$ is abelian by the induction hypothesis, hence so too is $G_n/Z(G_n)$. A second application of Theorem A.6.4 shows $G_n' \subseteq Z(G_n)$. But here too $\text{card}(Z(G_n)) = 2$ (Theorem 1.1.9, Part (b)) and also G_n' is nontrivial hence must equal $Z(G_n)$ establishing the claim.

It is easily seen that $\exp(D_4/Z(D_4)) = 2$ and an easy induction argument using Line (1) shows $\exp(G_n/Z(G_n)) = 2$ for all natural numbers n.

Finally to prove the result, let $L'x \in L/L'$. Then $x \in L$ so $x = h_n(y)$ for some n and some $y \in G_n$. Since $\exp(G_n/Z(G_n)) = 2, y^2 \in Z(G_n) = G_n'$ (by the second paragraph). Consequently $x^2 = h_n(y^2) \in L'$ completing the proof.

LEMMA *If A is a subdirectly irreducible abelian group having exponent p, then $A \cong Z(p)$.*
By subdirect irreducibility, there exists $a \in H(A) \setminus \{0\}$. Since $\exp(A) = p$, $o(a) = p$ so it suffices to show $A = \langle a \rangle$. Let $x \in A$. If $x = 0$, then $x \in \langle a \rangle$ as required. If $x \neq 0$, then $H(A) \subseteq \langle x \rangle$ so $a \in \langle x \rangle$. But $o(a) = o(x) = p$ implies that $\text{card}(\langle a \rangle) = \text{card}(\langle x \rangle) = p$, hence $\langle a \rangle = \langle x \rangle$ showing $x \in \langle a \rangle$.

RESULT 6.4.12 *L is coatomic.*
For convenience denote L/L' by A. Note that A is abelian. Also $\exp(A) = 2$ by Result 6.4.11. It is claimed that A is coatomic. Suppose H is a proper subgroup of A. Then some $a \subset A \setminus H$. Let $S = \{B \mid B$ is a subgroup of A, $H \subseteq B$, and $a \not\subseteq B\}$. Since $H \in S$, S is nonempty, and clearly $(S,\subseteq)$ is a partially ordered set. It will be shown that the hypothesis of Zorn's lemma holds for $(S,\subseteq)$.

Suppose C is a subchain of S. If C is empty, then any element of S will serve as an upper bound for C. Assume then C is nonempty. Let $J = \cup C$. Then J is a subgroup of A. Since $C \neq \emptyset$, there is some $\gamma_0 \in C$, hence $H \subseteq \gamma_0 \subseteq J$ so $H \subseteq J$. Finally if $a \in J$, then $a \in \gamma \in C$ for some γ which is false, hence $a \notin J$. Thus $J \in S$ and so J is the required upper bound for C and hence the Zorn's lemma hypothesis is met.

Apply Zorn's lemma to get a maximal element of S, call it M. Since $H \subseteq M$, to establish the claim that A is coatomic, it suffices to show M is a maximal subgroup of A. First note that $Ma \neq 1$. If F/M is any nontrivial subgroup of A/M, then $M \subset F$ and so $F \notin S$ by maximality of M. Therefore $a \in F$, and hence $Ma \in F/M$. This shows $Ma \in H(A/M) \setminus \{1\}$ and so A/M is subdirectly irreducible. Since $\exp(A) = 2$, $\exp(A/M) = 2$ also. Thus $A/M \cong Z(2)$ by the lemma. Conclude M is a maximal subgroup of A and hence A is coatomic.

Since $Z(L) = L'$ (Result 6.4.6), the above implies $L/Z(L)$ is coatomic. Therefore L is a central extension of a coatomic group $(Z(L) \cong Z(2))$ by a coatomic group $(A = L/Z(L))$. Conclude L is coatomic by the Lemma to Result 5.10.15.

EXERCISES

1. Show L is not finitely generated.
2. (Alternate proof of Result 6.4.7.) Show L is an FC group by proving that extensions of finite groups by FC groups are FC groups.

NOTES

1. The group L is of interest as an example of an infinite nilpotent group whose center is finite (cf Miscellaneous Exercise I.7). Moreover it shows that nilpotent groups with residually finite or finitely generated centers need not be residually finite or finitely generated.
2. As a counterexample, L shows indecomposable bounded nilpotent groups need not be finite. (This is in contrast to the abelian case. Bounded abelian groups are isomorphic to restricted direct products of finite cyclic groups by Prufer's theorem (Rotman [31], Corollary 9.32), hence the indecomposable ones are finite.)

□□□□□

Example 6.5 (The Infinite Dihedral Group)

Let $D_\infty = Dih(Z)$.

RESULT 6.5.1 *Let* $x = (1,0)$ *and* $y = (0,1)$. *Then* x *and* y *generate* D_∞.
For all $(i,j) \in D_\infty$, $(i,j) = (i,0)(0,j) = x^i y^j$.

RESULT 6.5.2 D_∞ *is coatomic.*
By finite generation (Result 6.5.1) and Theorem A.10.11.

RESULT 6.5.3 $Z(D_\infty) = \{1\}$.
By Corollary 3 to Theorem 1.2.5 and the fact that Z is torsion-free.

RESULT 6.5.4 D_∞ *is not*
(a) *hypercentral,*
(b) *nilpotent,*
(c) *locally nilpotent.*
Since $Z(D_\infty)$ is trivial (Result 6.5.3), D_∞ is clearly nonhypercentral and nonnilpotent proving Parts (a) and (b). Part (c) follows from Part (b) since D_∞ is fi-

nitely generated (Result 6.5.1).

LEMMA *If n is a positive integer, then there exists m such that n does not belong to the cyclic subgroup of Z generated by 2^m.*
Suppose to the contrary that $n \in \langle 2^m \rangle$ for all m. Let 2^a be the largest power of 2 dividing n. Thus $n = 2^a d$ where 2 and d are relatively prime. Then $n \in \langle 2^{a+1} \rangle$ implies $n = k2^{a+1}$ for some k and hence $k2^{a+1} = 2^a d$ implying $k2 = d$ so $2|d$ the desired contradiction.

RESULT 6.5.5 *D_∞ is both residually finite and residually nilpotent.*
By recourse to Theorem 1.3.9 it suffices to show that for any nontrivial $x \in D_\infty$, there exists a normal subgroup B such that $x \notin B$ and D_∞/B is both finite and nilpotent. If $x \notin Z^*$, then let $B = Z^*$, in which case $D_\infty/B \cong Z(2)$ which is finite and nilpotent. Assume then $x \in Z^*$. Thus $x = (i,0)$ for some integer i. Since x can be replaced by x^{-1} if necessary, i may be assumed positive. By the lemma, there exists m such that $i \notin \langle 2^m \rangle$. This clearly implies $x \notin \langle (2^m,0) \rangle$. By Theorem 1.2.6, $\langle (2^m,0) \rangle \lhd D_\infty$. Note that $\langle (2^m,0) \rangle = \langle 2^m \rangle^*$ so $Z^*/\langle (2^m,0) \rangle = Z^*/\langle 2^m \rangle^*$ which is clearly isomorphic to $Z/\langle 2^m \rangle$ and hence isomorphic to $Z(2^m)$. Thus $D_\infty/\langle (2^m,0) \rangle$ is an extension of an isomorphic copy of $Z(2^m)$ by $Z(2)$ and is thus a finite 2-group, hence is both finite and nilpotent (nilpotent by Theorem A.5.7).

RESULT 6.5.6 *For $3 \leqslant n$, D_n is a homomorphic image of D_∞.*
Let $e{:}Z \to Z/\langle n \rangle$ be the natural map. Then e is surjective so induces by Result 1.2.8 a surjective homomorphism $e^*{:}D_\infty \to Dih(Z/\langle n \rangle)$. But $Z/\langle n \rangle \cong Z(n)$ and by Result 4.1.1 $Dih(Z(n)) \cong D_n$. Conclude existence of a homomorphism from D_∞ onto D_n as required.

LEMMA *Quotient groups of* MN *groups are* MN *groups.*
Let A be an MN group, let $B \lhd A$, and suppose C/B is a maximal subgroup of A/B. Then C is a maximal subgroup of A so $C \lhd A$. Consequently $C/B \lhd A/B$.

RESULT 6.5.7 *D_∞ is not an* MN *group.*
By Result 6.5.6, there is a normal subgroup K such that $D_\infty/K \cong D_3$. But D_3 is not an MN group (Result 4.2.10) so by the lemma, D_∞ is not either.

RESULT 6.5.8 *D_∞ is solvable.*
By Theorem 1.2.10.

RESULT 6.5.9 *D_∞ is not an* FC *group.*
Since there are an infinite number of odd integers, it suffices to show (1,1) is conjugate to $(x,1)$ in D_∞ for each odd integer x. If x is odd, then $x - 1$ is even so $x - 1 = 2a$ for some integer a. Therefore $(a,0)(1,1)(a,0)^{-1} = (a + 1,1)(-a,0) = (2a + 1,1) = (x,1)$.

LEMMA *Let H be any nontrivial subgroup of Z. Then Z/H is finite.*
Since Z is cyclic so are all of its subgroups. Let $H = \langle n \rangle$. Note $n \neq 0$ because H is nontrivial. It is claimed that $Z/H \subseteq \{H, H+1, \ldots, H+n-1\}$. Let $H + z \in Z/H$. Then $z, z-1, \ldots, z-(n-1)$ are n consecutive integers so n divides one of them, say $n \mid z - i$. Thus $z - i \in H$ so $H + z = H + i$ proving the claim. Thus Z/H is finite.

RESULT 6.5.10 *D_∞ is just-infinite.*
Suppose K is a nontrivial normal subgroup and seek to show D_∞/K is finite. Note that $Z^* \subseteq Z^*K \subseteq D_\infty$. This and $D_\infty/Z^* \cong Z(2)$ imply $Z^*K = Z^*$ or D_∞. In the first case, $K \subseteq Z^*$. This and $K \neq \{1\}$ imply Z^*/K is finite by the lemma. Hence D_∞/K is an extension of a finite group (Z^*/K) by a finite group (D_∞/Z^*) and hence is finite. In the second case $D_\infty/K \cong Z^*/(Z^* \cap K)$. If $Z^* \cap K = \{1\}$, then $D_\infty = Z^* \oplus K$ implying $K \cong D_\infty/Z^* \cong Z(2)$ so K is abelian. Hence D_∞ is abelian by the Corollary to A.2.1. This contradiction (to Result 6.5.3) shows $Z^* \cap K$ is nontrivial. Hence $Z^*/(Z^* \cap K)$ is finite by the lemma. Hence D_∞/K is finite.

LEMMA *Just-infinite groups are indecomposable.*
Let G be a just-infinite group. Suppose $G = B \oplus C$ where B and C are both nontrivial and seek a contradiction. Thus G/B and G/C are finite. But $G/B \cong C$ so G is an extension of a finite group (C) by a finite group (G/C) implying G is finite, the desired contradiction.

RESULT 6.5.11 *D_∞ is indecomposable.*
By Result 6.5.10 and the lemma.

LEMMA *Just-infinite groups are hopfian.*
Let G be a just-infinite group and suppose $G/B \cong G$. Then G/B is infinite so B must be trivial by the just-infinite condition.

RESULT 6.5.12 *D_∞ is hopfian.*
By Result 6.5.10 and the lemma.

RESULT 6.5.13 *D_∞ is not cohopfian.*
Define $f: Z \to Z$ by $f(i) = 2i$. It is clear that f is a homomorphism and moreover that f is injective. By Theorem 1.2.8, Parts (a) and (b), $f^*: D_\infty \to D_\infty$ defined by $f^*(a,b) = (f(a),b)$ is also an injective homomorphism. But f^* is not surjective because there exists no $(a,b) \in D_\infty$ such that $f^*(a,b) = (1,0)$. (For such (a,b), $2a = 1$.) Conclude D_∞ is not cohopfian.

LEMMA *In a residually nilpotent group elements of relatively prime order commute.*
Suppose A is a residually nilpotent group and $x,y \in A$ are elements of order n and m respectively such that n and m are relatively prime. Assume $xy \neq yx$ and

seek a contradiction. Thus $x^{-1}y^{-1}xy \neq 1$. Therefore by Theorem 1.3.9, there exists a normal subgroup H such that $x^{-1}y^{-1}xy \notin H$ and A/H is nilpotent. Since $(Hx)^n = 1$, $o(Hx)|n$ and similarly $o(Hy)|m$. Thus $o(Hx)$ and $o(Hy)$ are relatively prime so $HxHy = HyHx$ by Theorem A.5.8. This implies $x^{-1}y^{-1}xy \in H$, the required contradiction.

RESULT 6.5.14 *Elements of D_∞ of relatively prime order commute.*
By residual nilpotency (Result 6.5.5) and the lemma.

RESULT 6.5.15 *D_∞ is hypercyclic.*
By Theorem 1.2.11.

EXERCISES

1. Show D_∞ is neither torsion nor torsion-free.
2. Show
   (a) D_∞ has no minimal normal subgroups.
   (b) D_∞ is not subdirectly irreducible.
3. Show $D_\infty \cong G(A,B \mid B^2 = 1, (AB)^2 = 1)$.
4. Let $f,g \subset \times\{D_n\}_{n\in N}$ be defined by $f(n) = x_n$ and $g(n) = y_n$ where x_n and y_n are the generators of D_n. Show $\langle f,g\rangle \cong D_\infty$.
5. Show D_∞ does not satisfy the descending chain condition on normal subgroups.
6. Show D_∞ satisfies the ascending chain condition.

NOTES

1. As a counterexample D_∞ separates the classes of nilpotent and residually nilpotent groups, and in addition shows
   (a) Extensions of FC groups by FC groups (even extensions by finite groups) need not be FC. (cf Example 6.4, Exercise 2.)
   (b) Direct products of locally nilpotent groups need not be locally nilpotent. This is seen as follows. By Result 6.5.5, D_∞ is residually nilpotent hence can be embedded in a direct product of nilpotent groups $G = \times\{A_i\}_{i\in I}$. Each A_i, being nilpotent, is surely locally nilpotent, but G is not locally nilpotent because local nilpotency would imply local nilpotency of D_∞ in contradiction to Result 6.5.4.
   (c) Groups in which elements of relatively prime order commute need not be locally nilpotent (Results 6.5.14 and 6.5.4, Part (c)).
   (d) The class of groups in which elements of relatively prime order commute is not quotient closed. (It is easily seen that D_3 does not belong to this class but D_3 is a homomorphic image of D_∞ (Result 6.5.6) and D_∞ does belong to the class (Result 6.5.14).)

Similarly the class of residually nilpotent groups is not quotient closed (D_3 is not residually nilpotent because finite residually nilpotent groups are nilpotent by Section 1.3, Exercise 12).

2. A presentation for D_∞ is given in Exercise 3.

□□□□□

Example 6.6

Let $H = Dih(Z(2^\infty))$.

RESULT 6.6.1 *H is a 2-group.*
The group H is an extension of the 2-group $Z(2^\infty)^*$ (a 2-group by Result 5.4.1) by $Z(2)$. Hence H is a 2-group (Lemma to Theorem 1.7.8).

RESULT 6.6.2 $H' = Z(2^\infty)^*$.
Since $Z(2^\infty)$ is 2-divisible (divisible in fact) by Result 5.4.2, the result follows from Corollary 2 to Theorem 1.2.5.

RESULT 6.6.3 *If F is a proper subgroup of H, then either $F = H'$ or F is finite.*
Either $F \cap H'$ is equal to H' or is a proper subgroup of H'. In the first case $H' \subseteq F$, and so $H' \subseteq F \subset H$. This and $H/H' \cong Z(2)$ imply $F = H'$. Consider now the second case. Since $H' \cong Z(2^\infty)$ (Result 6.6.2), $F \cap H'$ is finite since all proper subgroups of $Z(2^\infty)$ are finite (Result 5.4.3). Also $F/(F \cap H') \cong FH'/H'$ which is a subgroup of H/H' which, as noted before, is isomorphic to $Z(2)$. Therefore $F/(F \cap H')$ must be finite. This and finiteness of $F \cap H'$ show F is finite.

RESULT 6.6.4 *H is subdirectly irreducible.*
Let $B = \mathrm{H}(Z(2^\infty)^*)$. Since $Z(2^\infty)$ is subdirectly irreducible (Result 5.4.5), B is nontrivial. Thus it will suffice to show $B \subseteq F$ whenever F is a nontrivial normal subgroup. If $F = H$, then surely $B \subseteq F$ so assume F is a proper subgroup. Apply Theorem 1.2.7 to get $F \subseteq Z(2^\infty)^*$. Thus F is a nontrivial normal subgroup of $Z(2^\infty)^*$ and so $B \subseteq F$ as required.

RESULT 6.6.5 *H is indecomposable.*
By Result 6.6.4 and Theorem A.10.10.

RESULT 6.6.6 *H has a unique maximal subgroup.*
By Result 5.4.2, $Z(2^\infty)$ is divisible. Application of Theorem 1.2.13 then yields the result.

RESULT 6.6.7
(a) $Z(H) = (Z(2^\infty)[2])^*$.
(b) $Z(H) \cong Z(2)$.
(c) $H/Z(H) \cong H$.
Part (a) by Corollary 3 to Theorem 1.2.5. Next observe that by Result 5.4.9, $Z(2^\infty)[2] = \langle Z + \frac{1}{2}\rangle$ which is isomorphic to $Z(2)$ since $o(Z + \frac{1}{2}) = 2$. Therefore $Z(H) \cong Z(2)$ proving Part (b).

Now let T denote $Z(2^\infty)[2]$. Then $H/T^* \cong Dih(Z(2^\infty)/T)$ by Theorem 1.2.9. But $T^* = Z(H)$ as just shown and $Z(2^\infty)/T \cong Z(2^\infty)$ by Result 5.4.12, Part (a). Thus $H/Z(H) \cong H$.

RESULT 6.6.8 *H is not nilpotent.*
Note H is nontrivial, hence cannot be class-0 nilpotent. Suppose H is class-n nilpotent for some $n > 0$. Then (Theorem A.5.5) $H/Z(H)$ is class-$(n - 1)$ nilpotent. But as $H/Z(H) \cong H$ by Result 6.6.7, this implies $n - 1 = n$.

RESULT 6.6.9 *H is not residually nilpotent.*
The group H is subdirectly irreducible (Result 6.6.4) and nonnilpotent (Result 6.6.8) hence Theorem 1.3.10 shows it is not residually nilpotent.

RESULT 6.6.10 *H is hypercentral.*
It must be shown that any nontrivial quotient group H/B has a nontrivial center. Since $H/B \neq \{1\}$, B is a proper, normal subgroup of H so $B \subseteq Z(2^\infty)^*$ by Theorem 1.2.7. Let $K = \{x \mid (x,0) \in B\}$. It is clear that K is a subgroup of $Z(2^\infty)$ and $B = K^*$. Applying Theorem 1.2.9 yields $H/K^* \cong Dih(Z(2^\infty)/K)$. Now either B is equal to or is a proper subgroup of $Z(2^\infty)^*$. In the first case $H/B \cong Z(2)$ which is abelian so there is nothing else to prove. Assuming then that $B \subset Z(2^\infty)^*$ implies $K \subset Z(2^\infty)$. Hence $Z(2^\infty)/K \cong Z(2^\infty)$ by Result 5.4.12. Therefore $H/B = H/K^* \cong Dih(Z(2^\infty)) = H$. This and $Z(H)$ nontrivial (Result 6.6.7) show $Z(H/B)$ nontrivial as required.

RESULT 6.6.11
(a) *H is not coatomic.*
(b) *H is not finitely generated.*
Since $Z(2^\infty)^*$ has index 2 it is a maximal subgroup of H, hence is the only maximal subgroup by Result 6.6.6. Then H is not coatomic since $^*Z(2) \subset H$ but $^*Z(2) \not\subseteq Z(2^\infty)^*$.

Part (b) follows from Part (a) and Theorem A.10.11.

RESULT 6.6.12 *H is locally nilpotent.* (See Note 4.)
Let F be a finitely generated subgroup of H. Then F must be proper since H itself is not finitely generated (Result 6.6.11). Applying Result 6.6.3 shows $F = H'$ or F is finite. In the first case $F \cong Z(2^\infty)$ (Result 6.6.2) so F is abelian, hence nilpotent (actually the first case cannot occur as $Z(2^\infty)$ is not finitely generated, but

no matter). In the second case F is a finite 2-group (Result 6.6.1) hence is nilpotent by Theorem A.5.7.

RESULT 6.6.13 *H is an* MN *group.*
The subgroup $Z(2^\infty)^*$ has index 2 hence is a maximal subgroup of H. By Result 6.6.6 then, $Z(2^\infty)^*$ is the only maximal subgroup. Finally note that $Z(2^\infty)^* \lhd H$.

See Note 5 for an alternate proof.

RESULT 6.6.14 *H is hypercyclic.*
By Result 6.6.10 and Theorem A.7.2, or by Theorem 1.2.11.

RESULT 6.6.15 *H is solvable.*
By Theorem 1.2.10.

RESULT 6.6.16 *H is not residually finite.*
By Theorem 1.3.10, residually finite subdirectly irreducible groups are finite. But H is subdirectly irreducible (Result 6.6.4) and infinite, hence cannot be residually finite.

Alternate proof: Subgroups of residually finite groups must be residually finite. But H has a subgroup, $Z(2^\infty)^*$, which is not residually finite (Result 5.4.11).

RESULT 6.6.17 *The lattice of normal subgroups* N(H) *is a chain.*
Let $C, D \lhd H$. Four cases are distinguished. If both C and D are proper subgroups then $C, D \subseteq Z(2^\infty)^*$ by Theorem 1.2.7. This and Result 5.4.4 imply $C \subseteq D$ or $D \subseteq C$. As a second case suppose $C \subset H$ and $D = H$. Then $C \subseteq D$. The remaining two cases are just as easily handled.

RESULT 6.6.18 *H is not an* FC *group.*
First note that $Z(2^\infty)$ is 2-divisible (divisible even) by Result 5.4.2. Theorem 1.2.12 may be applied to show $(x,1)$ is conjugate to $(Z + ½,1)$ in H for all $x \in Z(2^\infty)$. Since $Z(2^\infty)$ is infinite, $(Z + ½,1)$ has an infinite number of conjugates.

RESULT 6.6.19 *H is nonhopfian.*
By Result 6.6.7, Parts (b) and (c).

RESULT 6.6.20 *H is cohopfian.*
Since $H' = Z(2^\infty)^*$ by Result 6.6.2, H/H' is isomorphic to $Z(2)$ hence is cohopfian. Also H' is cohopfian (Result 5.4.13). Finally H' is a fully invariant subgroup of H (Theorem A.6.2) so H is cohopfian by the Lemma to Result 5.12.7.

EXERCISES

1. Show all subnormal subgroups of H are normal.
2. Show all normal subgroups of H are fully invariant.

3. Show H does not satisfy the ascending chain condition on normal subgroups.
4. Let $n \in N$ such that n is odd. Define $g_n : H \to H$ by $g_n(b,a) = (nb,a)$. Show $g_n \in \mathrm{Aut}(H)$.
5. Show $Z(2^\infty)^*$ is the only proper subgroup of H having finite index.
6. Show the class of groups satisfying the descending chain condition is extension closed and conclude H satisfies the descending chain condition.

NOTES

1. The group H is of interest as an example of a hypercentral group that is not residually nilpotent, or more basically, as an example of a hypercentral group that is not nilpotent.
2. As a counterexample H shows that groups in which elements of relatively prime order commute need not be residually nilpotent. (Since H is a 2-group (Result 6.6.1), elements of relatively prime order commute vacuously. But H is not residually nilpotent by Result 6.6.9.)
3. Groups, such as H, that are extensions of abelian groups that satisfy the descending chain condition by finite groups are called *Chernikov*.
4. Result 6.6.12 follows immediately from the fact that hypercentral groups are locally nilpotent. This fact, however, will not be proven in this book, so an independent proof of Result 6.6.12 was necessary.
5. Result 6.6.13 also follows from the fact that locally nilpotent groups are MN. This will be proven later on (Lemma to Result 6.10.8).

□□□□□

Example 6.7

Let $H = Dih(Q)$.

RESULT 6.7.1 $H' = Q^*$.
Since Q is 2-divisible (divisible in fact) by Result 5.1.2, the result follows from Corollary 2 to Theorem 1.2.5.

RESULT 6.7.2 *H has a unique maximal subgroup.*
By Result 5.1.2, Q is divisible. Application of Theorem 1.2.13 then yields the result.

RESULT 6.7.3 *D_∞ can be embedded in H.*
By Theorem 1.2.8, the inclusion $i: Z \to Q$ induces an injection $i^*: D_\infty \to H$.

RESULT 6.7.4
(a) *H is not nilpotent.*
(b) *H is not locally nilpotent.*
Since D_∞ can be embedded in H (Result 6.7.3), H has a finitely generated, nonnilpotent subgroup (finitely generated by Result 6.5.1, nonnilpotent by Result 6.5.4, Part (b)). Thus H is not locally nilpotent and hence nonnilpotent.

RESULT 6.7.5 *H is an* MN *group.*
The subgroup Q^* has index 2 hence is a maximal subgroup of H. By Result 6.7.2 then, Q^* is the only maximal subgroup of H. Finally note $Q^* \lhd H$.

RESULT 6.7.6 *H is hypercyclic.*
By Theorem 1.2.11.

RESULT 6.7.7 *H is solvable.*
By Theorem 1.2.10.

RESULT 6.7.8 *H is indecomposable.*
Suppose $H = J \oplus K$ where J and K are nontrivial subgroups, and seek a contradiction. Then since Q is 2-divisible (Result 5.1.2), $J, K \subset H$ implies $J, K \subseteq Q^*$ by Theorem 1.2.7. Consequently $G = JK \subseteq Q^*$ which is false.

Alternately the result follows from Exercise 10 and the fact that H is not torsion (cf the Lemma to Result 6.5.11).

RESULT 6.7.9 *H is hopfian.*
Since $H' = Q^*$ by Result 6.7.1, H/H' is isomorphic to $Z(2)$, hence is hopfian. Also H' is hopfian (Result 5.1.12). Finally H' is a fully invariant subgroup of H (Theorem A.6.2) so H is hopfian by the Lemma to Result 5.12.6.

Alternately the result follows from Exercise 10 and the fact that H is not torsion (cf the Lemma to Result 6.5.12).

RESULT 6.7.10 *H is not residually nilpotent.*
Let $b \in Q$ be arbitrary for the moment and define a sequence of elements of H recursively by $x_1 = (b,0)$ and $x_{n+1} = x_n{}^{-1}(0,1)^{-1}x_n(0,1)$. It is claimed that $x_n = ((-1)^{n-1}2^{n-1}b,0)$ for all positive integers n. This will be shown by induction. When $n = 1$, $x_1 = (b,0)$ so induction begins. Assume the induction hypothesis, $x_{n-1} = ((-1)^{n-2}2^{n-2}b,0)$. Then (Theorem 1.2.5) $x_n = x_{n-1}{}^{-1}(0,1)^{-1}x_{n-1}(0,1)$ which equals $((-2)(-1)^{n-2}2^{n-2}b,0) = ((-1)^{n-1}2^{n-1}b,0)$ proving the claim.

Assume now that H is residually nilpotent and seek a contradiction. By Theorem 1.3.9, $(1,0) \neq 1$ implies existence of a normal subgroup N such that $(1,0) \notin N$ and H/N is nilpotent, say class-k nilpotent. Let $m = k + 1$ and let $x_1, x_2, \ldots, x_m$ be defined as in the first paragraph with $b = 1/(-1)^{m-1}2^{m-1}$. Thus $x_m = (1,0)$. Now $Nx_1 \in H/N = Z_k(H/N)$ so $(Nx_1)^{-1}(N(0,1))^{-1}Nx_1N(0,1)$ belongs to $Z_{k-1}(H/N)$. That is, $Nx_2 \in Z_{k-1}(H/N)$. Repeat this argument an additional

$k - 1$ times to show $Nx_{k+1} \in Z_0(H/N)$. That is, $Nx_m = 1$. Thus $x_m \in N$ so $(1,0) \in N$, the required contradiction.

RESULT 6.7.11 *Elements of H of relatively prime order commute.*
Suppose $x,y \in H$ such that $o(x) = n$ and $o(y) = m$ where n and m are relatively prime. Then either 2 does not divide n or 2 does not divide m. Assume the first case. Then 2 and n are relatively prime so there exist integers c and d such that $2c + nd = 1$. Consequently $x = x^{2c}x^{nd}$. But $o(x) = n$ implies $x^n = 1$ so $x = x^{2c}$. Now note that $H/Q^* \cong Z(2)$ so $x^2 \in Q^*$. Therefore $x = x^{2c} \in Q^*$. But $Q^* \cong Q$ so $x \in Q^*$ and $o(x)$ is finite imply $x = 1$ because Q is torsion-free (Result 5.1.1). Therefore x and y commute. The case 2 does not divide m shows $y = 1$ so again x and y commute.

RESULT 6.7.12 *The lattice of normal subgroups,* N(H), *is distributive.*
Let A, B, and C be normal subgroups and seek to show

$$A(B \cap C) = AB \cap AC. \tag{1}$$

CASE ONE, $A = H$. Then $A(B \cap C) = H$ while $AB \cap AC = H \cap H = H$, so Equation (1) holds.
CASE TWO, $B = H$. Then $A(B \cap C) = AC$ while $AB \cap AC = H \cap AC = AC$, so Equation (1) holds.
CASE THREE, $C = H$. Then $A(B \cap C) = AB$ while $AB \cap AC = AB \cap H = AB$, so Equation (1) holds.
CASE FOUR, $A,B,C \subset H$. Then $A,B,C \subseteq Q^*$ by Result 1.2.7. Therefore Equation (1) holds by Result 5.1.11.

EXERCISES

1. Show
   (a) H is not coatomic.
   (b) H is not finitely generated.
2. Show H has no minimal normal subgroups and conclude that H is not subdirectly irreducible.
3. Show H is not residually finite.
4. Show H is neither torsion nor torsion-free.
5. Show H is not an FC group.
6. Show
   (a) $Z(H) = \{1\}$.
   (b) H is not hypercentral.
7. Show
   (a) H does not satisfy the ascending chain condition on normal subgroups.
   (b) H does not satisfy the descending chain condition on normal subgroups.
8. Show H is cohopfian.

9. Show Q^* is the only proper subgroup of H having finite index.
10. Show H/N is torsion for all nontrivial normal subgroups N.
11. Show $H \cong G(\{X_i \mid i \in N, i \neq 0\} \cup \{Y\} \mid (X_{i+1})^{i+1}X_i^{-1} = 1$ for all i, $Y^2 = 1, (YX_i)^2 = 1$ for all i).
12. Show that the only characteristic subgroups of H are $\{1\}$, Q^*, and H.
13. Show all subnormal subgroups of H are normal.

NOTES

1. As a counterexample, H
   (a) separates the generalized nilpotent classes *locally nilpotent* and MN: By Result 6.7.5, H is an MN group, but H is not locally nilpotent by Result 6.7.4,
   (b) shows subgroups of MN groups need not be MN groups: H is an MN group by Result 6.7.5, but D_∞ can be embedded in H (Result 6.7.3) and D_∞ is not an MN group by Result 6.5.7.
2. A presentation for H is given in Exercise 11.

□□□□□

Example 6.8

Let G be the subgroup of $T(3, Q)$ generated by $UT(3, Q^2)$ and the matrix

$$A = \begin{pmatrix} 1 & 0 & 0 \\ 0 & 2 & 0 \\ 0 & 0 & 1 \end{pmatrix}$$

(see Note 1).

NOTATION Let B and C denote the matrices $\Theta_3(2,1,1)$ and $\Theta_3(3,2,1)$ respectively and let $U = UT(3, Q^2)$.

LEMMA
(a)

$$A^{-1} = \begin{pmatrix} 1 & 0 & 0 \\ 0 & \frac{1}{2} & 0 \\ 0 & 0 & 1 \end{pmatrix}.$$

(b) *For any unitriangular matrix M (with entries in Q)*

$$AMA^{-1} = \begin{pmatrix} 1 & 0 & 0 \\ 2M_{21} & 1 & 0 \\ M_{31} & M_{32}/2 & 1 \end{pmatrix}$$

and

$$A^{-1}MA = \begin{pmatrix} 1 & 0 & 0 \\ M_{21}/2 & 1 & 0 \\ M_{31} & 2M_{32} & 1 \end{pmatrix}.$$

Part (a) is shown simply by multiplying A and the given candidate for A^{-1} and noting that the product is I.

To show Part (b), $AMA^{-1} =$

$$\begin{pmatrix} 1 & 0 & 0 \\ 0 & 2 & 0 \\ 0 & 0 & 1 \end{pmatrix}\begin{pmatrix} 1 & 0 & 0 \\ M_{21} & 1 & 0 \\ M_{31} & M_{32} & 1 \end{pmatrix}\begin{pmatrix} 1 & 0 & 0 \\ 0 & ½ & 0 \\ 0 & 0 & 1 \end{pmatrix} = \begin{pmatrix} 1 & 0 & 0 \\ 2M_{21} & 2 & 0 \\ M_{31} & M_{32} & 1 \end{pmatrix}\begin{pmatrix} 1 & 0 & 0 \\ 0 & ½ & 0 \\ 0 & 0 & 1 \end{pmatrix}$$

$$= \begin{pmatrix} 1 & 0 & 0 \\ 2M_{21} & 1 & 0 \\ M_{31} & M_{32}/2 & 1 \end{pmatrix}$$

while $A^{-1}MA =$

$$\begin{pmatrix} 1 & 0 & 0 \\ 0 & ½ & 0 \\ 0 & 0 & 1 \end{pmatrix}\begin{pmatrix} 1 & 0 & 0 \\ M_{21} & 1 & 0 \\ M_{31} & M_{32} & 1 \end{pmatrix}\begin{pmatrix} 1 & 0 & 0 \\ 0 & 2 & 0 \\ 0 & 0 & 1 \end{pmatrix} = \begin{pmatrix} 1 & 0 & 0 \\ M_{21}/2 & ½ & 0 \\ M_{31} & M_{32} & 1 \end{pmatrix}\begin{pmatrix} 1 & 0 & 0 \\ 0 & 2 & 0 \\ 0 & 0 & 1 \end{pmatrix}$$

$$= \begin{pmatrix} 1 & 0 & 0 \\ M_{21}/2 & 1 & 0 \\ M_{31} & 2M_{32} & 1 \end{pmatrix}.$$

RESULT 6.8.1

(a) *U is a normal subgroup of G.*

(b) $G = U\langle A\rangle$.

(c) *Every $M \in G$ is of the form*

$$M = \begin{pmatrix} 1 & 0 & 0 \\ M_{21} & 2^i & 0 \\ M_{31} & M_{32} & 1 \end{pmatrix}.$$

Clearly U is a subgroup of G. Using Theorem A.10.3, it need only be shown that $AUA^{-1} \subseteq U$ and $A^{-1}UA \subseteq U$ to demonstrate normality of U. This, however, is clear from the lemma.

Part (b) follows immediately from Part (a) and the definition of G.

To prove Part (c) all that needs to be shown is that for $M \in G$, $M_{11} = M_{33} = 1$, $M_{12} = M_{13} = M_{23} = 0$, and $M_{22} = 2^i$. Let $M \in G$. Then $M = SA^i$ where $S \in U$ and i is an integer by Part (b). An easy induction proof shows

$$A^i = \begin{pmatrix} 1 & 0 & 0 \\ 0 & 2^i & 0 \\ 0 & 0 & 1 \end{pmatrix}.$$

thus

$$M = \begin{pmatrix} 1 & 0 & 0 \\ S_{21} & 1 & 0 \\ S_{31} & S_{32} & 1 \end{pmatrix} \begin{pmatrix} 1 & 0 & 0 \\ 0 & 2^i & 0 \\ 0 & 0 & 1 \end{pmatrix} = \begin{pmatrix} 1 & 0 & 0 \\ S_{21} & 2^i & 0 \\ S_{31} & 2^i S_{22} & 1 \end{pmatrix}$$

as required.

LEMMA *If R is a commutative ring with identity, then UT(3.R) is generated by transvections of the form $\Theta_3(2,1,x)$ and of the form $\Theta_3(3,2,y)$.*

Let $M \in UT(3,R)$. Then by Theorem 3.1.9, M can be expressed as the product $M = \Theta_3(2,1,M_{21})\Theta_3(3,1,M_{31})\Theta_3(3,2,M_{32})$. Since the first and third transvections are of the special form required in the statement of the lemma, it suffices to show that the middle transvection $\Theta_3(3,1,M_{31})$ can be expressed as a product of transvections of the required form.

First note that for any a and b

$$\Theta_3(2,1,a)\Theta_3(3,2,b) = \begin{pmatrix} 1 & 0 & 0 \\ a & 1 & 0 \\ 0 & 0 & 1 \end{pmatrix} \begin{pmatrix} 1 & 0 & 0 \\ 0 & 1 & 0 \\ 0 & b & 1 \end{pmatrix} = \begin{pmatrix} 1 & 0 & 0 \\ a & 1 & 0 \\ 0 & b & 1 \end{pmatrix}.$$

Two applications of the above equation show $[\Theta_3(2,1,1-M_{31})\Theta_3(3,2,1)] \cdot [\Theta_3(2,1,M_{31})\Theta_3(3,2,-1)] \cdot \Theta_3(2,1,-1)$

$$= \begin{pmatrix} 1 & 0 & 0 \\ 1-M_{31} & 1 & 0 \\ 0 & 1 & 1 \end{pmatrix} \begin{pmatrix} 1 & 0 & 0 \\ M_{31} & 1 & 0 \\ 0 & -1 & 1 \end{pmatrix} \begin{pmatrix} 1 & 0 & 0 \\ -1 & 1 & 0 \\ 0 & 0 & 1 \end{pmatrix}$$

$$= \begin{pmatrix} 1 & 0 & 0 \\ 1 & 1 & 0 \\ M_{31} & 0 & 1 \end{pmatrix} \begin{pmatrix} 1 & 0 & 0 \\ -1 & 1 & 0 \\ 0 & 0 & 1 \end{pmatrix} = \begin{pmatrix} 1 & 0 & 0 \\ 0 & 1 & 0 \\ M_{31} & 0 & 1 \end{pmatrix} = \Theta_3(3,1,M_{31}).$$

RESULT 6.8.2 *$G = \langle A,B,C \rangle$. Thus G is finitely generated.*

That $A \in \langle A,B,C \rangle$ is, of course, immediate so it suffices to show $U \subseteq \langle A,B,C \rangle$. Using the lemma it need only be shown that $\Theta_3(2,1,x)$ and $\Theta_3(3,2,y)$ belong to $\langle A,B,C \rangle$ for all $x,y \in Q^2$. Let $x = m/2^n$. Note that (Corollary to Theorem 3.1.3)

$$B^m = \Theta_3(2,1,m) = \begin{pmatrix} 1 & 0 & 0 \\ m & 1 & 0 \\ 0 & 0 & 1 \end{pmatrix}.$$

Apply the Lemma to Result 6.8.1 n times to the above matrix to get

$$A^{-n}B^m A^n = \begin{pmatrix} 1 & 0 & 0 \\ m/2^n & 1 & 0 \\ 0 & 0 & 1 \end{pmatrix} = \Theta_3(2,1,x)$$

so $\Theta_3(2,1,x) \in \langle A,B,C \rangle$. A similar argument shows $\Theta_3(3,2,y) = A^k C^j A^{-k} \in \langle A,B,C \rangle$ when $y = j/2^k$.

RESULT 6.8.3 *G is coatomic.*
By Result 6.8.2 and Theorem A.10.11.

RESULT 6.8.4
(a) $Z(G) = Z(U)$.
(b) $Z(G) \cong Q^2$.
Let $M \in Z(U)$. To show $M \in Z(G)$ it suffices to show M commutes with all generators of G. Hence it suffices to show M commutes with A. By Theorem 3.2.2, $M \in Z(U)$ implies $M \in [Z](U)$, hence $M_{21} = M_{32} = 0$. Thus

$$M = \begin{pmatrix} 1 & 0 & 0 \\ 0 & 1 & 0 \\ M_{31} & 0 & 1 \end{pmatrix},$$

so

$$MA = \begin{pmatrix} 1 & 0 & 0 \\ 0 & 2 & 0 \\ M_{31} & 0 & 1 \end{pmatrix} \quad \text{and} \quad AM = \begin{pmatrix} 1 & 0 & 0 \\ 0 & 2 & 0 \\ M_{31} & 0 & 1 \end{pmatrix}$$

showing M and A commute. Thus $Z(U) \subseteq Z(G)$. To complete the proof of Part (a) it suffices to show $Z(G) \subseteq U$. Let $M \in Z(G)$. By Result 6.8.1, Part (c),

$$M = \begin{pmatrix} 1 & 0 & 0 \\ M_{21} & 2^i & 0 \\ M_{31} & M_{32} & 1 \end{pmatrix}$$

for some integer i. Multiplying M by B yields $(MB)_{21} = M_{21} + 2^i$. Multiplication in the reverse order yields $(BM)_{21} = 1 + M_{21}$. But $M \in Z(G)$ so M and B commute. Hence the last two equations combine to show $2^i = 1$. Consequently $M \in U$. This completes the proof of Part (a). Part (b) follows from Part (a) and Theorem 3.2.6.

RESULT 6.8.5 *G is solvable.*
By Result 3.2.15, $T(3,Q)$ is solvable. Hence so too is its subgroup G (Theorem A.6.5).

RESULT 6.8.6
(a) *G is nonhypercyclic.*
(b) *G is nonhypercentral.*
(c) *G is nonnilpotent.*
(d) *G is not locally nilpotent.*
Since quotient groups of hypercyclic groups are easily seen to be hypercyclic, to prove Part (a) it suffices to show $G/Z(G)$ is not hypercyclic. Assume then that $G/Z(G)$ is hypercyclic and seek a contradiction. Since $Z(G) = Z(U)$ (Result 6.8.4), $Z(G) \subset U$, so $U/Z(G)$ is nontrivial. Also $U \lhd G$ (Result 6.8.1) which implies $U/Z(G) \lhd G/Z(G)$. Apply Theorem A.7.3 to get existence of $R/Z(G)$ such that $R/Z(G)$ is a nontrivial cyclic normal subgroup of $G/Z(G)$ and $R/Z(G) \subseteq$

$U/Z(G)$. Let the coset $Z(G)M$ generate $R/Z(G)$. Since $M \in U$, M is unitriangular. Since $R/Z(G) \lhd G/Z(G)$, $(Z(G)A)^{-1}Z(G)MZ(G)A$ belongs to $R/Z(G)$ so is equal to $(Z(G)M)^n$ for some integer n. Consequently $A^{-1}MA = DM^n$ for some $D \in Z(G)$. Similarly $AMA^{-1} = EM^m$ for some m and some $E \in Z(G)$. By Result 6.8.4 again, $D \in Z(G)$ implies $D \in Z(U)$. Thus D is unitriangular so $D_{12} = D_{13} = D_{23} = 0$. Also $D_{21} = D_{32} = 0$ (Result 3.2.2). Similarly $E_{12} = E_{13} = E_{23} = E_{21} = E_{32} = 0$.

Now

$$A^{-1}MA = \begin{pmatrix} 1 & 0 & 0 \\ M_{21}/2 & 1 & 0 \\ M_{31} & 2M_{32} & 1 \end{pmatrix} \quad \text{and} \quad AMA^{-1} = \begin{pmatrix} 1 & 0 & 0 \\ 2M_{21} & 1 & 0 \\ M_{31} & M_{32}/2 & 1 \end{pmatrix}$$

by the Lemma to Result 6.8.1.

Next note that $(DM^n)_{21} = \Sigma_{j=1}^3 D_{2j}(M^n)_{j1} = (M^n)_{21}$ because $D_{21} = 0, D_{22} = 1$, and $D_{23} = 0$. It is easily seen that if k is any integer, then $(M^k)_{21} = kM_{21}$ and $(M^k)_{32} = kM_{32}$. Thus $(DM^n)_{21} = nM_{21}$. Similarly $(EM^m)_{32} = mM_{32}$.

Equate $(A^{-1}MA)_{21}$ and $(DM^n)_{21}$ to get $M_{21}/2 = nM_{21}$. Then $M_{21} = 2nM_{21}$ implying $M_{21} = 0$ or $2n = 1$. However n being an integer precludes $2n = 1$ so $M_{21} = 0$. Equate $(AMA^{-1})_{32}$ and $(EM^m)_{32}$ to get $M_{32}/2 = mM_{32}$ which implies $M_{32} = 0$. Thus

$$M = \begin{pmatrix} 1 & 0 & 0 \\ 0 & 1 & 0 \\ M_{31} & 0 & 1 \end{pmatrix}$$

and so $M \in Z(G)$ showing $R/Z(G)$ is trivial, the required contradiction.

This completes the proof of Part (a). Parts (b) and (c) follow from Part (a) because of Theorem A.7.2 and its corollary. Part (d) follows from Part (c) because G is finitely generated (Result 6.8.2).

LEMMA *If A is a group and $x, y \in A$ such that $x \neq 1$ and $x^{-1}y^{-1}xy = x$, then A is not residually nilpotent.*

Suppose A is residually nilpotent and seek a contradiction. Then $x \neq 1$ implies by Theorem 1.3.9 that there exists a normal subgroup H such that $x \notin H$ and A/H is nilpotent, say class-k nilpotent. Then $Hx \in A/H = Z_k(A/H)$ implies that $Hx^{-1}Hy^{-1}HxHy \in Z_{k-1}(A/H)$. This and $x^{-1}y^{-1}xy = x$ imply $Hx \in Z_{k-1}(A/H)$. Repeat this argument $k - 1$ additional times to conclude $Hx \in Z_0(A/H)$. That is, $Hx = 1$ and so $x \in H$, the required contradiction.

RESULT 6.8.7 *G is not residually nilpotent.*

By the Lemma to Result 6.8.1

$$A^{-1}CA = \begin{pmatrix} 1 & 0 & 0 \\ 0 & 1 & 0 \\ 0 & 2 & 1 \end{pmatrix}.$$

Therefore

$$C^{-1}A^{-1}CA = \begin{pmatrix} 1 & 0 & 0 \\ 0 & 1 & 0 \\ 0 & -1 & 1 \end{pmatrix}\begin{pmatrix} 1 & 0 & 0 \\ 0 & 1 & 0 \\ 0 & 2 & 1 \end{pmatrix} = \begin{pmatrix} 1 & 0 & 0 \\ 0 & 1 & 0 \\ 0 & 1 & 1 \end{pmatrix} = C.$$

Application of the lemma now yields the result.

EXERCISES

1. Show
   (a) $G' = U.$
   (b) G' is nilpotent.
2. Show $U \cap \langle A \rangle = \{1\}$.
3. Show G is indecomposable.
4. Show G is torsion-free.
5. (a) Let $M \in G$ and let n be a positive integer. Suppose $M_{22} = 2^i$. Show
   (α) $(M^n)_{21} = (1 + 2^i + \ldots + 2^{(n-1)i})M_{21}$.
   (β) $(M^n)_{22} = 2^{ni}$.
   (γ) $(M^n)_{32} = (1 + 2^i + \ldots + 2^{(n-1)i})M_{32}$.
   (b) Show G is an R-group.
6. Show G is hopfian.
7. Show G is not an FC group.
8. For positive integers n define $g_n : G \to G$ by

$$g_n\begin{pmatrix} 1 & 0 & 0 \\ a & 2^i & 0 \\ b & c & 1 \end{pmatrix} = \begin{pmatrix} 1 & 0 & 0 \\ na & 2^i & 0 \\ nb & c & 1 \end{pmatrix}.$$

   Show
   (a) g_n is an injective homomorphism.
   (b) g_2 is an automorphism.
   (c) G is noncohopfian.
9. Show G is not an MN group.
10. Show G satisfies neither the ascending chain condition on normal subgroups nor the descending chain condition on normal subgroups.

NOTES

1. The notation $UT(3, Q^2)$ presupposes Q^2 is a ring. It is easily seen that Q^2 is a subring of the rational field Q since $(a/2^i)(b/2^j) = ab/2^{i+j}$.
2. As a counterexample G shows that neither the class of finitely generated groups nor the class of coatomic groups is subgroup closed. This is because $Z(G) \cong Q^2$ (Result 6.8.4) so $Z(G)$ is noncoatomic by Miscellaneous Exercise V.2 at the end of this chapter. Theorem A.10.11 then shows $Z(G)$ is not finitely generated either.
3. The group G is also of interest as an example of a finitely generated solvable

group that does not satisfy the ascending chain condition on normal subgroups (see Exercise 10). In view of the following theorem this demonstrates an interesting difference between length-2 solvable groups and groups (such as G) that are length-n solvable where $3 \leqslant n$. **THEOREM**: *Finitely generated length-2 solvable groups satisfy the ascending chain condition on normal subgroups.* (See Hall [17].)

4. By the theorem quoted in Note 2 to Example 6.9, G is residually finite.
5. See Miscellaneous Exercise V.3 at the end of this chapter.

□□□□□

Example 6.9

Let K be the subgroup of $T(3,Q)$ generated by the matrices

$$A = \begin{pmatrix} 1 & 0 & 0 \\ 0 & 2 & 0 \\ 0 & 0 & 1 \end{pmatrix} \text{ and } B = \begin{pmatrix} 1 & 0 & 0 \\ 1 & 1 & 0 \\ -4 & 1 & 1 \end{pmatrix}.$$

It will be shown that

$$C = \begin{pmatrix} 1 & 0 & 0 \\ 0 & 1 & 0 \\ 3 & 0 & 1 \end{pmatrix}$$

belongs to K and moreover that the subgroup $H = \langle C \rangle$ is normal. Let $G = K/H$.

RESULT 6.9.1

(a) *Every matrix $M \in K$ has the form*

$$M = \begin{pmatrix} 1 & 0 & 0 \\ M_{21} & 2^j & 0 \\ M_{31} & M_{32} & 1 \end{pmatrix}$$

for some integer j.

(b)
$$A^{-1} = \begin{pmatrix} 1 & 0 & 0 \\ 0 & \tfrac{1}{2} & 0 \\ 0 & 0 & 1 \end{pmatrix} \textit{ and } B^{-1} = \begin{pmatrix} 1 & 0 & 0 \\ -1 & 1 & 0 \\ 5 & -1 & 1 \end{pmatrix}.$$

Let L denote the group of Example 6.8, that is, $L = UT(3,Q^2)\langle A \rangle$. Note that $B \in UT(3,Q^2) \subseteq L$. This and $A \in L$ imply K is a subgroup of L. Therefore Part (a) follows from Result 6.8.1, Part (c).

Part (b) is shown by calculation. That the product of A and the given candidate for A^{-1} is I is easily seen. The calculation for B is

$$\begin{pmatrix} 1 & 0 & 0 \\ 1 & 1 & 0 \\ -4 & 1 & 1 \end{pmatrix}\begin{pmatrix} 1 & 0 & 0 \\ -1 & 1 & 0 \\ 5 & -1 & 1 \end{pmatrix} = \begin{pmatrix} 1 & 0 & 0 \\ 0 & 1 & 0 \\ 0 & 0 & 1 \end{pmatrix} = I.$$

RESULT 6.9.2 $C \in K$.

First note that

$$B^2 = \begin{pmatrix} 1 & 0 & 0 \\ 1 & 1 & 0 \\ -4 & 1 & 1 \end{pmatrix}\begin{pmatrix} 1 & 0 & 0 \\ 1 & 1 & 0 \\ -4 & 1 & 1 \end{pmatrix} = \begin{pmatrix} 1 & 0 & 0 \\ 2 & 1 & 0 \\ -7 & 2 & 1 \end{pmatrix}.$$

Therefore

$$A^{-1}B^2A = \begin{pmatrix} 1 & 0 & 0 \\ 1 & 1 & 0 \\ -7 & 4 & 1 \end{pmatrix}$$

by the Lemma to Result 6.8.1. Inverting yields

$$A^{-1}B^{-2}A = \begin{pmatrix} 1 & 0 & 0 \\ -1 & 1 & 0 \\ 11 & -4 & 1 \end{pmatrix}.$$

Therefore $B(A^{-1}B^{-2}A)B^{-1}(A^{-1}B^2A) =$

$$\begin{pmatrix} 1 & 0 & 0 \\ 1 & 1 & 0 \\ -4 & 1 & 1 \end{pmatrix}\begin{pmatrix} 1 & 0 & 0 \\ -1 & 1 & 0 \\ 11 & -4 & 1 \end{pmatrix}\begin{pmatrix} 1 & 0 & 0 \\ -1 & 1 & 0 \\ 5 & -1 & 1 \end{pmatrix}\begin{pmatrix} 1 & 0 & 0 \\ 1 & 1 & 0 \\ -7 & 4 & 1 \end{pmatrix}$$

$$= \begin{pmatrix} 1 & 0 & 0 \\ 0 & 1 & 0 \\ 6 & -3 & 1 \end{pmatrix}\begin{pmatrix} 1 & 0 & 0 \\ 0 & 1 & 0 \\ -3 & 3 & 1 \end{pmatrix} = \begin{pmatrix} 1 & 0 & 0 \\ 0 & 1 & 0 \\ 3 & 0 & 1 \end{pmatrix} = C.$$

RESULT 6.9.3

(a) $C \in Z(K)$.

(b) $H \lhd K$.

To prove Part (a) it suffices to show that C commutes with both generators of K. That C commutes with B is immediate because $B \in UT(3, Q)$ and C belongs to the center of $UT(3, Q)$ by Theorem 3.2.2. As for commutation with A, by the Lemma to Result 6.8.1,

$$ACA^{-1} = \begin{pmatrix} 1 & 0 & 0 \\ 0 & 1 & 0 \\ 3 & 0 & 1 \end{pmatrix} = C$$

hence $AC = CA$. This completes the proof of Part (a). Part (b) follows.

RESULT 6.9.4

(a) *G is finitely generated.*

(b) *G is coatomic.*

Since K is finitely generated, so is its quotient group G, proving Part (a). Part (b) follows from Part (a) and Theorem A.10.11.

LEMMA *Let F denote the transvection* $F = \Theta_3(3,2,-3/2)$. *Then* $F \in K$.

By the Lemma to Result 6.8.1,

$$ABA^{-1} = \begin{pmatrix} 1 & 0 & 0 \\ 2 & 1 & 0 \\ -4 & \frac{1}{2} & 1 \end{pmatrix}$$

Therefore $B^{-2}(ABA^{-1})C^{-1} =$

$$\begin{pmatrix} 1 & 0 & 0 \\ -1 & 1 & 0 \\ 5 & -1 & 1 \end{pmatrix}\begin{pmatrix} 1 & 0 & 0 \\ -1 & 1 & 0 \\ 5 & -1 & 1 \end{pmatrix}\begin{pmatrix} 1 & 0 & 0 \\ 2 & 1 & 0 \\ -4 & \frac{1}{2} & 1 \end{pmatrix}\begin{pmatrix} 1 & 0 & 0 \\ 0 & 1 & 0 \\ -3 & 0 & 1 \end{pmatrix}$$

$$= \begin{pmatrix} 1 & 0 & 0 \\ -2 & 1 & 0 \\ 11 & -2 & 1 \end{pmatrix}\begin{pmatrix} 1 & 0 & 0 \\ 2 & 1 & 0 \\ -7 & \frac{1}{2} & 1 \end{pmatrix} = \begin{pmatrix} 1 & 0 & 0 \\ 0 & 1 & 0 \\ 0 & -3/2 & 1 \end{pmatrix} = F.$$

Thus $F \in K$.

RESULT 6.9.5

(a) *G is not nilpotent.*

(b) *G is not residually nilpotent.*

(c) *G is not locally nilpotent.*

Let F be the matrix in the above lemma. Then

$$A^{-1}FA = \begin{pmatrix} 1 & 0 & 0 \\ 0 & 1 & 0 \\ 0 & -3 & 1 \end{pmatrix}$$

by the Lemma to Result 6.8.1. Therefore

$$F^{-1}A^{-1}FA = \begin{pmatrix} 1 & 0 & 0 \\ 0 & 1 & 0 \\ 0 & 3/2 & 1 \end{pmatrix}\begin{pmatrix} 1 & 0 & 0 \\ 0 & 1 & 0 \\ 0 & -3 & 1 \end{pmatrix} = \begin{pmatrix} 1 & 0 & 0 \\ 0 & 1 & 0 \\ 0 & -3/2 & 1 \end{pmatrix} = F.$$

Let $\bar{A}$ and $\bar{F}$ denote the cosets (relative to the normal subgroup H) determined by A and F respectively. Thus $\bar{A}, \bar{F} \in G$ and $\bar{F}^{-1}\bar{A}^{-1}\bar{F}\bar{A} = \bar{F}$. Finally note that $\bar{F} \neq 1$ (equality would imply $F \in H$ and hence $F = C^n$ for some integer n, so

$$F = \begin{pmatrix} 1 & 0 & 0 \\ 0 & 1 & 0 \\ 3n & 0 & 1 \end{pmatrix}$$

would hold, which is false). Applying the Lemma to Result 6.8.7 shows G is not residually nilpotent, proving Part (b).

Part (a) follows from Part (b). Part (c) follows from Part (a) since G is finitely generated (Result 6.9.4).

RESULT 6.9.6 *G is solvable.*
By Theorem 3.2.15, $T(3,Q)$ is solvable. But G is a quotient group of K which is a subgroup of $T(3,Q)$. Conclude G is solvable by Theorem A.6.5.

RESULT 6.9.7 *G is nonhopfian.*
Let

$$D = \begin{pmatrix} 1 & 0 & 0 \\ 0 & 1 & 0 \\ 0 & 0 & 4 \end{pmatrix}$$

(D belongs to $T(3,Q)$ but not to K). It is easily seen that

$$D^{-1} = \begin{pmatrix} 1 & 0 & 0 \\ 0 & 1 & 0 \\ 0 & 0 & \frac{1}{4} \end{pmatrix}.$$

It is first claimed that for any matrix

$$M = \begin{pmatrix} 1 & 0 & 0 \\ M_{21} & 2^i & 0 \\ M_{31} & M_{32} & 1 \end{pmatrix}$$

belonging to K,

$$D^{-1}MD = \begin{pmatrix} 1 & 1 & 0 \\ M_{21} & 2^i & 0 \\ M_{31}/4 & M_{32}/4 & 1 \end{pmatrix}.$$

This is seen as follows;

$$\begin{pmatrix} 1 & 0 & 0 \\ 0 & 1 & 0 \\ 0 & 0 & \frac{1}{4} \end{pmatrix}\begin{pmatrix} 1 & 0 & 0 \\ M_{21} & 2^i & 0 \\ M_{31} & M_{32} & 1 \end{pmatrix}\begin{pmatrix} 1 & 0 & 0 \\ 0 & 1 & 0 \\ 0 & 0 & 4 \end{pmatrix} = \begin{pmatrix} 1 & 0 & 0 \\ M_{21} & 2^i & 0 \\ M_{31}/4 & M_{32}/4 & \frac{1}{4} \end{pmatrix}\begin{pmatrix} 1 & 0 & 0 \\ 0 & 1 & 0 \\ 0 & 0 & 4 \end{pmatrix}$$

$$= \begin{pmatrix} 1 & 0 & 0 \\ M_{21} & 2^i & 0 \\ M_{31}/4 & M_{32}/4 & 1 \end{pmatrix}.$$

Now let $f: T(3,Q) \to T(3,Q)$ be defined by $f(M) = D^{-1}MD$ and let g be f restricted to K. Clearly f is an injective homomorphism, hence so is g. It is claimed that $\operatorname{Im} g \subseteq K$. To show this it suffices to show $D^{-1}AD \in K$ and $D^{-1}BD \in K$. Since

$$D^{-1}AD = \begin{pmatrix} 1 & 0 & 0 \\ 0 & 2 & 0 \\ 0 & 0 & 1 \end{pmatrix} = A$$

by the claim in the previous paragraph, $D^{-1}AD \in K$ holds. To show $D^{-1}BD \in K$, it will be shown that $D^{-1}BD = BAFA^{-1}C$ where $F = \Theta_3(3,2,-3/2)$ ($F \in K$ by the Lemma to Result 6.9.5). First note that

$$D^{-1}BD = \begin{pmatrix} 1 & 0 & 0 \\ 1 & 1 & 0 \\ -1 & \frac{1}{4} & 1 \end{pmatrix}$$

by the claim in the previous paragraph. Also

$$AFA^{-1} = \begin{pmatrix} 1 & 0 & 0 \\ 0 & 1 & 0 \\ 0 & -\frac{3}{4} & 1 \end{pmatrix}$$

by the Lemma to Result 6.8.1. Therefore $B(AFA^{-1})C =$

$$\begin{pmatrix} 1 & 0 & 0 \\ 1 & 1 & 0 \\ -4 & 1 & 1 \end{pmatrix}\begin{pmatrix} 1 & 0 & 0 \\ 0 & 1 & 0 \\ 0 & -\frac{3}{4} & 1 \end{pmatrix}\begin{pmatrix} 1 & 0 & 0 \\ 0 & 1 & 0 \\ 3 & 0 & 1 \end{pmatrix} = \begin{pmatrix} 1 & 0 & 0 \\ 1 & 1 & 0 \\ -4 & \frac{1}{4} & 1 \end{pmatrix}\begin{pmatrix} 1 & 0 & 0 \\ 0 & 1 & 0 \\ 3 & 0 & 1 \end{pmatrix}$$

$$= \begin{pmatrix} 1 & 0 & 0 \\ 1 & 1 & 0 \\ -1 & \frac{1}{4} & 1 \end{pmatrix} = D^{-1}BD.$$

Thus $D^{-1}BD \in K$ completing the demonstration that $\operatorname{Im} g \subseteq K$.

Thus far it has been shown that $g:K \to K$ is an injective homomorphism. Next it is claimed that g is onto K. It suffices to show both generators of K belong to $\operatorname{Im} g$. As noted above, $D^{-1}AD = A$ so $A \in \operatorname{Im} g$. Now let

$$R = \begin{pmatrix} 1 & 0 & 0 \\ 1 & 1 & 0 \\ -16 & 4 & 1 \end{pmatrix}.$$

Then by the claim in the second paragraph,

$$D^{-1}RD = \begin{pmatrix} 1 & 0 & 0 \\ 1 & 1 & 0 \\ -4 & 1 & 1 \end{pmatrix} = B.$$

From which it follows that $R = DBD^{-1}$. Now note that

$$C^3 = \begin{pmatrix} 1 & 0 & 0 \\ 0 & 1 & 0 \\ 9 & 0 & 1 \end{pmatrix}$$

by the Corollary to Theorem 3.1.3 so

$$DBD^{-1}C^3 = \begin{pmatrix} 1 & 0 & 0 \\ 1 & 1 & 0 \\ -16 & 4 & 1 \end{pmatrix}\begin{pmatrix} 1 & 0 & 0 \\ 0 & 1 & 0 \\ 9 & 0 & 1 \end{pmatrix} = \begin{pmatrix} 1 & 0 & 0 \\ 1 & 1 & 0 \\ -7 & 4 & 1 \end{pmatrix}.$$

Next note that

$$B^2 = \begin{pmatrix} 1 & 0 & 0 \\ 1 & 1 & 0 \\ -4 & 1 & 1 \end{pmatrix}\begin{pmatrix} 1 & 0 & 0 \\ 1 & 1 & 0 \\ -4 & 1 & 1 \end{pmatrix} = \begin{pmatrix} 1 & 0 & 0 \\ 2 & 1 & 0 \\ -7 & 2 & 1 \end{pmatrix}$$

so by the Lemma to Result 6.8.1,

$$A^{-1}B^2A = \begin{pmatrix} 1 & 0 & 0 \\ 1 & 1 & 0 \\ -7 & 4 & 1 \end{pmatrix} = DBD^{-1}C^3$$

from which it follows that $A^{-1}B^2AC^{-3} = DBD^{-1}$ so $B = D^{-1}(A^{-1}B^2AC^{-3})D$ and hence $B \in \operatorname{Im} g$.

This proves $g: K \to K$ is an automorphism.

Since the center of a group is a characteristic subgroup, $C \in Z(K)$ (Result 6.9.3) implies $g(C) \in Z(K)$, hence $\langle g(C) \rangle \lhd K$. But $\langle g(C) \rangle = g[H]$ so $g[H] \lhd K$. Apply Theorem A.10.1 to get a surjective homomorphism $g': K/H \to K/g[H]$. It is claimed that g' is also injective. If $g'(Hx) = 1$, then $g[H]g(x) = 1$ implying $g(x) \in g[H]$. Thus $g(x) = g(h)$ for some $h \in H$. But then $x = h$ by injectivity of g and so $x \in H$ implying $Hx = 1$ and showing g' is injective and hence is an isomorphism.

Now

$$D^{-1}CD = \begin{pmatrix} 1 & 0 & 0 \\ 0 & 1 & 0 \\ \frac{3}{4} & 0 & 1 \end{pmatrix}$$

by the claim in the second paragraph. Let the last matrix be denoted by W. Thus $W \in g[H]$. Also (Corollary to Theorem 3.1.3) $W^4 = C$. Hence $C \in g[H]$ implying $H \subseteq g[H]$ and hence $(K/H)/(g[H]/H) \cong K/g[H]$. This and the previously established isomorphism effected by g' yield $(K/H)/(g[H]/H) \cong K/H$.

Thus to prove K/H is nonhopfian it suffices to show $g[H]/H$ is nontrivial. As noted in the previous paragraph, $W \in g[H]$. If $W \in H$, however, then $W = C^i$ for some integer i. But all the entries of C^i are integers ($C \in UT(3,Z)$ hence $C^i \in UT(3,Z)$) so this cannot happen. Thus $W \in g[H] \setminus H$ so $g[H]/H$ is nontrivial.

LEMMA (A.I. Malcev) *Finitely generated residually finite groups are hopfian.* Let A be a finitely generated residually finite group, suppose $A/H \cong A$ where H is nontrivial, and seek a contradiction. By nontriviality there exists $h \in H$, $h \neq 1$. Applying Theorem 1.3.9 yields a normal subgroup N such that $h \notin N$ and A/N is finite. Let $\operatorname{card}(A/N) = n$. Let $B = \{J \mid J$ is a subgroup of A and $[A:J] = n\}$, let $C = \{J/H \mid J/H$ is a subgroup of A/H and $[A/H:J/H] = n\}$, and let $D = \{J \mid J \in B$ and $H \subseteq J\}$. Clearly $\operatorname{card}(C) = \operatorname{card}(D)$. Since A and A/H are isomorphic, they each have the same number of subgroups of index n. That is, $\operatorname{card}(B) = \operatorname{card}(C)$. Thus $\operatorname{card}(B) = \operatorname{card}(D)$. Finally note that Theorem A.10.9 guarantees the set B to be finite. The facts $D \subseteq B$, $\operatorname{card}(B) = \operatorname{card}(D)$, and B finite imply $D = B$. Then $N \in B$ implies $N \in D$ hence $H \subseteq N$ so $h \in N$, the required contradiction.

RESULT 6.9.8 *G is not residually finite.*
By Results 6.9.4 and 6.9.7, and the lemma.

EXERCISES

1. Show G is not an FC group.
2. Show G is neither torsion nor torsion-free.
3. Show that elements of G of finite order belong to $Z(G)$.
4. Show that elements of G of relatively prime order commute.
5. Let $U = UT(3, Q) \cap K$. Show
   (a) $U \lhd K$.
   (b) $Z(K) = Z(U)$.
6. Show
   (a) K is not hypercyclic.
   (b) G is not hypercyclic.
7. Show G' is nilpotent.
8. Define $g:K \to T(3,Q)$ by

$$g\begin{pmatrix} 1 & 0 & 0 \\ a & 2^i & 0 \\ b & c & 1 \end{pmatrix} = \begin{pmatrix} 1 & 0 & 0 \\ 7a & 2^i & 0 \\ 7b & c & 1 \end{pmatrix}.$$

   Show
   (a) g is an injective homomorphism.
   (b) $g(B) = BA^2B^2A^{-1}B^{-1}A^{-1}C^{-9}$ and consequently that $g:K \to K$.
   (c) K is noncohopfian.
   (d) G is noncohopfian.
9. If W is a group satisfying the ascending chain condition on normal subgroups, show W is hopfian. Conclude G does not satisfy the ascending chain condition on normal subgroups, and consequently neither does K.

NOTES

1. The question of the existence of a finitely generated nonhopfian group was an unsolved problem for a number of years before a counterexample was published around 1950. The example given here is due to M.F. Newman and was published in 1966.

2. A group is called *linear* if and only if it can be embedded in a group of matrices over some field. By a theorem of Malcev, finitely generated linear groups are residually finite. The group G then shows the class of linear groups is not quotient closed.

□□□□□

Example 6.10

Let p be a prime. Let $G = Z(p^\infty)$ wr $Z(p^\infty)$.

RESULT 6.10.1 *G is a p-group.*
Since $Z(p^\infty)$ is a p-group (Result 5.4.1), so is G by Theorem 1.7.8.

RESULT 6.10.2 *G is semidivisible.*
Since $Z(p^\infty)$ is divisible (Result 5.4.2), G is semidivisible by Theorem 1.7.7.

RESULT 6.10.3 $Z(G) = \{1\}$.
By Theorem 1.7.5.

RESULT 6.10.4
(a) *G is not hypercentral.*
(b) *G is not nilpotent.*
Since $Z(G)$ is trivial (Result 6.10.3), Part (a) is clearly true. Part (b) follows from Part (a) and Corollary 2 to Theorem A.5.6.

LEMMA 1 *Torsion abelian groups are locally finite.*
Let A be a torsion abelian group and let $B = \langle x_1, x_2, \ldots, x_n \rangle$ be a finitely generated subgroup. Then since B is abelian, each $\langle x_i \rangle \lhd B$. Hence $B = \langle x_1 \rangle \langle x_2 \rangle \ldots \langle x_n \rangle$. Since A is torsion, $o(x_i)$ is finite for each i and hence card($\langle x_i \rangle$) is finite. Consequently B is finite since $B = \langle x_1 \rangle \langle x_2 \rangle \ldots \langle x_n \rangle$ implies card(B) $\leqslant$ card($\langle x_1 \rangle$) card($\langle x_2 \rangle$) . . . card($\langle x_n \rangle$).

LEMMA 2 *Extensions of locally finite groups by locally finite groups are locally finite.*
Suppose A/H and H are locally finite, let $F = \langle b_1, b_2, \ldots, b_n \rangle$ be a finitely generated subgroup of A, and seek to show F is finite. Assume for convenience that the set $S = \{b_1, b_2, \ldots, b_n\}$ is closed under inverses. (If it is not, add inverses to it as necessary and relabel. It will still be finite.) Let $L/H = \langle Hb_1, Hb_2, \ldots Hb_n \rangle$. Note that L/H is a finitely generated subgroup of A/H, so it must be fi-

nite, say of cardinality t. Thus $L/H = \{Hb_1, Hb_2, \ldots, Hb_n, Hb_{n+1}, \ldots, Hb_t\}$.

Since L/H is a group, for all $i,j \leqslant t$, $Hb_iHb_j \in L/H$ so $Hb_iHb_j = Hb_k$ for some $k \leqslant t$. Hence there exists $h_{i,j} \in H$ such that $b_ib_j = h_{i,j}b_k$.

Let $C = \langle h_{i,j} \mid 1 \leqslant i,j \leqslant t\rangle$. Then C is a finitely generated subgroup of H hence is finite. Let D denote that set $\{cb_i \mid c \in C \text{ and } 1 \leqslant i \leqslant t\}$. Since C is finite, so is D, hence to complete the proof it suffices to show $F \subseteq D$. Let $x \in F$. Thus x can be expressed as a product $x = y_1y_2 \ldots y_s$ where each $y_i \in S$ and repititions are allowed. It will be shown that $x \in D$ by induction on s. If $s = 1$, then $x = b_i$ for some i, hence $x \in D$ so induction begins.

Assume now the induction hypothesis $w \in D$ when w can be expressed as a product of $s - 1$ elements of the set S and seek to show $x = y_1y_2 \ldots y_s \in D$. Relabel the elements y_1 and y_2 as $y_1 = b_i$ and $y_2 = b_j$. Then by the second paragraph $x = h_{i,j}b_ky_3y_4 \ldots y_s$ for some k. Then $b_ky_3y_4 \ldots y_s \in D$ by the induction hypothesis, so $b_ky_3y_4 \ldots y_s = cb_q$ for some q, and hence $x = (h_{i,j}c)b_q \in D$.

RESULT 6.10.5
(a) *G is locally finite.*
(b) *G is locally nilpotent.*
The group $\mathrm{F}[Z(p^\infty), Z(p^\infty)]$ is abelian and torsion (torsion by Result 5.4.1 and Theorem 1.4.6), hence is locally finite by Lemma 1. Consequently G is an extension of a locally finite group ($\mathrm{F}[Z(p^\infty), Z(p^\infty)]^*$) by a locally finite group ($Z(p^\infty)$) hence G is locally finite by Lemma 2.

As for Part (b), let H be a finitely generated subgroup of G. Then H is finite by Part (a), hence must be a finite p-group since G is a p-group (Result 6.10.1). conclude H is nilpotent by Theorem A.5.7.

LEMMA *If A and B are groups such that A is nontrivial and B is infinite, then A wr B is nonhypercyclic.*
Let $W = A$ wr B. Since A is nontrivial, so is $\mathrm{F}[A,B]^*$ so by Theorem A.7.3 it suffices to show there does not exist a nontrivial cyclic subgroup H such that $H \lhd W$ and $H \subseteq \mathrm{F}[A,B]^*$. Suppose $\langle(f,1)\rangle \lhd W$. It will be shown that $f = 1$ and hence that $(f,1)$ is trivial. To do this it must be shown that $f(b) = 1$ for all $b \in B$. Since $\mathrm{S}(f)$ is finite and B is infinite, there exists $c \in B \backslash \mathrm{S}(f)$. Thus $f(c) = 1$. Let $x = c^{-1}b$. Then $(1,x^{-1})^{-1}(f,1)(1,x^{-1}) \in \langle(f,1)\rangle$ by normality. Consequently $(1,x^{-1})^{-1}(f,1)(1,x^{-1}) = (f,1)^n$ for some integer n. Hence $(f^x,1) = (f^n,1)$ where f^x is the special wreath-product notation ($f^x(t) = f(tx)$ for all $t \in B$) and where f^n is the usual notation for powers of the group element f. The last equation implies $f^x = f^n$. Thus $f(b) = f(cx) = f^x(c) = f^n(c) = f(c)^n = 1$.

RESULT 6.10.6 *G is not hypercyclic.*
By the lemma.

RESULT 6.10.7 *G is solvable.*
Since $Z(p^\infty)$ is abelian it is solvable, hence G is solvable by Theorem 1.7.9.

LEMMA (D.H. McLain) *Locally nilpotent groups are* MN *groups.*
Let A be a locally nilpotent group and let M be a maximal subgroup. It is first assumed that $A' \not\subseteq M$. Thus there exists $x \in A' \setminus M$. Because $x \in A'$, $x = c_1 c_2 \ldots c_n$ where each $c_i = a_i^{-1} b_i^{-1} a_i b_i$. Now since M is a maximal subgroup of A and $x \notin M$ then $A = \langle M \cup \{x\} \rangle$. Thus each $a_i \in \langle M \cup \{x\} \rangle$ and so there exist $m(i_1), m(i_2), \ldots, m(i_{h(i)}) \in M$ such that a_i can be expressed as some product of the elements of $\{m(i_1), m(i_2), \ldots, m(i_{h(i)}), x\}$. In other words $a_i \in \langle m(i_1), m(i_2), \ldots, m(i_{h(i)}), x\rangle$. Similarly there exist $m(i_1)', m(i_2)', \ldots, m(i_{k(i)})' \in M$ such that $b_i \in \langle m(i_1)', m(i_2)', \ldots, m(i_{k(i)})', x\rangle$. Let H denote the subgroup of A generated by x, every $m(i_s)$, and every $m(i_s)'$. Let K denote the subgroup generated by all the above generators of H with the exception of x.

Now $c_i \in H'$ for each i because $a_i, b_i \in H$. Recall $x = c_1 c_2 \ldots c_n$. Thus $x \in H'$. Note that every generator of K is an element of M so $K \subseteq M$. This and $x \notin M$ imply $x \notin K$ so K must be a proper subgroup of H. Now by definition of H, H is finitely generated. Apply Theorem A.10.11 to get existence of a maximal subgroup B of H such that $K \subseteq B$. Now local nilpotence of A implies that the finitely generated subgroup H is nilpotent. Thus Theorem A.10.12 can be applied to yield $B \lhd H$, and hence $H/B \cong Z(p)$ for some prime p. By Theorem A.6.4 then, $H' \subseteq B$. This and $x \in H'$ imply $x \in B$. This and $K \subseteq B$ imply $H \subseteq B$ by definition of H, thus contradicting the fact that B is a maximal subgroup of H. Thus the original assumption $A' \not\subseteq M$ has lead to a contradiction.

Consequently $A' \subseteq M$. Then M/A' is a subgroup of A/A' and furthermore is a normal subgroup since A/A' is abelian. This implies $M \lhd A$ as was required.

RESULT 6.10.8 *G has no maximal subgroups.*
Suppose M is a maximal subgroup. Then $M \lhd G$ by Result 6.10.5 and the lemma. This and the fact that M is maximal imply $G/M \cong Z(p)$ for some prime p. However G is semidivisible (Result 6.10.2) so a contradiction to Theorem A.8.2 has been reached.

RESULT 6.10.9 *G is indecomposable.*
Suppose $G = C \oplus D$ where both C and D are nontrivial and seek a contradiction. For convenience denote $F[Z(p^\infty), Z(p^\infty)]^*$ by F^*.
CASE ONE, $G = F^*D$. Then $C \cong G/D = F^*D/D \cong F^*/(F^* \cap D)$. But F^* is abelian, hence so too is C. It is claimed that $C \subseteq Z(G)$. Let $c \in C$ and $g \in G$. Then $G = C \oplus D$ implies $g = c_1 d_1$ where $c_1 \in C$ and $d_1 \in D$. Then $cg = cc_1 d_1 = c_1 c d_1$ (because C is abelian) $= c_1 d_1 c$ (by Theorem A.2.1) $= gc$ hence $c \in Z(G)$ as claimed. However $Z(G)$ is trivial by Result 6.10.3. Thus $C = \{1\}$, the desired contradiction.
CASE TWO, $G = F^*C$. Similar to Case One.
CASE THREE, $F^*D \subset G$ and $F^*C \subset G$. Since $G/F^* \cong Z(p^\infty)$, Result 5.4.4 shows that either $F^*D/F^* \subseteq F^*C/F^*$ or $F^*C/F^* \subseteq F^*D/F^*$. In the first case $D \subseteq F^*C$ and hence $CD \subseteq F^*C$. But $CD = G$ so $G = F^*C$ the desired contradiction. The second case yields the contradiction $G = F^*D$.

EXERCISES

1. Show G is not finitely generated.
2. Show G is not residually finite.
3. Show G is nonhopfian.
4. Show G does not satisfy the ascending chain condition on normal subgroups.

NOTES

1. The group G illustrates one of the most celebrated applications of wreath products, namely the construction of infinite p-groups having trivial centers. This is in contrast to the finite case, as shown in Theorem A.1.1.
2. As a counterexample G separates the generalized nilpotent classes *locally nilpotent* and *hypercentral.*
3. By Result 6.10.8, $\Phi(G) = G$ so G is an example of an infinite group whose Frattini subgroup is not nilpotent. This is in contrast to the finite case as shown in Scott [33], Theorem 7.3.14.
4. See Lemma 9.21 in Robinson [30], vol II for a proof that G is not divisible. Note this shows the class of divisible groups is not extension closed.
5. See Miscellaneous Exercises V.1 and V.2 at the end of this chapter.

□□□□□

Example 6.11

Let p be a prime. Let $G = UT(\infty, Z(p))$.

RESULT 6.11.1 *G is a p-group.*
By Theorem 3.2.8, each group in the direct system $(\{UT(k_n, Z(p))\}_{n \in N}, \{\pi_n^m \mid n \leqslant m\})$ used to define G is a p-group. Thus G is a p-group by Theorem 1.5.10, Part (b).

RESULT 6.11.2 *G is subdirectly irreducible.*
For convenience let U_n denote $UT(k_n, Z(p))$, the nth group of the direct system used to define G. Theorem 1.5.11 can be applied to yield the result provided it is shown that the direct system satisfies the three conditions; (a) all bonding maps are injective, (b) the group U_0 is subdirectly irreducible, and (c) for $n \leqslant m$, $\pi_n^m [\mathrm{H}(U_n)] \subseteq \mathrm{H}(U_m)$.

By Theorem 3.2.9, $d_n : U_n \to U_{n+1}$ is injective. But a bonding map π_n^m is either the identity map or a composition $\pi_n^m = d_{m-1} \circ d_{m-2} \circ \ldots \circ d_n$. In either case π_n^m is injective, proving (a).

By Theorem 3.2.6, $Z(U_n) \cong Z(p)$ for all n. But each U_n is nilpotent by Theorem 3.2.3. Hence each U_n is subdirectly irreducible by Theorem A.5.10. In particular U_0 is subdirectly irreducible, proving Part (b).

Again note that $Z(U_n) \cong Z(p)$ for all n. But by subdirect irreducibility of U_n, $\{1\} \neq H(U_n) \subseteq Z(U_n)$ so $H(U_n) = Z(U_n)$. Thus to show (c) it suffices to show

$$\pi_n^m [Z(U_n)] \subseteq Z(U_m) \quad \text{for } n \leqslant m. \tag{1}$$

If $n = m$, then π_n^m is the identity and so Inclusion (1) is clearly true. Assume then that $n < m$. Then $\pi_n^m = d_{m-1} \circ d_{m-2} \circ \ldots \circ d_n$ and so Inclusion (1) follows from successive applications of Theorem 3.2.10.

RESULT 6.11.3 *G is indecomposable.*
By Result 6.11.2 and Theorem A.10.10.

RESULT 6.11.4 *G is*
(a) *perfect,*
(b) *nonsolvable,*
(c) *nonnilpotent,*
(d) *nonhypercyclic,*
(e) *nonhypercentral.*
By Theorem 3.2.13, G is perfect proving Part (a). Part (b) follows from Part (a). Part (c) follows from Part (b) and Theorem A.6.7. Part (d) follows from Part (a) and Theorem A.7.4. Part (e) follows from Part (d) and Theorem A.7.2.

RESULT 6.11.5 *G is*
(a) *locally nilpotent,*
(b) *locally solvable.*
By Theorem 3.2.3, each group in the direct system used to define G is nilpotent. But the class of nilpotent groups is closed under subgroups and homomorphic images (Theorem A.5.2). Consequently G is locally nilpotent by Theorem 1.5.12 proving Part (a). Part (b) follows from Part (a) and Theorem A.6.7.

RESULT 6.11.6 *G is not*
(a) *residually finite,*
(b) *residually solvable,*
(c) *residually nilpotent.*
Since G is subdirectly irreducible (Result 6.11.2), if it were residually finite it would also be finite by Theorem 1.3.10. Thus G is not residually finite which proves Part (a). Parts (b) and (c) are similar, depending on the facts that G is neither solvable nor nilpotent (Result 6.11.4).

RESULT 6.11.7
(a) *G has no maximal subgroups.*
(b) *G is an* MN *group.*
If M is a maximal subgroup of G, then $M \lhd G$ by Result 6.11.5, Part (a), and the

Lemma to Result 6.10.8. Hence G/M is a finite cyclic group and consequently abelian. Applying Theorem A.6.4 then yields $G' \subseteq M$. But $G' = G$ (Result 6.11.4, Part (a)) so $G' \subseteq M$ implies $G = M$ in contradiction to the assumption that M is a maximal subgroup. This shows G has no maximal subgroups, proving Part (a). Part (a) implies Part (b) vacuously.

EXERCISES

1. Show G is not finitely generated.
2. Show $G/Z(G)$ is a nonsolvable p-group with trivial center.

NOTES

1. The group G is of interest as an example of a locally solvable group that is not residually solvable, or more basically, as an example of a locally solvable group that is not solvable.
2. Secondly, G provides an example of a locally nilpotent group that is not hypercyclic (cf Theorem A.7.2).

□□□□□

Example 6.12

Let p be a prime. Let $U_0 = U_1 = Z(p)$ and let $U_n = UT(n,Z(p))$ for natural numbers $n \geqslant 2$. Let e_0 and e_1 be the identity isomorphism on $Z(p)$ and let $e_n{:}Z(p) \to Z(U_n)$ be the isomorphism described in Theorem 3.2.6 for $n \geqslant 2$. Thus $(Z(p), \{U_n\}_{n\in N}, \{e_n\}_{n\in N})$ is a countable central pushout system. Let L be its central pushout.

RESULT 6.12.1 *L is subdirectly irreducible.*
Recall that L is the direct limit of groups $\{G_i\}_{i\in N}$ such that $G_0 = U_0 = Z(p)$ and G_i is the central pushout of $(Z(p), G_{i-1}, U_i, k_{i-1}, e_i)$ for $1 \leqslant i$. In order to apply Theorem 1.5.11 and conclude the result, the following must be shown: (a) all bonding maps are injective, (b) G_0 is subdirectly irreducible, and (c) $i \leqslant j$ implies $\pi_i^j[\mathrm{H}(G_i)] \subseteq \mathrm{H}(G_j)$.

Condition (a) is true by Theorem 1.6.1. Next note that since $e_i{:}Z(p) \to Z(U_i)$ is an isomorphism for all i, so is each $k_i{:}Z(p) \to Z(G_i)$ by Theorem 1.6.5. By Theorem 3.2.8, each U_i is a finite p-group, hence each G_i is a finite p-group (Theorem 1.1.6) so each G_i is nilpotent by Theorem A.5.7. Application of Theorem A.5.10 then shows each G_i is subdirectly irreducible. In particular G_0 is subdirectly irreducible, proving condition (b).

Again note that $Z(G_i) \cong Z(p)$ for all i. But by subdirect irreducibility of G_i, $\{1\} \neq H(G_i) \subseteq Z(G_i)$ so $H(G_i) = Z(G_i)$. Thus to show condition (c) it suffices to show $i \leqslant j$ implies $\pi_i^j[Z(G_i)] \subseteq Z(G_j)$. This, however, is true by Theorem 1.6.6.

RESULT 6.12.2 *L is indecomposable.*
By Result 6.12.1 and Theorem A.10.10.

RESULT 6.12.3 *L is not solvable.*
Suppose to the contrary that L is length-m solvable. Then since $UT(2^m + 1, Z(p))$ can be embedded in L (Theorem 1.6.1), $UT(2^m + 1, Z(p))$ must be solvable of length less than or equal to m by Theorem A.6.5. This contradicts Theorem 3.2.11.

RESULT 6.12.4
(a) *L is hypercentral.*
(b) *L is hypercyclic.*
(c) *L is hyperabelian.*
For $2 \leqslant n$, the group U_n is nilpotent by Theorem 3.2.3. Clearly U_0 and U_1 are nilpotent as well, so L is hypercentral by Theorem 1.6.10. This proves Part (a). Part (b) follows from Part (a) and Theorem A.7.2. Part (c) follows from Part (b).

RESULT 6.12.5
(a) *L is not residually solvable.*
(b) *L is not residually nilpotent.*
(c) *L is not residually finite.*
Part (a) by Theorem 1.3.10 and Results 6.12.1 and 6.12.3. Parts (b) and (c) are similar.

EXERCISES

1. Show L is a p-group.
2. Show L is locally nilpotent.
3. Show L is not finitely generated.

NOTE

1. The group L is of interest as an example of a hyperabelian group that is not residually solvable, or more basically, as an example of a hyperabelian group that is not solvable.

□□□□□

Example 6.13 (The Homogeneous* Modular Group)

Let $G = SL(2,Z)$.

RESULT 6.13.1 *G is neither torsion nor torsion-free.*
Since $\Theta_2(1,2,1)$ has infinite order (its nth power is $\Theta_2(1,2,n)$) while $o(-I) = 2$.

RESULT 6.13.2 *Let* $A = \left(\begin{smallmatrix} 1 & 1 \\ 0 & 1 \end{smallmatrix}\right)$ *and let* $B = \left(\begin{smallmatrix} 0 & 1 \\ -1 & 0 \end{smallmatrix}\right)$. *Then* $G = \langle A,B \rangle$.
Denote $\langle A,B \rangle$ by H. First note that for all integers i, $\Theta_2(1,2,i) \in H$ because $\Theta_2(1,2,i) = A^i$. It will be shown that $M \in H$ for all $M \in G$ by induction on $|M_{21}|$. If $|M_{21}| = 0$, then $M_{21} = 0$. Consequently $1 = \det(M) = M_{11}M_{22}$ so $M_{11} = M_{22} = \pm 1$. Thus either $M = \left(\begin{smallmatrix} 1 & i \\ 0 & 1 \end{smallmatrix}\right)$ or $M = \left(\begin{smallmatrix} -1 & i \\ 0 & -1 \end{smallmatrix}\right)$ for some integer i. In the first case $M \in H$ as previously noted. In the second case $B^2M = \left(\begin{smallmatrix} 0 & 1 \\ -1 & 0 \end{smallmatrix}\right)\left(\begin{smallmatrix} 0 & 1 \\ -1 & 0 \end{smallmatrix}\right)\left(\begin{smallmatrix} -1 & i \\ 0 & -1 \end{smallmatrix}\right) = \left(\begin{smallmatrix} -1 & 0 \\ 0 & -1 \end{smallmatrix}\right)\left(\begin{smallmatrix} -1 & i \\ 0 & -1 \end{smallmatrix}\right) = \left(\begin{smallmatrix} 1 & -i \\ 0 & 1 \end{smallmatrix}\right) \in H$ so $M \in H$ as well. Hence induction begins.

Assume now the induction hypothesis, $|M_{21}| < n$ implies $M \in H$, and suppose $|M_{21}| = n$. As the case $|M_{21}| = 0$ has been treated above, it may be assumed that $|M_{21}| \neq 0$. Thus $M_{21} \neq 0$ so the division algorithm may be applied to yield existence of integers d and r such that $M_{11} = dM_{21} + r$ where $r = 0$ or $|r| < |M_{21}|$. Hence $B^{-1}A^{-d}M =$

$$\begin{pmatrix} 0 & -1 \\ 1 & 0 \end{pmatrix}\begin{pmatrix} 1 & -d \\ 0 & 1 \end{pmatrix}\begin{pmatrix} M_{11} & M_{12} \\ M_{21} & M_{22} \end{pmatrix} = \begin{pmatrix} 0 & -1 \\ 1 & -d \end{pmatrix}\begin{pmatrix} M_{11} & M_{12} \\ M_{21} & M_{22} \end{pmatrix}$$

$$= \begin{pmatrix} -M_{21} & -M_{22} \\ M_{11} - dM_{21} & M_{12} - dM_{22} \end{pmatrix}.$$

Thus $(B^{-1}A^{-d}M)_{21} = M_{11} - dM_{21} = r$. If $r = 0$, $B^{-1}A^{-d}M \in H$ as shown in the first paragraph. If $|r| < |M_{21}|$, then $|(B^{-1}A^{-d}M)_{21}| = |r| < |M_{21}| = n$ so that $B^{-1}A^{-d}M \in H$ by the induction hypothesis. Thus in either case $B^{-1}A^{-d}M \in H$ which further implies $M \in H$.

RESULT 6.13.3 *G is coatomic.*
By Result 6.13.2 and Theorem A.10.11.

RESULT 6.13.4 *G has a finite nonabelian simple homomorphic image.*
Let J be the ideal of Z generated by 5. Then $Z/J \cong Z(5)$ and so $SL(2,Z/J) \cong SL(2,5)$. Define $f: SL(2,Z) \to SL(2,Z/J)$ by $f(M)_{ij} = J + M_{ij}$ for $1 \leqslant i,j \leqslant 2$. It is easily seen that f is a homomorphism. By Theorem 3.2.18, to show f is onto $SL(2,Z/J)$, it suffices to show all transvections of $SL(2,Z/J)$ belong to Im f. Let A be a transvection. Thus either $A = \Theta_2(1,2,J+x)$ or $A = \Theta_2(2,1,J+x)$ for some $x \in Z$. In the first case $\Theta_2(1,2,x) \in SL(2,Z)$ and $f(\Theta_2(1,2,x)) = A$. The second

---

*The group $PSL(2,Z) = SL(2,Z)/Z(SL(2,Z))$ is called the Inhomogeneous Modular Group.

case is similar. Thus f is onto $SL(2,Z/J)$ and so $SL(2,Z)/\ker f \cong SL(2,Z/J)$. (Actually $\ker f = C(J)$ but this is incidental to the proof.) Thus $SL(2,5)$ is a homomorphic image of $SL(2,Z)$. But by definition, $PSL(2,5)$ is a homomorphic image of $SL(2,5)$, hence one of $SL(2,Z)$ as well. Finally $PSL(2,5)$ is finite, nonabelian, and is simple by Result 4.11.2.

RESULT 6.13.5

(a) *G is not solvable.*

(b) *G is not locally solvable.*

If G were solvable all of its quotient groups would be likewise by Theorem A.6.5. But G has a finite nonabelian simple quotient group (Result 6.13.4), so G must be nonsolvable, proving Part (a).

Part (b) follows from Part (a) since G is finitely generated (Result 6.13.2).

RESULT 6.13.6 *G is not hyperabelian.*

By Result 6.13.4, there is a normal subgroup, K, of G such that G/K is finite, nonabelian, and simple. Thus G/K has no nontrivial normal abelian subgroups proving G is not hyperabelian.

LEMMA *Let $Z(3^k)$ denote the ring of integers* mod 3^k *where $1 \leqslant k$. Then the group $SL(2,Z(3^k))$ is solvable.*

Let J be the principal ideal of $Z(3^k)$ generated by 3; thus $J = \{3t \mid t \in Z(3^k)\}$. It is seen without difficulty that $\text{card}(J) = \text{card}(Z(3^k))/3 = 3^{k-1}$.

First it is claimed that if a is any element of J, then $a + 1$ is a unit in $Z(3^k)$. Suppose $a = 3t$. Then $a + 1 = 3t + 1$. Since $3t + 1$ and 3 are relatively prime, so are $3t + 1$ and 3^k. Therefore there exist integers b and c such that $(3t + 1)b + 3^k c = 1$. That is, $(3t + 1)b = 1$ in $Z(3^k)$ establishing the claim. This justifies the notation $1/(a + 1)$ for $a \in J$. Let

$$F = \left\{ \begin{pmatrix} 1 + a & b \\ c & (bc + 1)/(a + 1) \end{pmatrix} \middle| \, a,b,c \in J \right\} .$$

The determinant of any matrix in F is $(1 + a)((bc + 1)/(a + 1)) - bc = 1$ so $F \subseteq SL(2,Z(3^k))$. To show F is a subset of the principal congruence subgroup $C(J)$ it need only be shown that $((bc + 1)/(a + 1)) - 1 \in J$ whenever $a,b,c \in J$. But $((bc + 1)/(a + 1)) - 1 = (bc + 1 - (a + 1))/(a + 1) = (bc - a)/(a + 1) = (bc - a)\cdot(1/(a + 1))$ which belongs to J since b,c, and a do and because J is an ideal. Thus $F \subseteq C(J)$. Conversely, let $A \in C(J)$. Then $A_{11} = 1 + a$, $A_{12} = b$, and $A_{21} = c$ where $a,b,c \in J$. Then $1 = \det(A) = (1 + a)A_{22} - bc$ which implies A_{22} is equal to $(bc + 1)/(a + 1)$ and so $A \in F$. Conclude $F = C(J)$. But clearly $\text{card}(F) = \text{card}(J)^3$. Therefore $\text{card}(C(J)) = (3^{k-1})^3 = 3^{3k-3}$ which implies $C(J)$ is nilpotent (Theorem A.5.7), hence solvable (Theorem A.6.7).

Next consider the ring $R = Z(3^k)/J$. Since this ring has cardinality $3^k/3^{k-1} = 3$, it is the 3-element field. By Theorem 3.2.17, therefore, $SL(2,Z(3^k))/C(J)$ can be embedded in $SL(2,3)$. But $SL(2,3)$ is solvable (Result 4.10.6), hence so is

$SL(2,Z(3^k))/C(J)$ (Theorem A.6.5). Finally since $SL(2,Z(3^k))/C(J)$ and $C(J)$ are both solvable, $SL(2,Z(3^k))$ is solvable by Theorem A.6.6.

RESULT 6.13.7
(a) *G is residually finite.*
(b) *G is residually solvable.*
Let J_k be the ideal of Z generated by 3^k for all natural numbers k. Thus $Z/J_k \cong Z(3^k)$. Let $\mathcal{E} = \{C(J_k) \mid k \in N\}$. It is claimed that $\cap \{J_k \mid k \in N\} = \{0\}$. Let m belong to this intersection. If $m \neq 0$, then $m = 3^i d$ for some i and d where 3 does not divide d. But $m \in J_{i+1}$ so $m = e3^{i+1}$ for some integer e which implies $3^i d = e3^{i+1}$ and hence $3^i(d - 3e) = 0$. Thus either $3^i = 0$ or $d = 3e$, false in either case. Hence $m = 0$ proving the claim. This implies $\cap\mathcal{E} = \{I\}$ because if $M \in \cap\mathcal{E}$, then $M \in C(J_k)$ for all natural numbers k, hence $M_{11} - 1, M_{12}, M_{21}, M_{22} - 1 \in J_k$ for all natural numbers k implying $M_{11} - 1 = 0$, $M_{12} = 0$, $M_{21} = 0$, and $M_{22} - 1 = 0$ by the claim, so $M = I$.

By the lemma each $SL(2,Z(3^k))$ and hence each $SL(2,Z/J_k)$ is solvable. It is clearly also finite. Hence by Theorem 3.2.17, $G/C(J_k)$ is finite and solvable for all $k \in N$. This and $\cap\mathcal{E} = \{I\}$ imply G is residually finite and residually solvable by Theorem 1.3.9.

RESULT 6.13.8 *G is hopfian.*
Since G is finitely generated (Result 6.13.2) and residually finite (Result 6.13.7), it is hopfian by the Lemma to Result 6.7.6.

RESULT 6.13.9
(a) *G has noncommuting elements of relatively prime order.*
(b) *G is not residually nilpotent.*
Let $B = \begin{pmatrix} 0 & 1 \\ -1 & 0 \end{pmatrix}$ and $C = \begin{pmatrix} 0 & -1 \\ 1 & -1 \end{pmatrix}$. Then $B^2 = \begin{pmatrix} 0 & 1 \\ -1 & 0 \end{pmatrix}\begin{pmatrix} 0 & 1 \\ -1 & 0 \end{pmatrix} = \begin{pmatrix} -1 & 0 \\ 0 & -1 \end{pmatrix}$ so $B^4 = \begin{pmatrix} -1 & 0 \\ 0 & -1 \end{pmatrix}\begin{pmatrix} -1 & 0 \\ 0 & -1 \end{pmatrix} = I$ and hence $o(B) = 4$. Also $C^3 = \begin{pmatrix} 0 & -1 \\ 1 & -1 \end{pmatrix}\begin{pmatrix} 0 & -1 \\ 1 & -1 \end{pmatrix}\begin{pmatrix} 0 & -1 \\ 1 & -1 \end{pmatrix} = \begin{pmatrix} -1 & 1 \\ -1 & 0 \end{pmatrix}\begin{pmatrix} 0 & -1 \\ 1 & -1 \end{pmatrix} = I$ so $o(C) = 3$. Thus B and C have relatively prime orders. But $BC = \begin{pmatrix} 0 & 1 \\ -1 & 0 \end{pmatrix}\begin{pmatrix} 0 & -1 \\ 1 & -1 \end{pmatrix} = \begin{pmatrix} 1 & -1 \\ 0 & 1 \end{pmatrix}$ while $CB = \begin{pmatrix} 0 & -1 \\ 1 & -1 \end{pmatrix}\begin{pmatrix} 0 & 1 \\ -1 & 0 \end{pmatrix} = \begin{pmatrix} 1 & 0 \\ 1 & 1 \end{pmatrix}$. This proves Part (a). Part (b) follows from Part (a) and the Lemma to Result 6.5.14.

EXERCISES

1. Show G is not an FC group.
2. Show G is not subdirectly irreducible.
3. Show G does not satisfy the descending chain condition on normal subgroups.
4. Show $G = \langle B, C \rangle$ where B and C denote the matrices of Result 6.13.9.
5. Show
   (a) D_3 is isomorphic to a quotient group of G.
   (b) $PSL(2,3)$ is isomorphic to a quotient group of G.

In Exercises 6–8, let G^m denote $\langle x^m \mid x \in G \rangle$.

6. Show
   (a) $G/G^2 \cong Z(2)$.
   (b) G^2 is the only subgroup of G having index 2.
7. Show $G/G^3 \cong Z(3)$.
8. If $m \in N$ such that m and 6 are relatively prime, show $G = G^m$.

NOTES

1. The group G is of interest as an example of a residually solvable group that is neither locally solvable nor hyperabelian. More basically, it separates the classes of solvable and residually solvable groups.
2. As a counterexample G shows
   (a) Direct products of locally solvable groups need not be locally solvable. This is seen as follows. By Result 6.13.7, G is residually solvable and can thus be embedded in a direct product of solvable groups $\times\{A_i\}_{i \in I}$. Since each A_i is solvable it is surely locally solvable. But $\times\{A_i\}_{i \in I}$ is not locally solvable because if it were then G would be also, contradicting Result 6.13.5.

   The same argument can be applied to show direct products of hyperabelian groups need not be hyperabelian provided it is first proven that the class *hyperabelian* is subgroup closed. An easy modification of the proof that the class *hypercyclic* is subgroup closed (Baer [1], Lemma 1) yields this result.

   (b) The class of residually solvable groups is not quotient closed (G is residually solvable but has a finite nonsolvable quotient group G/H by Result 6.13.4: that G/H is not residually solvable follows from Section 1.3, Exercise 12.)
3. Considerably more material on G including a presentation ($G \cong \mathrm{G}(X, Y \mid X^4 = 1, X^2Y^{-3} = 1)$) and information on the principal congruence subgroups, normal subgroups, commutator subgroup, and subgroups of the form G^m (see Exercises 6–8) can be found in Newman [28] or Rankin [29].

□□□□□

Example 6.14

Let $G = PSL(2,Q)$.

RESULT 6.14.1 *G is simple.*
By Theorem 3.2.22.

RESULT 6.14.2
(a) *G is subdirectly irreducible.*
(b) *G is indecomposable.*
Since G is simple (Result 6.14.1), $\mathrm{H}(G) = G$ proving Part (a). Part (b) follows from Part (a) and Theorem A.10.10.

RESULT 6.14.3 *G is neither torsion nor torsion-free.*
Since $\Theta_2(2,1,1)^m = \Theta_2(2,1,m)$ for all positive integers m (Corollary to Theorem 3.1.3), it follows that $[\Theta_2(2,1,1)]^m = [\Theta_2(2,1,m)] \neq 1$. Thus $[\Theta_2(2,1,1)]$ has infinite order showing G is not torsion. Next note that $[(\begin{smallmatrix} 1 & 1 \\ -1 & 0 \end{smallmatrix})]$ has order 3 by Theorem 3.2.25 so G is not torsion-free.

RESULT 6.14.4 *G is not divisible.*
Let $A = (\begin{smallmatrix} 2 & 0 \\ 1 & 1/2 \end{smallmatrix})$. It suffices to show there is no $[M] \in G$ such that $[M]^2 = [A]$. Suppose to the contrary that $[M]^2 = [A]$. For convenience subscripts on the entries of M will be avoided by letting $M = (\begin{smallmatrix} a & b \\ c & d \end{smallmatrix})$. Then $M^2 = (\begin{smallmatrix} a & b \\ c & d \end{smallmatrix})(\begin{smallmatrix} a & b \\ c & d \end{smallmatrix})$

$$= \begin{pmatrix} a^2 + bc & b(a+d) \\ c(a+d) & bc + d^2 \end{pmatrix} \tag{1}$$

so $[M]^2 = [A]$ implies $M^2 = CA$ for some $C \in \mathrm{Z}(SL(2,Q))$ hence the matrix in Line (1) is either $(\begin{smallmatrix} 2 & 0 \\ 1 & 1/2 \end{smallmatrix})$ or

$$\begin{pmatrix} -1 & 0 \\ 0 & -1 \end{pmatrix}\begin{pmatrix} 2 & 0 \\ 1 & ½ \end{pmatrix} = \begin{pmatrix} -2 & 0 \\ -1 & -½ \end{pmatrix}.$$

In either case, $b(a + d) = 0$ so $b = 0$ or $a + d = 0$. If $b = 0$, then in the first case $2 = a^2 + bc = a^2$ while in the second case $-2 = a^2 + bc = a^2$. Thus a contradiction has been reached as neither 2 nor -2 have square roots in Q. Assume then that $a + d = 0$. Then in the first case $1 = c(a + d) = 0$ while in the second case $-1 = c(a + d) = 0$, both being contradictions.

RESULT 6.14.5 *G is semidivisible.*
By Theorem A.8.1 it suffices to show $SL(2,Q)$ is semidivisible. Let $A \in SL(2,Q)$ and let n be a positive integer. Then (Result 3.2.18) $A = B_1B_2 \ldots B_k$ where each B_i is a transvection. If $B_i = \Theta_2(2,1,x)$, then (Corollary to Theorem 3.1.3) $B_i = C_i^n$ for $C_i = \Theta_2(2,1,x/n)$. On the other hand if $B_i = \Theta_2(1,2,x)$, then let $C_i = \Theta_2(1,2,x/n)$ and again $B_i = C_i^n$. Thus $A = C_1^nC_2^n \ldots C_k^n$.

EXERCISE

1. Show G is not locally solvable.

NOTES

1. As a counterexample G
   (a) separates the classes of divisible and semidivisible groups,
   (b) shows that neither the class of groups satisfying the ascending chain condition on normal subgroups, nor the class of groups satisfying the descending chain condition on normal subgroups is subgroup closed: G satisfies both chain conditions, being simple, but Q can be embedded in G and Q satisfies neither condition (Miscellaneous Exercises IV.5 and IV.6 at the end of Chapter 5),
   (c) shows that infinite subgroups of just-infinite groups need not be just-infinite. Since G is infinite and simple, it is just-infinite but Q can be embedded in G and Q is not just-infinite (Q/Z is not finite).
2. Exercise 1 can be generalized; no nonabelian simple group is locally solvable.

□□□□□

Example 6.15

Let $F_0 = D_3$. For all $n \in N$, let F_{n+1} be the group of permutations of the set F_n (the symmetric group of F_n). Let $d_n : F_n \to F_{n+1}$ be the left regular representation: that is, for $x \in F_n$, $d_n(x) : F_n \to F_n$ is defined by $d_n(x)(y) = xy$. For $n \leqslant m$ let π_n^m be the identity map on F_n if $n = m$ and let $\pi_n^m = d_{m-1} \circ d_{m-2} \circ \ldots \circ d_n$ if $n < m$. Clearly $(\{F_n\}_{n\in N}, \{\pi_n^m \mid n \leqslant m\})$ is a direct system. Let G be its direct limit.

RESULT 6.15.1 *Every finite group A can be embedded in G.*
If k is a natural number such that $2 < k$, then 2 and k are distinct divisors of $k!$, and hence $2k \leqslant k!$. Also $2 < k$ implies $1 \leqslant k$ so $1 + k \leqslant 2k$. Thus $1 + k \leqslant k!$ for all k such that $2 < k$.

It is claimed that $n \leqslant \text{card}(F_n)$ for all $n \in N$. This will be shown by induction. When $n = 0$, $\text{card}(F_n) = 6$ so induction begins. Assume now the induction hypothesis, $n - 1 \leqslant \text{card}(F_{n-1})$. Since the case $n = 0$ has just been handled, it may also be assumed that $1 \leqslant n$. Thus $\pi_0^{n-1} : F_0 \to F_{n-1}$ is clearly injective being the composition of injections (that each d_k is injective is easily seen). Thus $\text{card}(F_0) \leqslant \text{card}(F_{n-1})$. Hence $2 < 6 \leqslant \text{card}(F_{n-1})$ and so $1 + \text{card}(F_{n-1}) \leqslant$

card$(F_{n-1})!$ by the first paragraph. This and the induction hypothesis imply $n \leqslant$ card$(F_{n-1})! = $ card(F_n) proving the claim.

Now let card$(A) = m$. By the above, $m \leqslant$ card(F_m) so S_m can be embedded in F_{m+1} by Theorem A.10.8. By Cayley's theorem, A can be embedded in S_m, thus A can be embedded in F_{m+1}. Finally it is clear that all bonding maps of the direct system are injective, hence so too are the associated maps by Theorem 1.5.6. Thus F_{m+1} can be embedded in G, so A can be embedded in G.

RESULT 6.15.2 *G is torsion.*
By Theorem 1.5.10.

LEMMA *Let A be a finite group and let $d{:}A \to S_A$ be the left regular representation. Then if J and K are isomorphic subgroups of A, $d[J]$ and $d[K]$ are conjugate in S_A.*
Let H be the subgroup of A generated by $J \cup K$. Let T be a transversal for H in A, let W be a transversal for J in H, and let W^* be a transversal for K in H. Since $J \cong K$, card(J) = card(K) so by Theorem A.10.14, card(W) = card(H)/card(J) = card(H)/card(K) = card(W^*). Let $e{:}W \to W^*$ be an equivalence (bijection). Let $f{:}J \to K$ be an isomorphism.

By Theorem A.10.14, for all $x \in A$, x is uniquely determined as $x = ht$ where $h \in H$ and $t \in T$. Similarly h is uniquely determined as $h = jw$ where $j \in J$ and $w \in W$. Thus $x = jwt$ and this representation is unique. Define the function $g{:}A \to A$ by $g(jwt) = f(j)e(w)t$. To show g is injective, suppose $x_1 = j_1w_1t_1$ and $x_2 = j_2w_2t_2$ are elements of A such that $g(x_1) = g(x_2)$. Thus $f(j_1)e(w_1)t_1$ and $f(j_2)e(w_2)t_2$ are equal, hence so are the cosets (with respect to H) that they determine. But $f(j_1)$, $f(j_2)$, $e(w_1)$, and $e(w_2)$ all belong to H so the coset equality above implies $Ht_1 = Ht_2$. This and the definition of transversal imply $t_1 = t_2$. This and $f(j_1)e(w_1)t_1 = f(j_2)e(w_2)t_2$ further imply $f(j_1)e(w_1) = f(j_2)e(w_2)$ so $Kf(j_1)e(w_1) = Kf(j_2)e(w_2)$ and hence $e(w_1) = e(w_2)$. Injectivity of e then yields $w_1 = w_2$. Hence $f(j_1) = f(j_2)$ so $j_1 = j_2$ by injectivity of f. Thus $x_1 = j_1w_1t_1 = j_2w_2t_2 = x_2$ proving g is injective. An even easier proof shows g is also surjective, or the proof may be dispensed with by use of the pigeonhole principle. Thus $g{:}A \to A$ is a bijective function, so $g \in S_A$.

It is claimed that $gd[J]g^{-1} = d[K]$, which will complete the proof. Let $j_0 \in J$. Then for all $x \in A$, $x = ht$ where $h \in H$ and $t \in T$, and moreover $h \in H$ implies $h = kw^*$ where $k \in K$ and $w^* \in W^*$. Since f and e are onto K and W^* respectively, there exist $j \in J$ and $w \in W$ such that $f(j) = k$ and $e(w) = w^*$. Then $x = kw^*t = f(j)e(w)t$. Since $g(jwt) = x$, $jwt = g^{-1}(x)$. Consequently $d(j_0)\circ g^{-1}(x) = d(j_0)(jwt) = j_0jwt$, and so $g\circ d(j_0)\circ g^{-1}(x) = g(j_0jwt) = f(j_0j)e(w)t = f(j_0)f(j)e(w)t = f(j_0)x = d(f(j_0))(x)$. This recall is for all $x \in A$, so $g\circ d(j_0)\circ g^{-1} = d(f(j_0)) \in d[K]$ so $gd[J]g^{-1} \subseteq d[K]$. A comparison of cardinalities now shows $gd[J]g^{-1} = d[K]$ as claimed.

RESULT 6.15.3
(a) *Any two finite isomorphic subgroups of G are conjugate in G. More partic-*

ularly,

(b) *If B and C are finite isomorphic subgroups of G and $\phi: B \to C$ is an isomorphism, then there exists $h \in G$ such that $\phi(b) = h^{-1}bh$ for all $b \in B$.*

(c) *If $x,y \in G$ such that $o(x) = o(y)$, then x and y are conjugate in G.*

Let J and K be finite isomorphic subgroups of G. Suppose $J = \{j_1, j_2, \ldots, j_n\}$ and $K = \{k_1, k_2, \ldots, k_n\}$. Then since G is a direct limit, for each $j_i \in J$, $j_i = [(x_i, m_i)]$ for some m_i and x_i while for each $k_i \in K$, $k_i = [(y_i, q_i)]$ for some q_i and y_i. Let $m = \max\{m_1, m_2, \ldots, m_n, q_1, q_2, \ldots, q_n\}$. Then for all $i \leqslant n$,

$$j_i = h_{m_i}(x_i) = h_m \circ \pi_{m_i}^m(x_i)$$

by Theorem 1.5.4. Thus $J \subseteq \operatorname{Im} h_m$ and similarly $K \subseteq \operatorname{Im} h_m$. Let $J^* = h_m^{-1}[J]$ and $K^* = h_m^{-1}[K]$, so that J^* and K^* are subgroups of F_m. The homomorphism h_m is injective by Theorem 1.5.6 because all bonding maps are clearly injective. Let h_m' be h_m restricted to J^* and let h_m'' be h_m restricted to K^*. Then $h_m': J^* \to J$ is an isomorphism (surjective because $J \subseteq \operatorname{Im} h_m$ as shown above). Similarly $h_m'': K^* \to K$ is an isomorphism so $J^* \cong K^*$. Since J^* and K^* are isomorphic subgroups of the finite group F_m, $\pi_m^{m+1}[J^*]$ is conjugate to $\pi_m^{m+1}[K^*]$ in F_{m+1} by the lemma. Thus there exists $y \in F_{m+1}$ such that $\pi_m^{m+1}[J^*] = y^{-1}\pi_m^{m+1}[K^*]y$. Let ν denote $h_{m+1}(y)$. It is claimed that $J = \nu^{-1}K\nu$. Let $j \in J$. Then since h_m' is an isomorphism, $j = h_m(j^*)$ for some $j^* \in J^*$. Therefore $\pi_m^{m+1}(j^*) = y^{-1}\pi_m^{m+1}(k^*)y$ for some $k^* \in K^*$. Applying the function h_{m+1} to this equation, and using the fact that $h_{m+1} \circ \pi_m^{m+1} = h_m$ (Theorem 1.5.4) results in $\nu^{-1}h_m(k^*)\nu = h_m(j^*) = j$. Note that $k^* \in K^*$ implies $h_m(k^*) \in K$ so $j \in \nu^{-1}K\nu$. Thus it has been shown that $J \subseteq \nu^{-1}K\nu$. A cardinality comparison shows equality completing the proof of Part (a).

Now let B and C be finite isomorphic subgroups of G with an isomorphism $\phi: B \to C$. Since $\operatorname{Hol}(B)$ is a finite group, there exists an embedding $e: \operatorname{Hol}(B) \to G$ by Result 6.15.1. Let $R = \operatorname{Im} e$. As usual for holomorphs, let B^* denote $\{(b,1) \mid b \in B\}$. Let $F = e[B^*]$. Thus $F \cong B^* \cong B$ so by Part (a) there exists $w \in G$ such that $F = w^{-1}Bw$. Also $F \cong B$ implies $F \cong C$ so there exists $y \in G$ such that $F = y^{-1}Cy$. For all $x \in F$, $wxw^{-1} \in B$ implying $\phi(wxw^{-1}) \in C$ and so $y^{-1}\phi(wxw^{-1})y$ belongs to F again. Thus β defined by $\beta(x) = y^{-1}\phi(wxw^{-1})y$ is a function from F into itself. It is clearly both a homomorphism and injective, hence application of the pigeonhole principle shows $\beta \in \operatorname{Aut}(F)$. Let α denote the isomorphism $\alpha: B \to B^*$ defined by $\alpha(b) = (b,1)$. Then $e \circ \alpha: B \to F$ is an isomorphism. This and $\beta \in \operatorname{Aut}(F)$ imply $(e \circ \alpha)^{-1} \circ \beta \circ e \circ \alpha \in \operatorname{Aut}(B)$. Apply Theorem 1.2.21 to produce an inner automorphism of $\operatorname{Hol}(B)$, call it f, such that $f \circ \alpha = \alpha \circ (e \circ \alpha)^{-1} \circ \beta \circ e \circ \alpha$, so $\beta = e \circ f \circ e^{-1}$. It is clear that f is an inner automorphism of $\operatorname{Hol}(B)$ implies β must be an inner automorphism of F. Thus there exists $k \in F$ such that for all $x \in F$, $\beta(x) = k^{-1}xk$. Let $h = wky^{-1}$. Then for all $b \in B$, $y\beta(w^{-1}bw)y^{-1} = y(k^{-1}w^{-1} \cdot bwk)y^{-1} = h^{-1}bh$. On the other hand, by definition of β, $y\beta(w^{-1}bw)y^{-1} = \phi(w(w^{-1}bw)w^{-1}) = \phi(b)$. Thus $\phi(b) = h^{-1}bh$ for all $b \in B$ completing Part (b).

To prove Part (c), suppose $o(x) = o(y) = n$. The function $\phi: \langle x \rangle \to \langle y \rangle$ defined by $\phi(ix) = iy$ is clearly an isomorphism. Application of Part (b) then yields $h \in G$ such that $h^{-1}xh = \phi(x) = y$.

RESULT 6.15.4 *G is simple.*

Suppose H is a nontrivial normal subgroup. Let $h \in H$ such that $h \neq 1$ and let $o(h) = m$ (every element has finite order by Result 6.15.2). Note that $m \neq 1$. It must be shown that $G = H$. So let $g \in G$ and suppose $o(g) = n$.

By Theorem 1.7.11, there exists a finite group A and two elements $a,b \in A$ such that $o(a) = o(b) = m$ and $o(ab) = n$. By Result 6.15.1, there is an embedding $e:A \to G$. Thus $e(ab)$ and g are two elements of G of order n. Applying Result 6.15.3, Part (c) yields $x^{-1}e(ab)x = g$ for some $x \in G$. Hence $g = (x^{-1}e(a)x)\cdot(x^{-1}e(b)x)$. But $o(x^{-1}e(a)x) = o(e(a)) = m = o(h)$, and similarly $o(x^{-1}e(b)x) = o(h)$. Two more applications of Result 6.15.3, Part (c) show $x^{-1}e(a)x = y^{-1}hy$ and $x^{-1}e(b)x = t^{-1}ht$ for some $y, t \in G$. It follows then from normality of H that $x^{-1}e(a)x$ and $x^{-1}e(b)x$ belong to H, hence so too does their product g.

RESULT 6.15.5

(a) *G is subdirectly irreducible.*

(b) *G is indecomposable.*

Since G is simple (Result 6.15.4), $H(G) = G$ proving Part (a). Part (b) follows from Part (a) and Theorem A.10.10.

RESULT 6.15.6 *G is divisible.*

Let $x \in G$ and let n be a positive integer. Let W denote the group $\langle x \rangle$ wr $Z(n)$. By Theorem 1.7.10, there exists an injection $e:\langle x \rangle \to W$ such that $e(x) = y^n$ for some $y \in W$. Since $\langle x \rangle$ is finite (G is torsion by Result 6.15.2), so is W. Consequently there is an embedding $j: W \to G$ by Result 6.15.1. Thus $j \circ e(x) = j(y)^n$. Since x and $j \circ e(x)$ are elements of G having the same order, they are conjugate in G by Result 6.15.3, Part (c). Thus $j \circ e(x) = t^{-1}xt$ for some $t \in G$. Consequently $x = t(j \circ e(x))t^{-1} = tj(y)^n t^{-1} = (tj(y)t^{-1})^n$ completing the proof that G is divisible.

EXERCISES

1. Suppose B is a finite group and A is a subgroup of B. Show every embedding $g:A \to G$ can be extended to an embedding $f:B \to G$.
2. Show how the result quoted in Note 1 implies G is noncohopfian.
3. Show G is not locally solvable.

NOTES

1. Our explication of most of the properties of G was based on Kegel and Wehrfritz [22], the authors of which in turn acknowledge Hall [16] as the original source. See Kegel and Wehrfritz [22] (Theorem 6.1, Parts (b) and (c)) for proofs of Exercise 1 and the following result: *Every countable locally finite group can be embedded in G.*
2. Note that G satisfies the descending chain condition on normal subgroups

(being simple) but is noncohopfian (Exercise 2). Cf the dual situation, Exercise 9 of Example 6.9.

□□□□□

MISCELLANEOUS EXERCISES

I Nilpotent Groups

I.1. Show $M(Z)$ is
(a) nilpotent,
(b) torsion-free,
(c) finitely generated and moreover satisfies the ascending chain condition.

I.2. Show that torsion-free class-2 nilpotent groups are R-groups and conclude that Examples 6.1 and 6.3 and Miscellaneous Exercise I.1 are R-groups.

I.3. Let p be a prime. Let

$$H = \left\{ \begin{pmatrix} 1 & 0 & 0 \\ 0 & 1 & 0 \\ x & 0 & 1 \end{pmatrix} \middle| \, x \in Z \right\}.$$

Show $H \lhd UT(3, Q)$ and let $G = UT(3, Q)/H$. Show
(a) G is nilpotent.
(b) G is divisible.
(c) The set of elements of G of finite order is a subgroup and this subgroup is not a direct factor of G.

Conclude, in contrast to the abelian case, that the torsion subgroup of a divisible nilpotent group need not be a direct factor.

I.4. Let $H = Z(2) \times Z(2)$. Define $d{:}H \to H$ by $d(x,y) = (y,x)$. Show $d \in \mathrm{Aut}(H)$ and define $g{:}Z \to \mathrm{Aut}(H)$ by $g_i = 1$ for even i and $g_i = d$ for odd i. Show (Z, g, H) is a semidirect system and let $G = H \times_g Z$. Show
(a) $G' \subseteq Z(G)$ and hence G is class-2 nilpotent.
(b) $\mathrm{card}(G') = 2$.
(c) G is an FC group.
(d) G is finitely generated and moreover satisfies the ascending chain condition.
(e) $H^* = \{t \mid t \in G \text{ and } o(t) \text{ is finite}\}$.
(f) H^* is not a direct factor of G.

Conclude, in contrast to the abelian case, that the torsion subgroup of a finitely generated nilpotent group need not be a direct factor.

I.5. Show $UT(3, Z_3[x])$ has exponent 3.

I.6. Let R be a commutative ring with identity, and let A be an R-module. Let $G(A,R) = A \times A \times R$ with a multiplication defined by $(a,b,i)(a',b',j) = (a + a' + ib', b + b', i + j)$. Show

(a) $G(A,R)$ is a group and moreover inverses are described by $(a,b,i)^{-1} = (ib - a, -b, -i)$.

(b) $Z(G(A,R)) = \{(a,0,0) \mid a \in A\}$ and $Z(G(A,R)) \cong A$.

(c) $G(A,R)' \subseteq Z(G(A,R))$ so $G(A,R)$ is class-2 nilpotent.

(d) $G(A,R)/Z(G(A,R)) \cong A \times R$.

Note that $G(A,R)$ is a generalization of the construction $M(A)$ (every abelian group A is a Z-module).

I.7. Let p be a prime. First show $Z(p)$ is a Q_p-module. Let $G = G(Z(p), Q_p)$ as described in Exercise I.6. Show

(a) $Z(G) \cong Z(p)$.

(b) G is an FC group.

(c) G is noncoatomic.

Conclude that nilpotent groups with coatomic centers need not be coatomic.

I.8. (a) Give an example of an uncountable nilpotent group with a finite center.

(b) Give an example of a torsion-free uncountable nilpotent group with a countable center.

II Hypercyclic Groups

II.1. Let G be the restricted direct product of all dihedral groups D_m such that $m = 2^n$ for some $n \geqslant 2$. Show G is

(a) solvable,

(b) hypercentral,

(c) nonnilpotent,

(d) an FC group.

II.2. Let $G = G(R, S \mid R^2 = 1, (RS)^2(SR)^{-2} = 1)$. Show G

(a) is infinite,

(b) is solvable,

(c) is hypercyclic,

(d) satisfies the ascending chain condition.

II.3. (a) Give an example of a nonsolvable torsion-free hypercentral group.

(b) Give an example of a nonsolvable divisible hypercentral group.

II.4. Let $G = Z \text{ wr } Z(2)$. Show G is

(a) hypercyclic,

(b) residually nilpotent and residually finite,

(c) finitely generated and moreover satisfies the ascending chain condition.

III Solvable Groups

III.1. Let $G = Z \text{ WR } Z$ (see Miscellaneous Exercises, Chapter 1). Show
(a) G is solvable.
(b) G is torsion-free.
(c) $(g,1)^2 = (h,1)^2$ where g and h are defined by

$$g(i) = \begin{cases} -1 & \text{if } i \text{ is even,} \\ 1 & \text{if } i \text{ is odd,} \end{cases} \quad \text{and} \quad h(i) = \begin{cases} 1 & \text{if } i \text{ is even,} \\ -1 & \text{if } i \text{ is odd.} \end{cases}$$

Thus G is not an R-group.
This example separates the classes of torsion-free and R-groups and also shows the class of R-groups is not extension closed.

III.2. (a) Give an example of a finitely generated solvable torsion-free group that is not an R-group.
(b) Give an example of a semidivisible solvable torsion-free group that is not an R-group.

III.3. Show $Z(2^\infty)$ wr $Dih(Z(2^\infty))$ is indecomposable.

III.4. Define $d{:}Z \times Z \to Z \times Z$ by $d(x,y) = (-y,x)$. Show $d \in \text{Aut}(Z \times Z)$ and has order 4. Define $g{:}Z(4) \to \text{Aut}(Z \times Z)$ by $g_i = d^i$ for $i = 0,1,2,3$. Let $G = (Z \times Z) \times_g Z(4)$. Show
(a) If $H \lhd G$, $H \subseteq (Z \times Z)^*$ and H is cyclic, then $H = \{1\}$.
(b) G is not hypercyclic.
(c) G is residually nilpotent.
Conclude residually nilpotent groups need not be hypercyclic (cf Theorem A.7.2).
(d) $(Z \times Z)^* = C_G((Z \times Z)^*)$.
(e) G is just-infinite.

III.5. Let p be a prime. Give an example of a solvable bounded p-group with trivial center.

III.6. Show Z wr Z is finitely generated.

IV Nonhypercyclic Nonsolvable Groups

IV.1. Give an example of a torsion-free locally nilpotent perfect group.

IV.2. Give an example of an uncountable simple group.

IV.3. Let $G = S_Z$. For any integer n, let $F_n = \{f \mid f \in S_Z \text{ and } f(n) = n\}$. Show
(a) F_n is a maximal subgroup of G.
(b) $[G{:}F_n]$ is infinite.

IV.4. Let G be the subgroup of $SL(2,Q)$ generated by $\Theta_2(2,1,1/2)$, $\Theta_2(2,1,1/3)$, $\Theta_2(1,2,1/2)$, and $\Theta_2(1,2,1/3)$. Show G is perfect.

IV.5. Let $G = G(X, Y \mid X^2 = 1, Y^3 = 1, (XY)^7 = 1)$. Show G is perfect.

V General

V.1. Show that if G is an FC group, then $G/Z(G)$ is residually finite. Conclude that Examples 6.1, 6.2, and 6.10 are not FC groups.

V.2. Show that if G is a group having a nontrivial divisible abelian quotient group, then G is noncoatomic. Conclude Examples 5.1–5.6, 5.9, 5.11, 5.12, 5.14, 5.15, 6.1, 6.2, and 6.10 are noncoatomic.

V.3. Show that the center of a subdirectly irreducible group is either trivial or subdirectly irreducible. Use this to show that Examples 6.1, 6.3, and 6.8 are not subdirectly irreducible.

SUGGESTIONS FOR FURTHER READING

See Robinson [30], volume I, Section 6.2 for a family of examples called the McLain groups. These are locally nilpotent groups whose construction essentially generalizes the family of matrix groups $UT(n,F)$ where F is a field. The McLain groups have a variety of applications as counterexamples in infinite group theory and are particularly useful for separating classes of generalized solvable groups.

A very pleasant wreath-product construction for just-infinite groups is given in McCarthy [26]. Some examples of infinite simple groups are found in Higman [20]. See Scott [33], Chapter 11, for a thorough discussion of infinite permutation groups.

Appendixes

A.1 Referenced Theorems

The following theorems are used in several places in the text and in the other appendixes. It is assumed that the reader is familiar with these theorems and hence their proofs will be omitted but referenced.

THEOREM A.1.1 *Nontrivial finite p-groups have nontrivial centers.*
Baumslag and Chandler [2], Theorem 5.11; Hall [14], Theorem 4.3.1; Herstein [18], Theorem 2.I; Rotman [31], Theorem 5.4; Scott [33], Theorem 6.3.2.

THEOREM A.1.2 *If G is a group of cardinality p^2 where p is a prime, then G is abelian.*
Baumslag and Chandler [2], Problem 5.19; Hall [14], page 49; Herstein [18], Corollary to Theorem 2.I; Rotman [31], Corollary 5.5; Scott [33], Theorem 6.3.5.

THEOREM A.1.3 *If G is a group and H is a subgroup of finite index n, then $G/\mathrm{Core}_G(H)$ can be embedded in S_n.*
Baumslag and Chandler [2], Theorem 7.2; Hall [14], Theorem 5.3.2; Herstein [18], Theorem 2.G; Rotman [31], Theorem 3.17.

THEOREM A.1.4 *If G is a group and B is a subgroup of G, then B has exactly $[G : \mathrm{N}_G(B)]$ conjugates in G.*
Baumslag and Chandler [2], Lemma 5.4; Hall [14], Theorem 1.6.1; Schenkman [32], Proposition III.1.p; Scott [33], Theorem 3.3.2.

THEOREM A.1.5 (Cauchy's theorem) *If p is a prime and G is a finite group such that p divides* card(G), *then G has an element of order p.*
Baumslag and Chandler [2], Proposition 5.9; Hall [14], Theorem 4.1.1; Herstein [18], Theorem 2.J; Rotman [31], Theorem 5.2; Schenkman [32], Corollary IV.1.f.

THEOREM A.1.6 (Sylow's theorem) *If p is a prime and G is a finite group, then*
(a) *p-Sylow subgroups of G exist.*
(b) *Any two p-Sylow subgroups of G are conjugate in G.*
(c) $n_p(G) \equiv 1 \bmod p$ *and* $n_p(G) \mid \mathrm{card}(G)$.
Baumslag and Chandler [2], Theorems 5.1 and 5.3; Hall [14], Theorems 4.2.1–4.2.3; Rotman [31], Theorem 5.9 and Corollary 5.11; Schenkman [32], Theorems IV.1.e and IV.1.i; Scott [33], Theorem 6.1.11.

A.2 Direct Factors and the ⊕ Notation

NOTATION $G = A \oplus B$ will be used to represent the following situation: A and B are normal subgroups of G, $G = AB$, and $A \cap B = \{1\}$.

If $G = A \oplus B$, then A and B are called **direct factors** of G.

THEOREM A.2.1 *If* $G = A \oplus B$, *then* $ab = ba$ *for all* $a \in A$ *and* $b \in B$.
Proof: Since B is normal in G, $a^{-1}ba \in B$. This and $b^{-1} \in B$ imply $b^{-1}a^{-1}ba \in B$. Since A is normal in G, $b^{-1}a^{-1}b \in A$. This and $a \in A$ imply $b^{-1}a^{-1}ba \in A$. Thus $b^{-1}a^{-1}ba$ belongs to $A \cap B$ which is trivial. Conclude $ba = ab$.

COROLLARY *If* $G = A \oplus B$ *and both A and B are abelian, then G is abelian.*
Proof: Let $x, x^* \in G$. Then $x = ab$ and $x^* = a^*b^*$ for some $a, a^* \in A$ and $b, b^* \in B$. Hence $xx^* = aba^*b^* = aa^*bb^*$ (by Theorem A.2.1) $= a^*ab^*b$ (since A and B are abelian) $= a^*b^*ab$ (Theorem A.2.1 again) $= x^*x$.

THEOREM A.2.2 *If* $G = A \oplus B$, *then every* $g \in G$ *has a unique representation* $g = ab$ *where* $a \in A$ *and* $b \in B$.
Proof: That $g = ab$ for some $a \in A$ and $b \in B$ follows immediately from the meaning of the ⊕ notation ($G = AB$). It need only be shown, therefore, that a and b are unique. Suppose $g = a_2b_2$ where $a_2 \in A$ and $b_2 \in B$. Then $ab = a_2b_2$ hence ${a_2}^{-1}a = b_2b^{-1}$ implying ${a_2}^{-1}a$ belongs to $A \cap B$ which is trivial. Thus $a_2 = a$. Similarly $b_2 = b$.

Because of Theorem A.2.2 we may define functions $\pi_A : G \to A$ and $\pi_B : G \to B$ as follows: $\pi_A(g) = a$ and $\pi_B(g) = b$ if and only if $g = ab$ is the unique representation for g.

THEOREM A.2.3 *If $G = A \oplus B$, then π_A and π_B are homomorphisms.*
Proof: Suppose g and g^* belong to G, and have unique representations $g = ab$ and $g^* = a^*b^*$. Then $gg^* = aba^*b^* = aa^*bb^*$ by Theorem A.2.1. Therefore $\pi_A(gg^*) = aa^* = \pi_A(g)\pi_A(g^*)$ and so π_A is a homomorphism. Similarly so is π_B.

THEOREM A.2.4 *If $G = A \times B$, then there are subgroups J and K of G such that $J \cong A$, $K \cong B$, and $G = J \oplus K$.*
Proof: Let $J = \{(a,1) \mid a \in A\}$ and let $K = \{(1,b) \mid b \in B\}$. Clearly J and K are subgroups of G. Define $j:A \to J$ and $k:B \to K$ by $j(a) = (a,1)$ and $k(b) = (1,b)$. It is easily seen that j and k are isomorphisms. Next note that $(a,1) \in J$ and $(x,y) \in G$ imply $(x,y)^{-1}(a,1)(x,y) = (x^{-1}ax,1) \in J$ so $J \lhd G$. Similarly $K \lhd G$. If $(x,y) \in J \cap K$, then $x = 1$ and $y = 1$ so $J \cap K$ is trivial. And finally if $(x,y) \in G$, then $(x,y) = (x,1)(1,y) \in JK$ so $G = JK$. Conclude $G = J \oplus K$.

THEOREM A.2.5 *If $G = A \oplus B$, then $G \cong A \times B$.*
Proof: Define $d:A \times B \to G$ by $d(a,b) = ab$. To show d is a homomorphism let (a,b), $(a^*,b^*) \in A \times B$. Then $d((a,b)(a^*,b^*)) = d(aa^*,bb^*) = aa^*bb^* = aba^*b^*$ (by Theorem A.2.1) $= d(a,b)d(a^*,b^*)$. To show injectivity of d, suppose that $d(a,b) = 1$. Then $ab = 1$ implying $a = b^{-1}$ and so $a \in A \cap B$ hence $a = 1$. Also $b = a^{-1}$ so $b = 1$. Thus $(a,b) = (1,1)$, the identity element of $A \times B$ proving d is injective. Finally to show d is surjective, let $g \in G$. Then $g = ab$ for some $a \in A$ and $b \in B$ because $G = AB$. Thus $g = d(a,b)$ and so d is surjective.

THEOREM A.2.6 *If $G = A \oplus B$ and $H \lhd A$, then $H \lhd G$.*
Proof: Let $h \in H$ and $g \in G$. Then $g = ab$ for some $a \in A$ and $b \in B$ because $G = AB$. Thus $g^{-1}hg = b^{-1}a^{-1}hab$. But $a^{-1}ha \in H$ by normality of H and therefore $a^{-1}hab = ba^{-1}ha$ by Theorem A.2.1. Thus $g^{-1}hg = b^{-1}ba^{-1}ha = a^{-1}ha \in H$.

THEOREM A.2.7 *If $G = A \oplus B$ and F is a fully invariant subgroup of G, then $F = (F \cap A) \oplus (F \cap B)$.*
Proof: Since A is a normal subgroup of G, $F \cap A$ is normal in F. Similarly $F \cap B \lhd F$. Since A and B intersect trivially, so do $F \cap A$ and $F \cap B$. Now all that must be shown is that $F = (F \cap A)(F \cap B)$. By Theorem A.2.3, both π_A and π_B are homomorphisms. Thus $\pi_A[F] \subseteq F$ and $\pi_B[F] \subseteq F$ by full invariance of F. Let $x \in F$. Then $x \in G$ so $x = ab$ for some $a \in A$ and $b \in B$. Hence $a = \pi_A(x)$. But $\pi_A[F] \subseteq F$ implies $a \in F$. Thus $a \in F \cap A$ and similarly $b \in F \cap B$. Then since $x = ab$, $x \in (F \cap A)(F \cap B)$ as required. This shows $F \subseteq (F \cap A)(F \cap B)$. The reverse inclusion is clear.

THEOREM A.2.8 *If $G = A \oplus B$ and F is a fully invariant subgroup of G, then $G/F = (FA/F) \oplus (FB/F)$.*
Proof: Since A and F are both normal subgroups of G, so is FA. Hence FA/F is normal in G/F. Similarly $FB/F \lhd G/F$. From $G = AB$ it follows that $G/F = (FA/F)(FB/F)$. Thus it suffices to show that FA/F and FB/F intersect trivially.

Let Fx belong to their intersection. Thus $x = fa = f^*b$ where $f, f^* \in F$, $a \in A$, and $b \in B$. By Theorem A.2.7, $F = (F \cap A) \oplus (F \cap B)$. Thus $f = cd$ and $f^* = c^*d^*$ where $c, c^* \in F \cap A$ and $d, d^* \in F \cap B$. Then $x = fa = f^*b$ implies $x = cda = c^*d^*b$. But $d \in B$ so d and a commute by Theorem A.2.1. Thus $x = (ca)d = c^*(d^*b)$. Theorem A.2.2 may now be applied to yield $ca = c^*$. Therefore $x = (ca)d = c^*d \in F$ showing $Fx = 1$ as required.

A.3 The Axiom of Choice and Zorn's Lemma

Both the axiom of choice and Zorn's lemma are set-theoretical propositions which guarantee the existence of something, a function in the first case and maximal elements in the second. These two propositions and a number of other set-theoretical propositions, among which are the well-ordering theorem and the Hausdorff maximal principle, enjoy a special position in mathematics. None of them are provable solely from the axioms of set theory* but they are all equivalent; that is, each may be proven from one of the others. Thus the working mathematician is forced into either wholesale acceptance or rejection of the lot of them. The "obvious" nature of some of these equivalents (in particular the axiom of choice seems obvious to most people) coupled with the fact that the equivalents are indispensable to many branches of mathematics has resulted in their all but universal acceptance.

No equivalence proof for the axiom of choice and Zorn's lemma is given here. For such a proof see Hewitt and Stromberg [19], Theorem 3.12. The reader encountering Zorn's lemma here for the first time is advised to accept it if it appears obvious, or to pursue a proof of its derivation from the axiom of choice if it does not.

Informally, the axiom of choice says that given a collection of nonempty sets, one element may be chosen from each of these sets. A formal statement of the axiom is as follows:

Axiom of choice If $\mathcal{R} = \{A_i \mid i \in I\}$ is a collection of nonempty sets, then there exists a function $f: I \to \cup\mathcal{R}$ such that $f(i) \in A_i$ for all $i \in I$.

When the index set I above is finite, the axiom of choice can be used, but is not necessary. For a discussion of why this is so, see Dugundji [9], p 21.

The statement of Zorn's lemma follows four preliminary definitions.

DEFINITION $(S, \leqslant)$ is a **partially ordered set** if and only if $\leqslant$ is a relation on S

*That they are unprovable from the axioms of set theory is, in fact, provable.

such that

(a) $x \leqslant x$ for all $x \in S$.

(b) $x \leqslant y$ and $y \leqslant x$ imply $x = y$.

(c) $x \leqslant y$ and $y \leqslant z$ imply $x \leqslant z$.

((a), (b), and (c) above are called the *reflexive, antisymmetric,* and *transitive* conditions respectively.)

DEFINITION If $(S, \leqslant)$ is a partially ordered set, then C is a **subchain** of $(S, \leqslant)$ if and only if

(a) C is a subset of S.

(b) For all $x, y \in C$, either $x \leqslant y$ or $y \leqslant x$.

If S happens to be a subchain of $(S, \leqslant)$, the partially ordered set $(S, \leqslant)$ is called a **chain.**

DEFINITION If $(S, \leqslant)$ is a partially ordered set and A is a subset of S, then x is an **upper bound** for A if and only if

(a) $x \in S$.

(b) $a \leqslant x$ for all $a \in A$.

DEFINITION If $(S, \leqslant)$ is a partially ordered set and $x \in S$, then x is a **maximal element** of $(S, \leqslant)$ if and only if $x \leqslant y$ for some $y \in S$ implies that $x = y$.

Zorn's lemma If $(S, \leqslant)$ is a nonempty partially ordered set every subchain of which has an upper bound, then S has (one or more) maximal elements.

A.4 Abelian Groups

THEOREM A.4.1 *If A is an abelian group and n is a positive integer, then both $A[n]$ and nA are fully invariant subgroups of A and $A/A[n] \cong nA$.*

Proof: Define $d{:}A \to A$ by $d(x) = nx$. Then $d(x + y) = n(x + y) = nx + ny = d(x) + d(y)$ so d is a homomorphism. It follows then from the first isomorphism theorem that $A/\ker d \cong \operatorname{Im} d$. Now $x \in \ker d$ if and only if $d(x) = 0$ which is true if and only if $nx = 0$, so $\ker d = A[n]$. Also $x \in \operatorname{Im} d$ if and only if $x = d(y)$ for some y which happens if and only if $x = ny$ for some y, so $\operatorname{Im} d = nA$. Thus $A[n]$ and nA are subgroups (they are the kernel and image of a homomorphism) and $A/A[n] \cong nA$.

To show $A[n]$ is fully invariant, let $x \in A[n]$ and let $f{:}A \to A$ be a homomorphism. Thus $nx = 0$ so $nf(x) = f(nx) = f(0) = 0$. Hence $f(x) \in A[n]$ as required. To show nA is fully invariant, let $x \in nA$ and let $f{:}A \to A$ be a homomorphism. Thus $x = ny$ for some y, so $f(x) = f(ny) = nf(y) \in nA$.

DEFINITION AND NOTATION If A is an abelian group, then $t(A)$ is the **torsion subgroup** of A if and only if $t(A)$ is the set of all elements of A having finite order.

THEOREM A.4.2 *If A is an abelian group, then*
(a) *$t(A)$ is a subgroup of A.*
(b) *$t(A)$ is a torsion group.*
(c) *$A/t(A)$ is a torsion-free group.*
Proof: Let $x,y \in t(A)$. Let $o(x) = m$ and $o(y) = n$. Then $mn(x - y) = mnx - mny = 0 - 0 = 0$, so $x - y$ has finite order, showing $t(A)$ is a subgroup. This proves Part (a) and Part (b) is obvious. Finally suppose $t(A) + x$ is an element of $A/t(A)$ of finite order m. Then $m(t(A) + x) = 0$ implies $t(A) + mx = 0$ so $mx \in t(A)$. Let k be the order of mx. Then $kmx = 0$ which implies x has finite order. Thus $x \in t(A)$ and so $t(A) + x = 0$ proving $A/t(A)$ is torsion-free.

THEOREM A.4.3 *If A is an abelian group and B is a subgroup of A such that $t(A) \subseteq B$, then $t(A) = t(B)$.*
Proof: Let $x \in t(A)$. Then x has finite order and $x \in B$ (since $t(A) \subseteq B$). Thus $x \in t(B)$ showing $t(A) \subseteq t(B)$. Conversely if $x \in t(B)$, then $x \in A$ and x has finite order so $x \in t(A)$.

THEOREM A.4.4 *If A is an abelian group, then $t(A)$ is a pure subgroup of A.*
Proof: Let $x \in nA \cap t(A)$ where n is a positive integer. Thus $x = ny$ for some y and $x \in t(A)$. Let $m = o(x)$. Then $0 = mx = mny = (mn)y$ so y has finite order hence $y \in t(A)$. This and $x = ny$ imply $x \in nt(A)$ proving $nA \cap t(A) \subseteq nt(A)$.

THEOREM A.4.5 *If A is an abelian group, then $t(A)$ is a fully invariant subgroup of A.*
Proof: Let $x \in t(A)$, let $f{:}A \to A$ be a homomorphism, and seek to show $f(x) \in t(A)$. Since $x \in t(A)$, x has finite order, say n. Then $nf(x) = f(nx) = f(0) = 0$ so $f(x)$ too has finite order as required.

THEOREM A.4.6 *If A and B are abelian groups, then $t(A \times B) = t(A) \times t(B)$.*
Proof: Let $(x,y) \in t(A \times B)$. Thus $n(x,y) = (0,0)$ for some n. Then $nx = 0$ and $ny = 0$. Consequently $x \in t(A)$ and $y \in t(B)$, so $(x,y) \in t(A) \times t(B)$. Conversely suppose $(x,y) \in t(A) \times t(B)$. Then $nx = 0$ and $my = 0$ for some n and m. Then $nm(x,y) = (0,0)$ which shows (x,y) has finite order proving $(x,y) \in t(A \times B)$.

THEOREM A.4.7 *If A and B are abelian groups, then $(A \times B)/t(A \times B) \cong [A/t(A)] \times [B/t(B)]$.*
Proof: Let G denote $A \times B$. Thus (Theorem A.2.4) $G = J \oplus K$ where $J \cong A$ and $K \cong B$. Theorems A.4.5 and A.2.8 imply $G/t(G) = [(J + t(G))/t(G)] \oplus [(K + t(G))/t(G)]$ and so $G/t(G) \cong [(J + t(G))/t(G)] \times [(K + t(G))/t(G)]$ by Theorem A.2.5. Now $(J + t(G))/t(G) \cong J/(J \cap t(G)) = J/t(J) \cong A/t(A)$. Similarly $(K + t(G))/t(G) \cong B/t(B)$. Therefore $G/t(G) \cong [A/t(A)] \times [B/t(B)]$.

A.5 The Ascending Central Series and Nilpotent Groups

Let G be a group. For all natural numbers n we define a subgroup $Z_n(G)$ as follows:

$$Z_0(G) = \{1\}, \qquad Z_n(G) = \{x \mid x^{-1}y^{-1}xy \in Z_{n-1}(G) \text{ for all } y \in G\}.$$

These subgroups constitute the **ascending central series.**

NOTE: The subgroup $Z_1(G)$ is just the center of G, that is, $Z(G)$. This is seen as follows: $x \in Z_1(G)$ if and only if $x^{-1}y^{-1}xy \in Z_0(G) = \{1\}$ for all $y \in G$ which happens if and only if $xy = yx$ for all $y \in G$ which happens if and only if $x \in Z(G)$.

THEOREM A.5.1 *Let G be a group. Then for all natural numbers n,*
(a) *$Z_n(G)$ is a normal subgroup of G.*
(b) *$Z_n(G) \subseteq Z_{n+1}(G)$.*
(c) *$Z_n(G/Z(G)) = Z_{n+1}(G)/Z(G)$.*

Proof: This will be shown by induction on n. When $n = 0$, $Z_n(G) = \{1\}$ which is surely a normal subgroup of G. Also $Z_{n+1}(G) = Z(G)$ by the above note so $Z_n(G) \subseteq Z_{n+1}(G)$. Finally $Z_{n+1}(G)/Z(G) = Z(G)/Z(G) = \{1\} = Z_n(G/Z(G))$ so Part (c) also holds. Thus induction begins.

Assume now the induction hypothesis, namely that Properties (a), (b), and (c) hold for $n - 1$ and seek to show that they hold for n. Let $a, b \in Z_n(G)$. Then for all $y \in G$, both $a^{-1}y^{-1}ay$ and $b^{-1}y^{-1}by$ belong to $Z_{n-1}(G)$. Then since $Z_{n-1}(G)$ is normal by induction, $b(a^{-1}y^{-1}ay)b^{-1}$ and $b(b^{-1}y^{-1}by)b^{-1} = y^{-1}byb^{-1}$ both belong to $Z_{n-1}(G)$. Again, since $Z_{n-1}(G)$ is a subgroup by the induction hypothesis, this implies $[b(a^{-1}y^{-1}ay)b^{-1}]\,[y^{-1}byb^{-1}]^{-1} \in Z_{n-1}(G)$. After simplification this becomes $(ab^{-1})^{-1}y^{-1}(ab^{-1})y \in Z_{n-1}(G)$ for all $y \in G$, which means that $ab^{-1} \in Z_n(G)$. Thus $Z_n(G)$ is a subgroup of G. To show normality, note that $Z_{n-1}(G/Z(G)) \lhd G/Z(G)$ by Part (a) of the induction hypothesis, and $Z_{n-1}(G/Z(G)) = Z_n(G)/Z(G)$ by Part (c) of the induction hypothesis. Thus $Z_n(G)/Z(G) \lhd G/Z(G)$ which implies $Z_n(G) \lhd G$.

To show Part (b), let $x \in Z_n(G)$. Then for all $y \in G$, $x^{-1}y^{-1}xy \in Z_{n-1}(G)$ which is a subset of $Z_n(G)$ by the induction hypothesis (Part (b)). Therefore $x^{-1}y^{-1}xy \in Z_n(G)$ for all $y \in G$, meaning $x \in Z_{n+1}(G)$.

Finally consider Part (c). Let $Z(G)x \in Z_n(G/Z(G))$. Then for all $y \in G$, $Z(G)y \in G/Z(G)$ so $[Z(G)x]^{-1}[Z(G)y]^{-1}Z(G)xZ(G)y \in Z_{n-1}(G/Z(G))$. Thus $Z(G)x^{-1}y^{-1}xy \in Z_n(G)/Z(G)$ by the induction hypothesis. That is, $x^{-1}y^{-1}xy \in Z_n(G)$ for all $y \in G$. Consequently $x \in Z_{n+1}(G)$ and so $Z(G)x \in Z_{n+1}(G)/Z(G)$. Reversing the preceding argument shows $Z_{n+1}(G)/Z(G) \subseteq Z_n(G/Z(G))$, completing the proof of Part (c).

With the aid of the ascending central series, nilpotency can be defined as follows.

DEFINITION The group G is class-n nilpotent if and only if $Z_n(G) = G$ and

$Z_{n-1}(G) \subset G$. The group G is **nilpotent** if and only if G is a class-n nilpotent group for some n.

LEMMA *If A and B are groups and $f:A \to B$ is a surjective homomorphism, then $f[Z_n(A)] \subseteq Z_n(B)$ for all natural numbers n.*
Proof: This will be shown by induction on n. When $n = 0, Z_n(A) = Z_n(B) = \{1\}$ so induction clearly begins. Assume the induction hypothesis, $f[Z_{n-1}(A)] \subseteq Z_{n-1}(B)$. For all $b \in B$, $b = f(y)$ for some y since f is onto. Thus $f(x)^{-1}b^{-1}f(x)b = f(x^{-1}y^{-1}xy)$. But $x \in Z_n(A)$ implies $x^{-1}y^{-1}xy \in Z_{n-1}(A)$ so $f(x^{-1}y^{-1}xy) \in Z_{n-1}(B)$ by induction. Thus $f(x)^{-1}b^{-1}f(x)b \in Z_{n-1}(B)$ for all b so $f(x) \in Z_n(B)$.

THEOREM A.5.2 *Subgroups and quotient groups of nilpotent groups are nilpotent. In particular, subgroups and quotient groups of class-n nilpotent groups are class-m nilpotent where $m \leqslant n$ (different values of m may, of course, be required for different subgroups and quotient groups).*
Proof: Let G be a class-n nilpotent group and let A be a subgroup of G. To show A is class-m nilpotent for some $m \leqslant n$ it suffices to show $Z_n(A) = A$. First it is claimed that $Z_k(G) \cap A \subseteq Z_k(A)$ for all k. This will be shown by induction. When $k = 0$, $Z_k(G)$ and $Z_k(A)$ are both trivial so induction begins. Assume now the induction hypothesis $Z_{k-1}(G) \cap A \subseteq Z_{k-1}(A)$ and let $x \in Z_k(G) \cap A$. Then for all $a \in A$, $x^{-1}a^{-1}xa \in Z_{k-1}(G)$ since $x \in Z_k(G)$. Also $x, a \in A$ implies that $x^{-1}a^{-1}xa \in A$ so $x^{-1}a^{-1}xa \in Z_{k-1}(G) \cap A$. Hence for all $a \in A$, $x^{-1}a^{-1}xa \in Z_{k-1}(A)$ by the induction hypothesis. Consequently $x \in Z_k(A)$ proving the claim. By the claim $Z_n(G) \cap A \subseteq Z_n(A)$. But $Z_n(G) = G$ and $G \cap A = A$. Thus $A = Z_n(A)$ as required.

As for quotient groups, to show G/H is class-m nilpotent for some $m \leqslant n$, it suffices to show $Z_n(G/H) = G/H$. Let $e:G \to G/H$ be the natural map. Then $e[Z_n(G)] \subseteq Z_n(G/H)$ by the lemma. But $Z_n(G) = G$, so $e[Z_n(G)] = G/H$. Therefore $G/H = Z_n(G/H)$ as required.

The following theorem says that central extensions by nilpotent groups are nilpotent.

THEOREM A.5.3 *If $H \subseteq Z(G)$ and G/H is nilpotent, then G is nilpotent. In particular, if G/H is class-n nilpotent, then G is either class-n or class-$(n + 1)$ nilpotent.*
Proof: It is claimed that $Hx \in Z_k(G/H)$ implies $x \in Z_{k+1}(G)$ for all k. This will be shown by induction. When $k = 0$, $Z_k(G/H)$ is trivial so $Hx \in Z_k(G/H)$ implies $x \in H \subseteq Z(G) = Z_1(G)$ so induction begins. Assume now the induction hypothesis, $Hy \in Z_{k-1}(G/H)$ implies $y \in Z_k(G)$, and let $Hx \in Z_k(G/H)$. Then for all $p \in G$, $Hp \in G/H$ so $(Hx)^{-1}(Hp)^{-1}HxHp \in Z_{k-1}(G/H)$. Thus $Hx^{-1}p^{-1}xp$ belongs to $Z_{k-1}(G/H)$ for all $p \in G$. Applying the induction hypothesis yields $x^{-1}p^{-1}xp \in Z_k(G)$ for all $p \in G$. Therefore $x \in Z_{k+1}(G)$ as required, thus proving the claim. This claim will be used to show $G = Z_{n+1}(G)$. Let $x \in G$. Then $Hx \in Z_n(G/H)$

since G/H is class-n nilpotent. Hence $x \in Z_{n+1}(G)$ by the claim, showing $G = Z_{n+1}(G)$ as noted.

It may or may not be true that $G = Z_n(G)$. If $Z_n(G) \neq G$, then G is class-$(n+1)$ nilpotent. If $G = Z_n(G)$, then to show G is class-n nilpotent, it need only be shown that $Z_{n-1}(G) \neq G$. But if $Z_{n-1}(G) = G$, then G is class-r nilpotent for some $r \leqslant n-1$, hence by Theorem A.5.2, G/H is class-m nilpotent for some $m \leqslant r$. But by hypothesis G/H is class-n nilpotent. Thus $n \leqslant r \leqslant n-1$. This contradiction shows $Z_{n-1}(G) \neq G$ completing the proof.

LEMMA *If $G = A \oplus B$, then $Z_n(G) = Z_n(A) \oplus Z_n(B)$ for all natural numbers n.*
Proof: This will be shown by induction. When $n = 0$, $Z_n(G)$, $Z_n(A)$, and $Z_n(B)$ are all trivial, so induction begins. Assume now the induction hypothesis, namely $Z_{n-1}(G) = Z_{n-1}(A) \oplus Z_{n-1}(B)$. First it must be shown that $Z_n(A)$ is a subgroup of $Z_n(G)$. Let $x \in Z_n(A)$ and let $g \in G = A \oplus B$. Hence $g = ab$ where $a \in A$ and $b \in B$. Then $x^{-1}g^{-1}xg = x^{-1}b^{-1}(a^{-1}xa)b = x^{-1}(a^{-1}xa)b^{-1}b$ (by Theorem A.2.1) $= x^{-1}a^{-1}xa$ which belongs to $Z_{n-1}(A)$ because $x \in Z_n(A)$. But $Z_{n-1}(A)$ is a subgroup of $Z_{n-1}(G)$ by the induction hypothesis. Consequently $x^{-1}g^{-1}xg$ belongs to $Z_{n-1}(G)$ for all $g \in G$ so $x \in Z_n(G)$, proving $Z_n(A) \subseteq Z_n(G)$. That $Z_n(B)$ is a subgroup of $Z_n(G)$ is similar. Next note that $Z_n(A) \lhd A$ (Theorem A.5.1) implies $Z_n(A) \lhd G$ by Theorem A.2.6. This and $Z_n(A) \subseteq Z_n(G)$ imply $Z_n(A) \lhd Z_n(G)$. Similarly $Z_n(B) \lhd Z_n(G)$.

Since it is obvious that $Z_n(A)$ and $Z_n(B)$ intersect trivially, all that remains to be shown is that $Z_n(G) = Z_n(A)Z_n(B)$. Let $z \in Z_n(G)$. Then $z \in G = A \oplus B$ implies $z = ab$ where $a \in A$ and $b \in B$. It will suffice to show $a \in Z_n(A)$ and $b \in Z_n(B)$. But $\pi_A : G \to A$ is clearly surjective so $\pi_A[Z_n(G)] \subseteq Z_n(A)$ by the Lemma to Theorem A.5.2. Therefore $a = \pi_A(z) \in Z_n(A)$. Similarly $b \in Z_n(B)$.

THEOREM A.5.4 *Suppose $G = A \oplus B$ where A and B are nilpotent. Then G is nilpotent. In particular, if A is class-n and B is class-m nilpotent, then G is nilpotent of class $k = \max\{n,m\}$.*
Proof: By the lemma, $Z_k(G) = Z_k(A) \oplus Z_k(B)$. Since A is class-n nilpotent, $Z_n(A) = A$. But $Z_n(A) \subseteq Z_k(A)$ so $Z_k(A) = A$. Similarly $Z_k(B) = B$. Therefore $Z_k(G) = A \oplus B = G$, so G is class-t nilpotent where $t \leqslant k$. However, A is a subgroup of G so $n \leqslant t$ by Theorem A.5.2, and similarly $m \leqslant t$ since B is a subgroup of G. Conclude $k \leqslant t$ so $k = t$, hence G is class-k nilpotent.

The following innocent-looking proposition is very important to the theory of nilpotent groups, and is, in fact, the key to proving many theorems about nilpotent groups by means of induction on the nilpotency class.

THEOREM A.5.5 *If G is a class-n nilpotent group where $1 \leqslant n$, then $G/Z(G)$ is class-$(n-1)$ nilpotent.*
Proof: By Theorem A.5.1, Part (c), $Z_{n-1}(G/Z(G)) = Z_n(G)/Z(G) = G/Z(G)$ because G is class-n nilpotent. This proves $G/Z(G)$ is class-m nilpotent for some

$m \leqslant n - 1$. But $m < n - 1$ cannot happen for then $m + 1 < n$. But by Theorem A.5.3, $G/Z(G)$ is class-m nilpotent implies G is class-m or class-$(m + 1)$ nilpotent. Both cases then violate the hypothesis (G is class-n nilpotent).

THEOREM A.5.6 *If G is nilpotent and H is a nontrivial normal subgroup of G, then $Z(G) \cap H$ is nontrivial.*
Proof: This will be shown by induction on the nilpotency class of G. If G is class-1 nilpotent (G cannot be class-0 nilpotent since it has a nontrivial subgroup), then $G = Z(G)$ so $Z(G) \cap H = H \neq \{1\}$, hence induction begins. As the induction hypothesis, assume the theorem holds for all class-k nilpotent groups. Let G be class-$(k + 1)$ nilpotent, assume the theorem does not hold for G, and seek a contradiction. Thus there exists a nontrivial normal subgroup H of G such that $Z(G) \cap H = \{1\}$. Consider the group $G/Z(G)$. Normality of H implies $Z(G)H \lhd G$, hence $Z(G)H/Z(G) \lhd G/Z(G)$. Also $Z(G)H/Z(G)$ is nontrivial for triviality would imply $Z(G)H = Z(G)$ which would further imply $H \subseteq Z(G)$ hence $H \cap Z(G) = H$ and so $\{1\} = H$ which is false. Now $G/Z(G)$ is class-k nilpotent by Theorem A.5.5 so by the induction hypothesis $Z(G/Z(G)) \cap (Z(G)H/Z(G))$ is nontrivial so has an element $Z(G)x$ such that $x \notin Z(G)$. Now $x \in Z(G)H$ implies $x = zh$ where $z \in Z(G)$ and $h \in H$. It is claimed that $h \in Z(G)$. For all $y \in G$, $h^{-1}y^{-1}hy = (z^{-1}x)^{-1}y^{-1}(z^{-1}x)y = x^{-1}zy^{-1}z^{-1}xy = x^{-1}y^{-1}xy$ since $z \in Z(G)$. But $Z(G)x \in Z(G/Z(G))$ so $Z(G)x$ commutes with $Z(G)y$. This implies that $x^{-1}y^{-1}xy \in Z(G)$, and so $h^{-1}y^{-1}hy \in Z(G)$. But $h \in H$ and $H \lhd G$ imply $y^{-1}hy$ belongs to H and hence $h^{-1}y^{-1}hy \in H$. Conclude $h^{-1}y^{-1}hy \in Z(G) \cap H$, and thus $h^{-1}y^{-1}hy = 1$, so $hy = yh$ proving $h \in Z(G)$ as claimed. Thus $h \in Z(G) \cap H$ so $h = 1$. This and $x = zh$ imply $x = z$ so $x \in Z(G)$, the required contradiction.

COROLLARY 1 *Nontrivial nilpotent groups have nontrivial centers.*
Proof: If G is a nontrivial nilpotent group, then $Z(G) \cap G$ is nontrivial by Theorem A.5.6.

COROLLARY 2 *Nilpotent groups are hypercyclic.*
Proof: Let G be a nilpotent group, and seek to show that all nontrivial quotient groups G/H have nontrivial centers. But G nilpotent implies G/H is nilpotent by Theorem A.5.2, so $Z(G/H)$ is nontrivial by Corollary 1.

THEOREM A.5.7 *Finite p-groups are nilpotent.*
Proof: This will be shown by induction on the cardinality of the group. If G is a group of cardinality 1, then $G = \{1\}$ so induction begins. Assume now the induction hypothesis that A is nilpotent whenever A is a finite p-group such that $\text{card}(A) < n$, and suppose G is a finite p-group such that $\text{card}(G) = n$. Then $Z(G)$ is nontrivial by Theorem A.1.1 so $\text{card}(G/Z(G)) < n$. Also $G/Z(G)$ is a p-group because G is. Thus $G/Z(G)$ is nilpotent by the induction hypothesis. Therefore G is nilpotent by Theorem A.5.3.

THEOREM A.5.8 *In a nilpotent group, elements of relatively prime order* *commute.*

Proof: This will be shown by induction on the nilpotency class. If G is class-0 nilpotent, then G is trivial so induction begins.

Assume now the induction hypothesis, elements of relatively prime order in class-$(n-1)$ groups commute, suppose G is class-n nilpotent, and suppose $x,y \in G$ such that $i = \mathrm{o}(x)$ and $k = \mathrm{o}(y)$ are relatively prime. Then $(\mathrm{Z}(G)x)^i = 1$ and $(\mathrm{Z}(G)y)^k = 1$ so $\mathrm{o}(\mathrm{Z}(G)x) | i$ and $\mathrm{o}(\mathrm{Z}(G)y) | k$ implying that $\mathrm{o}(\mathrm{Z}(G)x)$ and $\mathrm{o}(\mathrm{Z}(G)y)$ are relatively prime since i and k are. Since $G/\mathrm{Z}(G)$ is class-$(n-1)$ nilpotent by Theorem A.5.5, the induction hypothesis yields $\mathrm{Z}(G)x\mathrm{Z}(G)y = \mathrm{Z}(G)y\mathrm{Z}(G)x$. Thus $x^{-1}y^{-1}xy = z$ for some $z \in \mathrm{Z}(G)$. Then $y^{-1}xy = xz$ so that $y^{-1}x^iy = (y^{-1}xy)^i = (xz)^i = x^iz^i$ since $z \in \mathrm{Z}(G)$. But $x^i = 1$, hence the previous equation becomes $1 = z^i$ which implies $\mathrm{o}(z) | i$. Beginning with $x^{-1}y^{-1}xy = z$ again, $x^{-1}y^{-1}x = zy^{-1}$ so $x^{-1}y^{-k}x = (x^{-1}y^{-1}x)^k = (zy^{-1})^k = z^ky^{-k}$ and hence $1 = z^k$ which implies $\mathrm{o}(z)|k$. But $\mathrm{o}(z)|i$ and $\mathrm{o}(z)|k$ imply $\mathrm{o}(z) = 1$ since i and k are relatively prime. Therefore $z = 1$ so $xy = yx$.

THEOREM A.5.9 *If G is nilpotent and $\mathrm{Z}(G)$ is indecomposable, then G is indecomposable.*

Proof: Suppose $G = A \oplus B$ and seek to show either A or B is trivial. First note that $\mathrm{Z}(G) = \mathrm{Z}(A) \oplus \mathrm{Z}(B)$ by the Lemma to Theorem A.5.4. But $\mathrm{Z}(G)$ is indecomposable. Thus either $\mathrm{Z}(A)$ or $\mathrm{Z}(B)$ is trivial. Without loss of generality assume $\mathrm{Z}(A)$ is trivial. By Theorem A.5.2, nilpotency of G implies nilpotency of A. Therefore A is trivial by Corollary 1 to Theorem A.5.6.

THEOREM A.5.10

(a) *If G is a nontrivial nilpotent group and $\mathrm{Z}(G)$ is subdirectly irreducible, then G is subdirectly irreducible.*

(b) *If G is a nilpotent group and $\mathrm{Z}(G) \cong \mathrm{Z}(p^n)$ for some prime p and some nonzero n, then G is subdirectly irreducible.*

Proof: Suppose $\mathrm{Z}(G)$ is subdirectly irreducible. Then $\mathrm{H}(\mathrm{Z}(G))$ is nontrivial. Thus it suffices to show $\mathrm{H}(\mathrm{Z}(G)) \subseteq \mathrm{H}(G)$, and to do that it suffices to show that $\mathrm{H}(\mathrm{Z}(G)) \subseteq N$ for all nontrivial normal subgroups N of G. But if N is such a subgroup, then $\mathrm{Z}(G) \cap N$ is nontrivial by Theorem A.5.6. Clearly $\mathrm{Z}(G) \cap N \lhd \mathrm{Z}(G)$, hence $\mathrm{H}(\mathrm{Z}(G)) \subseteq \mathrm{Z}(G) \cap N \subseteq N$ as required. This proves Part (a).

To prove Part (b), it suffices to show that cyclic groups of cardinality p^n where $n \neq 0$ are subdirectly irreducible. Let $A = \langle a \rangle$ be such a group. Then $\mathrm{o}(a) = p^n$ so $p^{n-1}a \neq 0$ and hence it suffices to show $p^{n-1}a \in B$ for all nontrivial normal subgroups B. Let B be such a subgroup. By Cauchy's theorem (Theorem A.1.5) there exists $b \in B$ such that $\mathrm{o}(b) = p$. Since $b \in \langle a \rangle$, $b = ka$ for some integer k. Then $0 = pb = pka$ which implies $\mathrm{o}(a)|pk$. Thus $p^n | pk$ implying $p^{n-1} | k$ so $k = p^{n-1}i$ for some i. Then $b = p^{n-1}ia \in \langle p^{n-1}a \rangle$, hence $\langle b \rangle \subseteq \langle p^{n-1}a \rangle$. But $\langle p^{n-1}a \rangle$ has cardinality $\mathrm{o}(p^{n-1}a) = p = \mathrm{o}(b) = \mathrm{card}(\langle b \rangle)$ so $\langle b \rangle = \langle p^{n-1}a \rangle$. Thus $p^{n-1}a \in \langle b \rangle$ so $p^{n-1}a \in B$ as required.

A.6 The Derived Series and Solvable Groups

The **derived series** of a group is defined as follows. Let G be any group and define $G^{(0)} = G$. Then for all positive integers n, define $G^{(n)}$ to be the commutator subgroup of $G^{(n-1)}$. Thus $G^{(1)} = G'$, $G^{(2)} = G''$, etcetera.

THEOREM A.6.1 *If G is a group and H is a subgroup of G, then $H^{(n)} \subseteq G^{(n)}$ for all n.*
Proof: This will be shown by induction. When $n = 0, H^{(n)} = H$ and $G^{(n)} = G$ so induction begins. Assume now the induction hypothesis, $H^{(n-1)} \subseteq G^{(n-1)}$, and seek to show $H^{(n)} \subseteq G^{(n)}$. Let $x \in H^{(n)}$. Thus $x = y_1y_2\ldots y_k$ where each y_i is a commutator of elements of $H^{(n-1)}$. Thus by the induction hypothesis, each y_i is a commutator of elements of $G^{(n-1)}$ and hence $y_i \in (G^{(n-1)})' = G^{(n)}$ for all i. Therefore $x \in G^{(n)}$.

THEOREM A.6.2 *If G is a group, then $G^{(n)}$ is a fully invariant subgroup for all n.*
Proof: This will be shown by induction. When $n = 0, G^{(n)} = G$ so induction begins. Assume now the induction hypothesis, $G^{(n-1)}$ is fully invariant, and seek to show $G^{(n)}$ is likewise. Let $f{:}G \to G$ be a homomorphism and let $x \in G^{(n)}$. Then $x = y_1y_2\ldots y_k$ where each y_i is a commutator $y_i = t_i^{-1}s_i^{-1}t_is_i$ where t_i and s_i belong to $G^{(n-1)}$. Thus each $f(y_i) = f(t_i)^{-1}f(s_i)^{-1}f(t_i)f(s_i)$. By the induction hypothesis, $f(t_i)$ and $f(s_i)$ belong to $G^{(n-1)}$ so $f(y_i) \in G^{(n)}$ for each i. But $f(x) = f(y_1)f(y_2)\ldots f(y_k)$ implying $f(x) \in G^{(n)}$.

THEOREM A.6.3 *If $G = A \oplus B$, then $G^{(n)} = A^{(n)} \oplus B^{(n)}$ for all n.*
Proof: First note that both $A^{(n)}$ and $B^{(n)}$ are contained in $G^{(n)}$ by Theorem A.6.1. Since $A^{(n)}$ is normal in A, it is also normal in G by Theorem A.2.6. This and $A^{(n)} \subseteq G^{(n)}$ imply $A^{(n)} \lhd G^{(n)}$. Similarly $B^{(n)} \lhd G^{(n)}$. Since A and B intersect trivially, so do $A^{(n)}$ and $B^{(n)}$. All that remains to be shown is that $G^{(n)} = A^{(n)}B^{(n)}$. This will be done by induction. When $n = 0$, $G^{(n)} = G$, $A^{(n)} = A$, and $B^{(n)} = B$ so induction begins. Assume now the induction hypothesis, $G^{(n-1)} = A^{(n-1)}B^{(n-1)}$. Let $x \in G^{(n)}$. Thus $x = t_1t_2\ldots t_k$ where each t_i is a commutator of elements in $G^{(n-1)}$ say $t_i = p_i^{-1}q_i^{-1}p_iq_i$ for $p_i,q_i \in G^{(n-1)}$. Then by the induction hypothesis, $p_i = a_ib_i$ and $q_i = c_id_i$ where $a_i,c_i \in A^{(n-1)}$ and $b_i,d_i \in B^{(n-1)}$. Thus $t_i = b_i^{-1}a_i^{-1}d_i^{-1}c_i^{-1}a_ib_ic_id_i = (a_i^{-1}c_i^{-1}a_ic_i)(b_i^{-1}d_i^{-1}b_id_i)$ by Theorem A.2.1. Consequently letting $g_i = a_i^{-1}c_i^{-1}a_ic_i$ and letting $h_i = b_i^{-1}d_i^{-1}b_id_i$ yields $g_i \in A^{(n)}$, $h_i \in B^{(n)}$, and $t_i = g_ih_i$. Conclude $t_i \in A^{(n)}B^{(n)}$ for each i and hence $x \in A^{(n)}B^{(n)}$.

One simple result will be needed about the first derived subgroup.

THEOREM A.6.4 *If G is a group and H is a normal subgroup such that G/H is abelian, then $G' \subseteq H$.*
Proof: It suffices to show all commutators of elements of G belong to H. Let $x,y \in G$. Then $HxHy = HyHx$ by commutativity of G/H. Hence $x^{-1}y^{-1}xy \in H$.

With the aid of the derived series solvable groups are defined as follows:

DEFINITION The group G is length-n solvable if and only if $G^{(n)} = \{1\}$ and $G^{(n-1)} \neq \{1\}$. The group G is **solvable** if and only if it is length-n solvable for some n.

LEMMA *If A and B are groups and $g{:}A \to B$ is a surjective homomorphism, then $B^{(n)} \subseteq g[A^{(n)}]$ for all n.*

Proof: This will be shown by induction on n. When $n = 0$, $B^{(n)} = B$ and $A^{(n)} = A$. Induction then begins since $B \subseteq g[A]$ by surjectivity of g. Assume now the induction hypothesis, $B^{(n-1)} \subseteq g[A^{(n-1)}]$, let $x \in B^{(n)}$, and seek to show $x \in g[A^{(n)}]$.

Since $x \in B^{(n)}$, $x = t_1 t_2 \ldots t_k$ where $t_i = p_i^{-1} q_i^{-1} p_i q_i$ for $p_i, q_i \in B^{(n-1)}$. Therefore by the induction hypothesis, $p_i = g(c_i)$ and $q_i = g(d_i)$ where $c_i, d_i \in A^{(n-1)}$. Consequently $c_i^{-1} d_i^{-1} c_i d_i \in A^{(n)}$, so $t_i = g(c_i^{-1} d_i^{-1} c_i d_i) \in g[A^{(n)}]$ for each i, which implies $x \in g[A^{(n)}]$.

THEOREM A.6.5 *Subgroups and quotient groups of solvable groups are solvable. In particular subgroups and quotient groups of length-n solvable groups are length-m solvable where $m \leqslant n$ (different values of m may, of course, be required for different subgroups and quotient groups).*

Proof: Let H be a subgroup of a length-n solvable group G. By Theorem A.6.1, $H^{(n)} \subseteq G^{(n)}$. Therefore $H^{(n)}$ is trivial so H is length m solvable for some $m \leqslant n$.

Next let G/H be a quotient group of G. It suffices to show $(G/H)^{(n)}$ is trivial. Let $e{:}G \to G/H$ be the natural map. Then $(G/H)^{(n)} \subseteq e[G^{(n)}]$ by the lemma. But $G^{(n)}$ is trivial, hence so too is $e[G^{(n)}]$. Thus $(G/H)^{(n)}$ is trivial as required.

THEOREM A.6.6 *Extensions of solvable groups by solvable groups are solvable. In particular if G/H is length-n solvable and H is length-m solvable, then G is solvable of length-k for some $k \leqslant m + n$.*

Proof: It suffices to show that $G^{(n+m)}$ is trivial. This will be accomplished by induction on n. When $n = 0$, the hypothesis states that G/H is class-0 solvable, in other words trivial. Hence $G = H$. Consequently $G^{(n+m)} = H^{(m)} = \{1\}$ and induction begins. Assume now the following induction hypothesis. If A is any group and $B \lhd A$ such that $(A/B)^{(n-1)}$ and $B^{(m)}$ are trivial, then $A^{(m+n-1)}$ is trivial. Seek to show triviality of $G^{(m+n)}$. Let $A/H = (G/H)'$. Hence $(A/H)^{(n-1)} = (G/H)^{(n)} = \{1\}$. This and triviality of $H^{(m)}$ imply triviality of $A^{(m+n-1)}$ by the induction hypothesis. By definition of A/H, $(G/H)/(A/H)$ is abelian. Thus G/A is abelian, so $G' \subseteq A$ by Theorem A.6.4. This implies $(G')^{(m+n-1)} \subseteq A^{(m+n-1)}$ by Theorem A.6.1. Hence $G^{(m+n)}$ is trivial as required.

THEOREM A.6.7 *Nilpotent groups are solvable.*

Proof: This will be shown by induction on the nilpotency class noting that class-0 nilpotent groups are trivial hence solvable. Assume the induction hypothesis, class-$(n-1)$ nilpotent groups are solvable, and suppose G is a class-n nilpotent group. Then $G/Z(G)$ is class-$(n - 1)$ nilpotent by Theorem A.5.5, hence solvable by induction. Also $Z(G)$ is abelian, hence solvable. Conclude G is solvable by Theorem A.6.6.

A.7 Hypercyclic Groups

THEOREM A.7.1 *Extensions of cyclic groups by hypercyclic groups are hyper-*

Proof: Let G be a group with cyclic normal subgroup H such that G/H is hypercyclic. Let G/K be a nontrivial quotient group of G. A nontrivial cyclic normal subgroup of G/K must be produced. Either $H \not\subseteq K$ or $H \subseteq K$. In the first case, $K \subset HK$ so HK/K is nontrivial. Also $HK/K \cong H/(H \cap K)$, hence HK/K is cyclic because H is. Finally $H,K \lhd G$ implies $HK \lhd G$, so $HK/K \lhd G/K$. Thus HK/K is the required cyclic normal subgroup.

Assume now that $H \subseteq K$. Then $(G/H)/(K/H) \cong G/K$ which is nontrivial so $(G/H)/(K/H)$ is a nontrivial quotient group of the hypercyclic group G/H. Thus there exists N/H such that $(N/H)/(K/H)$ is a nontrivial cyclic normal subgroup of $(G/H)/(K/H)$. Therefore N/K is a nontrivial cyclic subgroup of G/K. It is easily seen that $N/K \lhd G/K$ completing the proof.

NOTE: See Miscellaneous Exercise IV.4 of Chapter 5 for a generalization of Theorem A.7.1.

THEOREM A.7.2 *Hypercentral groups are hypercyclic.*

Proof: Let G be hypercentral, and suppose G/H is nontrivial. A nontrivial cyclic normal subgroup of G/H must be produced. Since G is hypercentral, $Z(G/H) \neq \{1\}$. Let $Hx \in Z(G/H)$ such that $Hx \neq 1$. That is, $x \notin H$. Let $B/H = \langle Hx \rangle$. Then B/H is contained in $Z(G/H)$ hence is normal in G/H. Thus B/H is the required nontrivial cyclic normal subgroup.

COROLLARY *Nilpotent groups are hypercyclic.*

Proof: By Theorem A.7.2 and Corollary 2 to Theorem A.5.6.

THEOREM A.7.3 *If G is a hypercyclic group and N is a nontrivial normal subgroup, then G has a nontrivial normal cyclic subgroup R such that $R \subseteq N$.*

Proof: Let $S = \{H \mid H \lhd G$ and $H \cap N$ is trivial$\}$. Since $\{1\} \in S$, S is nonempty. It is claimed that S satisfies the hypothesis of Zorn's lemma. Let C be a subchain of S. If C is empty, then any element of S will serve as an upper bound for C. Assume then that C is nonempty. Let $K = \cup C$. Then $K \lhd G$ and $H \cap K = \cup\{H \cap B \mid B \in C\} = \{1\}$ so $K \in S$. Since K is clearly an upper bound for C, the hypothesis of Zorn's lemma is satisfied as claimed.

Applying Zorn's lemma produces a maximal element of S, call it B. Thus $B \lhd$

G and $B \cap N$ is trivial. Note that $B \neq G$ because equality would imply $\{1\} = B \cap N = G \cap N = N$ contrary to hypothesis. Then G/B is nontrivial so has, by the hypercyclic property, a nontrivial normal cyclic subgroup, call it A/B. Since $A/B \lhd G/B$, $A \lhd G$. Since B is proper in A, $A \notin S$ by maximality of B. Thus $A \cap N$ must be nontrivial. Let $R = A \cap N$. Thus $R \lhd G$, $R \subseteq N$, and R is nontrivial. It remains only to show that R is cyclic.

Since $R \subseteq N$, $R \cap B = \{1\}$. Thus $R \cong R/(R \cap B) \cong BR/B$. But BR/B is cyclic being a subgroup of the cyclic group A/B. Therefore R is cyclic.

Lemma 1 *If G is a group and H is a cyclic normal subgroup, then $G' \subseteq C_G(H)$.*
Proof: Suppose $H = \langle h \rangle$. Let $x,y \in G$ and seek to show their commutator belongs to $C_G(H)$. By normality of H,

$$xhx^{-1} = h^i \quad \text{and} \quad yhy^{-1} = h^k \tag{1}$$

for some integers i and k. Let h^n be an arbitrary element of H. Then $yh^ny^{-1} = (yhy^{-1})^n = h^{nk}$ implying $x(yh^ny^{-1})x^{-1} = xh^{nk}x^{-1} = (xhx^{-1})^{nk} = h^{nik}$. From Equation (1) it follows that $h = x^{-1}h^ix = y^{-1}h^ky$. Thus $y^{-1}(xyh^ny^{-1}x^{-1})y = y^{-1}h^{nik}y = (y^{-1}h^ky)^{ni} = h^{ni}$. Therefore $x^{-1}(y^{-1}xyh^ny^{-1}x^{-1}y)x = x^{-1}h^{ni}x = (x^{-1}h^ix)^n = h^n$. Postmultiplying by $x^{-1}y^{-1}xy$ shows $x^{-1}y^{-1}xyh^n = h^nx^{-1}y^{-1}xy$ for all $h^n \in H$. Conclude $x^{-1}y^{-1}xy \in C_G(H)$.

LEMMA 2 *If G is a perfect group, then $Z_2(G) = Z(G)$.*
Proof: Let $t \in Z_2(G)$ and seek to show $t \in Z(G)$. Define $f: G \to G$ by $f(y) = y^{-1}t^{-1}yt$. For $y, y^* \in G$ then, $f(yy^*) = y^{*-1}y^{-1}t^{-1}yy^*t = y^{*-1}(y^{-1}t^{-1}yt)t^{-1}y^*t = (y^{-1}t^{-1}yt)y^{*-1}t^{-1}y^*t$ because $y^{-1}t^{-1}yt \in Z(G)$ since $t \in Z_2(G)$. Thus $f(yy^*) = f(y)f(y^*)$ and so f is a homomorphism. Moreover $\text{Im } f \subseteq Z(G)$ again because $t \in Z_2(G)$. Thus $G/\ker f$ can be embedded in $Z(G)$ and hence is abelian which implies $G' \subseteq \ker f$ by Theorem A.6.4. This and G perfect imply $G = \ker f$. Hence for all $x \in G$, $f(x) = 1$. Thus $x^{-1}t^{-1}xt = 1$ and so $tx = xt$ for all $x \in G$. Conclude $t \in Z(G)$.

THEOREM A.7.4 *If G is a nontrivial hypercyclic group, then $G' \neq G$.*
Proof: Assume $G' = G$ and seek a contradiction. If G is abelian, then G' is trivial. This and $G \neq \{1\}$ yield and immediate contradiction to $G' = G$, so assume G is nonabelian. Thus $G/Z(G)$ is nontrivial so the hypercyclic hypothesis implies $G/Z(G)$ has a nontrivial normal cyclic subgroup $A/Z(G)$. Application of Lemma 1 shows $K' \subseteq C_K(A/Z(G))$ where K denotes the group $G/Z(G)$. But since G is identical to its commutator subgroup, so too is any quotient group of G, as is easily seen. Thus $K' = K$ and so $K \subseteq C_K(A/Z(G))$. This implies $A/Z(G) \subseteq Z(K)$ from which it follows that $Z(K)$ is nontrivial. Thus there exists $Z(G)t$ belonging to $Z(G/Z(G)) \setminus \{1\}$ implying $t \in Z_2(G) \setminus Z(G)$ in contradiction to Lemma 2.

A.8 Semidivisible, Divisible, and p-divisible Groups

THEOREM A.8.1 *Homomorphic images of semidivisible groups are semidivisible.*

Proof: Suppose $g:A \to B$ is a surjective homomorphism where A is semidivisible. Let n be a positive integer and let $b \in B$. Then $b = g(x)$ for some $x \in A$, hence by semidivisibility, $x = y_1{}^n y_2{}^n \ldots y_k{}^n$ where $y_1, y_2, \ldots, y_k \in A$. Consequently $b = g(y_1)^n g(y_2)^n \ldots g(y_k)^n$ proving that B is semidivisible.

THEOREM A.8.2 *If G is a semidivisible group, then G has no nontrivial finite quotient groups.*

Proof: Suppose to the contrary that G/H is a nontrivial finite quotient group, say of cardinality n. Let $e:G \to G/H$ be the natural map. Then G/H is semidivisible by Theorem A.8.1. Now G/H nontrivial implies there exists $g \notin H$. By semidivisibility $Hg = (Hy_1)^n(Hy_2)^n \ldots (Hy_k)^n$ where $y_1, y_2, \ldots, y_k \in G$. But $Hy_i \in G/H$ and $\text{card}(G/H) = n$ imply $(Hy_i)^n = 1$. Thus $Hg = 1$ so $g \in H$, the required contradiction.

THEOREM A.8.3 *Semidivisible abelian groups are divisible.*

Proof: Let A be a semidivisible abelian group. Let $x \in A$ and let n be a positive integer. Then $x = na_1 + na_2 + \ldots + na_k$ where $a_1, a_2, \ldots, a_k \in A$. Let $y = a_1 + a_2 + \ldots + a_k$. Then $x = ny$ because A is abelian. Conclude A is divisible.

THEOREM A.8.4 *If A is an abelian group, and D is a divisible subgroup, then D is a direct factor of A.*

Proof: Let $S = \{H \mid H$ is a subgroup of A and $H \cap D$ is trivial$\}$. Then $(S, \subseteq)$ is a partially ordered set and S is nonempty because $\{0\} \in S$. It is claimed that this partially ordered set satisfies the hypothesis of Zorn's lemma. Suppose C is a subchain of S. If C is empty, any element of S will serve as an upper bound for C, so assume C is nonempty. Then $(\cup C) \cap D = \{0\}$ because if $x \in (\cup C) \cap D$, then $x \in H \in C$ for some H. Therefore $x \in H \cap D$, and consequently $x = 0$. On the other hand since C is nonempty, there exists some $H_0 \in C$, so $0 \in \cup C$ implying $\{0\} \subseteq (\cup C) \cap D$ and hence that $\{0\} = (\cup C) \cap D$ as claimed. Hence $\cup C \in S$ and as $\cup C$ is clearly an upper bound for C, this proves the claim.

Zorn's lemma may now be applied. Doing so yields existence of a subgroup K which is a maximal element of S. Thus $K \cap D = \{0\}$. To complete the proof then, it suffices to show $A = D \oplus K$, and to do that it suffices to show $A = D + K$. So assume $D + K \neq A$ and seek a contradiction. Thus there exists $x \in A$ such that $x \notin D + K$. Hence $x \notin K$, so K is a proper subgroup of $K + \langle x \rangle$. This and maximality of K imply $K + \langle x \rangle \notin S$, hence $(K + \langle x \rangle) \cap D$ must be nontrivial. Thus there exists $t \in (K + \langle x \rangle) \cap D$ such that $t \neq 0$. Then $t = k + nx$ where $k \in K$ and n is an integer. Thus $nx \in D + K$. Note that n does not equal 0 for if it did it would imply $t = k$, hence $t \in K \cap D$, hence $t = 0$ which is false. It may furthermore be assumed that n is positive, because if it is not, then $-n$ is positive, and

nx can be replaced by $(-n)(-x)$. Next note that n does not equal 1 for if it did it would imply $t = k + x$ hence $x \in D + K$ which is false. Thus $2 \leqslant n$. Let n' be the smallest element of $\{m | m \in N, 2 \leqslant m$, and $mx \in D + K\}$. (Note this set is nonempty as n belongs to it.) Let p be any prime dividing n'. Thus $n' = pq$ for some $q \in N$ (possibly $q = 1$). Note $q < n'$ so $qx \notin D + K$ by definition of n'. Let $y = qx$. Then $py = pqx = n'x \in D + K$. Hence $py = d + k'$ for some $d \in D$ and some $k' \in K$.

Applying the hypothesis of divisibility of D yields existence of $d_1 \in D$ such that $d = pd_1$. Let $z = y - d_1$. Then $pz = py - pd_1 = d + k' - pd_1 = d + k' - d = k' \in K$. If $z \in K$, then $z + d_1 \in D + K$ which would imply $y \in D + K$, hence $qx \in D + K$ which is false. Thus $z \notin K$. Consequently K is a proper subgroup of $K + \langle z \rangle$ so $K + \langle z \rangle \notin S$ by maximality of K. Conclude $(K + \langle z \rangle) \cap D$ is nontrivial so there exists $t_2 \in (K + \langle z \rangle) \cap D$ such that $t_2 \neq 0$. This means $t_2 = k_2 + mz$ where $k_2 \in K$ and m is an integer. If p divides m, then $m = pc$ for some integer c implying $t_2 = k_2 + pcz = k_2 + ck' \in K$ so $t_2 \in D \cap K = \{0\}$. Thus $t_2 = 0$ which is false. Hence p does not divide m. But p is a prime, so then p and m must be relatively prime. Hence there exist integers a and b such that $am + bp = 1$. Then $z = (am + bp)z = amz + bpz$. But $mz = t_2 - k_2 \in D + K$ and $pz = k' \in K$. Thus $z \in D + K$. But $y = z + d_1$. Thus $y \in D + K$ and finally $y = qx$ so $qx \in D + K$, the required contradiction.

THEOREM A.8.5 *Pure subgroups of divisible abelian groups are divisible.*

Proof: Let A be a divisible abelian group, and let B be a pure subgroup. Suppose $x \in B$ and n is a positive integer. Then since $x \in A$, $x = ny$ for some $y \in A$. But then $x \in nA \cap B$ so $x \in nB$ by purity of B. Conclude B is divisible.

THEOREM A.8.6 *Let B be an abelian p-group. Then B is q-divisible for all primes $q \neq p$ and moreover if $b \in B$, then there is a unique $y \in B$ satisfying $b = qy$.*

Proof: Let $o(b) = p^n$. Since p and q are distinct primes they are relatively prime hence so too are p^n and q. Consequently $up^n + vq = 1$ for some integers u and v. Therefore $b = up^n b + vqb = vqb = qy$ for $y = vb$. To show that y is unique, suppose $b = qy = qy^*$. Then $q(y - y^*) = 0$ implies $o(y - y^*) | q$. But since B is a p-group, $o(y - y^*) = p^m$ for some m hence $p^m | q$ which implies $p^m = 1$. Then $y - y^* = 0$ as required.

THEOREM A.8.7 *If the abelian group A is p-divisible for all primes p, then A is divisible.*

Proof: If n is a positive integer, then n can be expressed as the product of (not necessarily distinct) primes $n = p_1 p_2 \ldots p_k$. Let $x \in A$. Then by hypothesis there exists y_1 such that $x = p_1 y_1$. Again $y_1 \in A$ so there exists y_2 such that $y_1 = p_2 y_2$. And in general for all y_i, $1 \leqslant i \leqslant k - 1$, there exists y_{i+1} such that $y_i = p_{i+1} y_{i+1}$. Hence $x = p_1 y_1 = p_1 p_2 y_2 = \ldots = (p_1 p_2 \ldots p_k) y_k = n y_k$. Thus A is divisible.

A.9 Lattices and Subgroup Lattices

The purpose of this appendix is to describe L(G) and N(G), respectively the lattice of subgroups and lattice of normal subgroups of a group G. Certain results in the text and a number of exercises give information about the lattice-theoretical structure of L(G) and N(G) for various groups.

First the concept of a lattice will be defined. This definition itself depends on several other definitions. Two of these (for the terms partially ordered set and upper bound) were given in Appendix A.3. The remainder follow.

DEFINITION If $(S,\leqslant)$ is a partially ordered set and $A \subseteq S$, then x is a **lower bound** for A if and only if
(a) $x \in S$.
(b) $x \leqslant a$ for all $a \in A$.

DEFINITION If $(S,\leqslant)$ is a partially ordered set and $A \subseteq S$, then x is a **least upper bound** for A if and only if
(a) x is an upper bound for A.
(b) $x \leqslant y$ for all y such that y is an upper bound for A.

DEFINITION If $(S,\leqslant)$ is a partially ordered set and $A \subseteq S$, then x is a **greatest lower bound** for A if and only if
(a) x is a lower bound for A.
(b) $y \leqslant x$ for all y such that y is a lower bound for A.

Note that in an arbitrary partially ordered set neither least upper bounds nor greatest lower bounds need exist. When they do, however, they are unique.

THEOREM A.9.1 *Let $(S,\leqslant)$ be a partially ordered set and let $A \subseteq S$. Suppose x and x^* are least upper bounds for A and that w and w^* are greatest lower bounds for A. Then*
(a) $x = x^*$.
(b) $w = w^*$.

Proof: Since x is a least upper bound for A and since x^* is an upper bound for A, $x \leqslant x^*$ by definition of least upper bound. Similarly $x^* \leqslant x$. Thus $x = x^*$ by the antisymmetric condition in the definition of partially ordered set. This proves Part (a). Part (b) is similar.

DEFINITION $(S,\leqslant)$ is a **lattice** if and only if $(S,\leqslant)$ is a partially ordered set and for each pair of elements $x,y \in S$, $\{x,y\}$ has a least upper bound and a greatest lower bound.

Because of Theorem A.9.1, notation can be given to the least upper bound and greatest lower bound of a set $\{x,y\}$ in a lattice.

NOTATION $x \vee y$ will denote the least upper bound of $\{x,y\}$ and $x \wedge y$ will denote the greatest lower bound.

NOTATION If G is a group, let $L(G)$ denote $\{H \mid H$ is a subgroup of $G\}$.

THEOREM A.9.2 *Let G be a group. Then* $(L(G),\subseteq)$ *is a lattice. Greatest lower bounds and least upper bounds in this lattice are described by* $H \wedge K = H \cap K$ *and* $H \vee K = \langle H \cup K \rangle$. *Moreover if G is abelian, then an alternate description for* $H \vee K$ *is* $H \vee K = H + K$.

Proof: It is easily seen that $(L(G),\subseteq)$ is a partially ordered set. That greatest lower bounds exist and are of the form described above will be shown as follows. Let $H,K \in L(G)$. Then $H \cap K$ is also a subgroup of G so belongs to $L(G)$. Now $H \cap K \subseteq H$ and $H \cap K \subseteq K$ showing $H \cap K$ is a lower bound of $\{H,K\}$. Suppose Y is any lower bound of $\{H,K\}$. Then $Y \subseteq H$ and $Y \subseteq K$ so $Y \subseteq H \cap K$. Thus $H \cap K$ is the greatest lower bound of $\{H,K\}$ as required.

Again suppose $H,K \in L(G)$. Then $\langle H \cup K \rangle$ is also a subgroup of G so belongs to $L(G)$. Since $H \subseteq \langle H \cup K \rangle$ and $K \subseteq \langle H \cup K \rangle$, $\langle H \cup K \rangle$ is an upper bound of $\{H,K\}$. Suppose Y is any upper bound of $\{H,K\}$. Then $H \subseteq Y$ and $K \subseteq Y$. Let $x \in \langle H \cup K \rangle$. Thus $x = c_1 c_2 \ldots c_n$ where for each i, $c_i \in H$ or $c_i \in K$. Then as $H \subseteq Y$ and $K \subseteq Y$, $c_i \in Y$ for each i. Consequently $x \in Y$ since Y is a subgroup. This shows $\langle H \cup K \rangle \subseteq Y$, thus $\langle H \cup K \rangle$ is the least upper bound of $\{H,K\}$ as required.

This completes the proof of the theorem except for the last sentence. This last result is easily seen by noting that if G is abelian, then $H \lhd G$ so $H + K$ is a subgroup, and then clearly $H + K = \langle H \cup K \rangle$.

NOTATION If G is a group, let $N(G)$ denote $\{H \mid H \lhd G\}$.

THEOREM A.9.3 *Let G be a group. Then* $(N(G),\subseteq)$ *is a lattice. Greatest lower bounds and least upper bounds in this lattice are described by* $H \wedge K = H \cap K$ *and* $H \vee K = HK$.

Proof: It is easily seen that $(N(G),\subseteq)$ is a partially ordered set. If $H,K \in N(G)$, then $H \cap K$ is normal in G so belongs to $N(G)$. The remainder of the demonstration that $H \cap K$ is the greatest lower bound for $\{H,K\}$ is identical to the corresponding proof for $L(G)$ (Theorem 9.2). Thus it suffices to consider least upper bounds.

If $H,K \in N(G)$, then HK is normal in G so belongs to $N(G)$. Since $H \subseteq HK$ and $K \subseteq HK$, HK is an upper bound of $\{H,K\}$. Suppose Y is any upper bound of $\{H,K\}$ and seek to show $HK \subseteq Y$. Let $x \in HK$. Then $x = hk$ for some $h \in H$ and $k \in K$. But then $h \in Y$ because $H \subseteq Y$ and $k \in Y$ because $K \subseteq Y$. Therefore $x = hk \in Y$.

We now define some conditions on lattices that will be discussed in the text and exercises. One of these conditions (the definition of a chain) has already been presented in Appendix A.3. A chain was defined there as a partially ordered set $(P,\leqslant)$ such that for all $x,y \in P$, either $x \leqslant y$ or $y \leqslant x$. In such a case, $(P,\leqslant)$ is clearly a lattice since $x \vee y$ and $x \wedge y$ can be defined by

$$x \vee y = \begin{cases} x & \text{if } y \leqslant x, \\ y & \text{if } x \leqslant y, \end{cases} \quad \text{and} \quad x \wedge y = \begin{cases} x & \text{if } x \leqslant y, \\ y & \text{if } y \leqslant x. \end{cases}$$

Other needed definitions are as follows.

DEFINITION The lattice L is **distributive** if and only if $x \vee (y \wedge z) = (x \vee y) \wedge (x \vee z)$ for all $x,y,z \in L$.

By Theorem A.9.3, the particularization of this definition to N(G) reads as follows: N(G) is distributive if and only if $A(B \cap C) = AB \cap AC$ for all normal subgroups A, B, and C of G. If a group G is abelian, then by Theorem A.9.2, the particularization of the definition to L(G) reads: L(G) is distributive if and only if $A + (B \cap C) = (A + B) \cap (A + C)$ for all subgroups A, B, and C of G. What form the definition takes for L(G) when G is nonabelian is a moot point since it can be shown that only abelian groups have distributive subgroup lattices.

NOTE: An exercise at the end of Chapter 5 shows that in an arbitrary lattice L, the condition $x \vee (y \wedge z) = (x \vee y) \wedge (x \vee z)$ for all x, y, and $z \in L$ is equivalent to the condition $x \wedge (y \vee z) = (x \wedge y) \vee (x \wedge z)$ for all x, y, and $z \in L$.

DEFINITION Let L_1 and L_2 be lattices. Then L_1 and L_2 are **isomorphic** if and only if there exists a function $f: L_1 \to L_2$ such that
(a) f is bijective
(b) for all $x,y \in L_1$, $x \leqslant y$ if and only if $f(x) \leqslant f(y)$.

Finite lattices can be represented by diagrams, called Hasse diagrams, wherein the elements of the lattice are depicted by circles and the relation $x \leqslant y$ is shown by drawing the circle representing x somewhat lower than the circle representing y and connecting the two circles with a straight line. The figures in Chapter 4 are all Hasse diagrams. This representation makes the above definition much clearer. Two (finite) lattices are isomorphic if and only if they have identical Hasse diagrams.

DEFINITION The lattice L is **self-dual** if and only if there exists a function $f: L \to L$ such that
(a) f is bijective
(b) for all $x,y \in L$, $x \leqslant y$ if and only if $f(y) \leqslant f(x)$.

Once again, in the finite case, the definition becomes much clearer when interpreted using Hasse diagrams. A (finite) lattice is self-dual if and only if its

Hasse diagram looks the same both right-side-up and upside-down.

We conclude this appendix with a subgroup-lattice characterization of the group $Z(2) \times Z(2)$. This result is needed for several other subgroup-lattice characterizations of finite groups given in the exercises in Chapter 4.

THEOREM A.9.4 *If A is a group such that* $L(A) \cong L(Z(2) \times Z(2))$, *then* $A \cong Z(2) \times Z(2)$.

Proof: First the subgroup lattice of $Z(2) \times Z(2)$ must be determined. Let G denote $Z(2) \times Z(2)$, and let $w = (1,0)$ and $x = (0,1)$ in G. Then it is easily seen that the list of subgroups for G is $\{0\}$, $\langle w \rangle$, $\langle x \rangle$, $\langle w + x \rangle$, and G. Thus $L(G)$ has the following Hasse diagram.

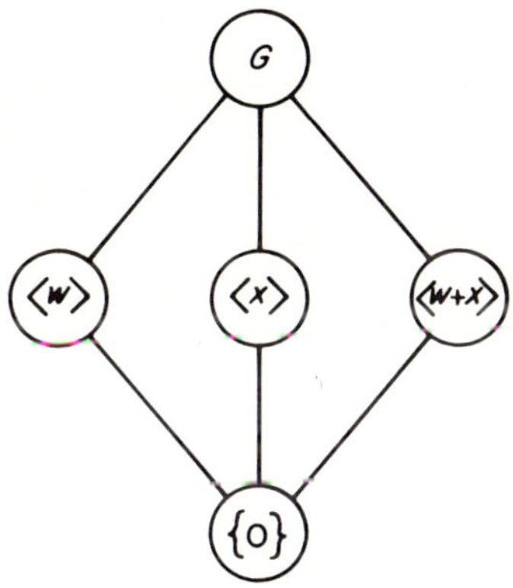

The problem may now be stated as follows: Given a group A whose subgroup lattice has the Hasse diagram

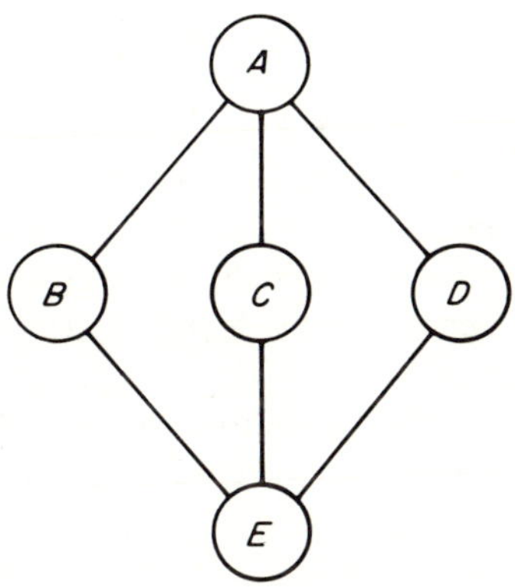

show $A \cong Z(2) \times Z(2)$.

It is clear that E must be the trivial subgroup. Since B has no nontrivial proper subgroups, it must be cyclic of prime cardinality, say $B = \langle b \rangle$ where $o(b) = p$. Similarly $C = \langle c \rangle$ where $o(c) = q$ and $D = \langle d \rangle$ where $o(d) = r$.

It is claimed that b and c commute. Suppose they do not and seek a contradiction. Consider the subgroup $\langle bc \rangle$. This subgroup cannot be A for then A would be cyclic, hence abelian, hence would not have a pair of noncommuting elements. Therefore $\langle bc \rangle = B, C, D,$ or E. If $\langle bc \rangle = E$, then $bc = 1$ which implies $b = c^{-1} \in C$. Hence $B \subseteq C$, a contradiction. If $\langle bc \rangle = B$, then $bc = b^i$ for some i so $c = b^{i-1} \in B$ and hence $C \subseteq B$, a contradiction. Similarly the contradiction $B \subseteq C$ is reached if $\langle bc \rangle = C$. Conclude then that $\langle bc \rangle = D$. A similar demonstration shows $\langle cb \rangle$ must equal D. Therefore $\langle bc \rangle = \langle cb \rangle$ so $cb = (bc)^t$ for some t. Thus $cb = b(cb)^{t-1}c$.

Next consider the subgroup $\langle cbc^{-1} \rangle$. Again this cannot be A so must be B, C, D, or E. If $\langle cbc^{-1} \rangle = E$, then $cbc^{-1} = 1$ which implies $b = 1$ and hence $B = E$, a contradiction. If $\langle cbc^{-1} \rangle = B$, then $cbc^{-1} = b^s$ for some s hence $cb = b^s c$. This and the last equation of the previous paragraph imply $b(cb)^{t-1}c = b^s c$ and hence $(cb)^{t-1} = b^{s-1}$. Recall $\langle cb \rangle = D$. Thus $b^{s-1} \in B \cap D = E$ so $b^{s-1} = 1$. Thus $b^s = b$. This and $cb = b^s c$ imply c and b commute, a contradiction. If $\langle cbc^{-1} \rangle = C$, then $cbc^{-1} = c^w$ for some w so $b = c^w$ and hence $B \subseteq C$, a contradiction. Finally if $\langle cbc^{-1} \rangle = D = \langle cb \rangle$, then $cb = (cbc^{-1})^k$ for some k so $cb = cb^k c^{-1}$, hence $c = b^{k-1}$ implying $C \subseteq B$, a contradiction. This establishes the claim that b and c commute.

Thus $c \in C_A(b)$ and so $C \subseteq C_A(b)$. Hence $C_A(b) = C$ or A. But $b \in C_A(b)$ precludes the first case. Thus $C_A(b) = A$ and so $b \in Z(A)$ implying $B \subseteq Z(A)$. A similar demonstration using $b \in C_A(c)$ shows $C \subseteq Z(A)$. Conclude $Z(A) = A$ so A is abelian.

This implies $B, C, D \lhd A$. Clearly then $A = B \oplus C$, $A = B \oplus D$, and $A = C \oplus D$. Application of Theorem A.2.5 yields $A \cong B \times C \cong B \times D \cong C \times D$ and hence $\text{card}(A) = \text{card}(B) \cdot \text{card}(C) = \text{card}(B) \cdot \text{card}(D) = \text{card}(C) \cdot \text{card}(D)$. Thus $pq = pr = qr$ and so $p = q = r$.

In summary, $A \cong B \times C \cong Z(p) \times Z(p)$. What must be shown to complete the proof, therefore, is that $p = 2$.

It will first be shown that $Z(p) \times Z(p)$, and hence A, is noncyclic. Suppose (x,y) generates $Z(p) \times Z(p)$. Then $(1,0) = n(x,y)$ for some n while $(0,1) = m \cdot (x,y)$ for some m. Therefore $1 = nx$, $0 = mx$, and $1 = my$. The first two equations imply $m = mnx = n(mx) = n \cdot 0 = 0$ which, in the group $Z(p)$, means $p \mid m$. Thus $m = pk$ for some k and hence $1 = my = pky = k(py) = k \cdot 0 = 0$. This contradiction shows A is noncyclic.

Since A is noncyclic, $A \neq \langle bc \rangle$. Therefore $\langle bc \rangle = B, C, D$, or E. That the first, second, and last possibilities cannot occur, however, has been shown in the fourth paragraph. Thus $\langle bc \rangle = D$.

In a similar fashion it is seen that $\langle bc^{-1} \rangle = D$. This, $\langle bc \rangle = D$, and commutativity of A imply $b^2 = bcbc^{-1} \in D$. Hence $b^2 \in B \cap D = E$ so $b^2 = 0$. Thus $o(b) \mid 2$. That is, $p \mid 2$. Conclude $p = 2$.

A.10 Miscellaneous

THEOREM A.10.1 *Suppose A and B are two groups, $H \lhd A$, $K \lhd B$, and $g:A \to B$ is a homomorphism such that $g[H] \subseteq K$. Define $g':A/H \to B/K$ by $g'(Hx) = Kg(x)$. Then g' is also a homomorphism. Moreover if g is surjective, so is g'.*

Proof: First it must be shown that g' is a function. Suppose $Hx = Hy$. Then $xy^{-1} \in H$ so $g(xy^{-1}) \in K$ by hypothesis. Thus $g(x)g(y)^{-1} \in K$, hence $Kg(x) = Kg(y)$ proving g' is a function. Now if $Hx, Hy \in A/H$, then $g'(HxHy) = g'(Hxy) = Kg(xy) = Kg(x)Kg(y) = g'(Hx)g'(Hy)$. Thus g' is a homomorphism.

To prove the surjectivity claim, suppose $Kb \in B/K$. Then $b = g(a)$ for some $a \in A$ by surjectivity of g. Therefore $g'(Ha) = Kg(a) = Kb$.

COROLLARY *Suppose A and B are two groups, $H \lhd A$, and $g:A \to B$ is a homomorphism such that $g[H] = \{1\}$. Define $g'':A/H \to B$ by $g''(Hx) = g(x)$. Then g'' is also a homomorphism. Moreover if g is surjective, so is g''.*

Proof: The function $\phi:B/\{1\} \to B$ defined by $\phi(\{1\}b) = b$ is clearly an isomorphism. It is also clear that $g'' = \phi \circ g'$. The results follow from the theorem.

THEOREM A.10.2 *If G is a group and H is a subgroup of index 2, then $H \lhd G$.*

Proof: There are two cosets of H, namely H and Hx for some $x \in G$. Thus $G = H \cup Hx$. Let $h \in H$ and $g \in G$, assume $g^{-1}hg \notin H$, and seek a contradiction. The assumption implies $g \notin H$ so $g \in Hx$ instead. Thus $g = h_2x$ for some $h_2 \in H$. Then $g^{-1}hg = x^{-1}h_2{}^{-1}hh_2x$. Now $g^{-1}hg \notin H$ implies $g^{-1}hg \in Hx$ so $g^{-1}hg = h_3x$ for some $h_3 \in H$. Therefore $x^{-1}h_2{}^{-1}hh_2x = h_3x$ implying $x = h_2{}^{-1}hh_2h_3{}^{-1} \in H$ which is false (H has index 2, not index 1).

THEOREM A.10.3 *Let G be a group and let H be a subgroup of G. Suppose $G = \langle X \rangle$ and $H = \langle Y \rangle$. Suppose further that*

(i) *$xyx^{-1} \in H$ and*

(ii) *$x^{-1}yx \in H$ for all $x \in X$ and all $y \in Y$.*

Then $H \lhd G$. Furthermore if every $x \in X$ has finite order, it is sufficient to show (i) *holds.*

Proof: First it is claimed that

$$xHx^{-1} \subseteq H \quad \text{for all } x \in X. \tag{1}$$

Let $h \in H$. Then $h = y_1{}^{e_1}y_2{}^{e_2}\ldots y_n{}^{e_n}$ where each $y_i \in Y$. Consequently $xhx^{-1} = (xy_1x^{-1})^{e_1}(xy_2x^{-1})^{e_2}\ldots(xy_nx^{-1})^{e_n}$ which belongs to H because each $xy_ix^{-1} \in H$ by hypothesis and because H is a subgroup. This proves the claim. A similar demonstration, using (ii) shows

$$x^{-1}Hx \subseteq H \quad \text{for all } x \in X. \tag{2}$$

Now since every $g \in G$ can be expressed as the product of elements of X and their inverses, repeated applications of (1) and (2) yield $g^{-1}Hg \subseteq H$ for all $g \in G$. Thus $H \lhd G$.

To establish the last sentence of the theorem note that, as in the first paragraph, Condition (i) implies (1). It suffices to show that this implies (2) when x has finite order. Suppose $o(x) = n$. Then $x^{n-1}H(x^{-1})^{n-1} \subseteq H$ by $n - 1$ applications of (1). But $x^{-1} = x^{n-1}$ since x has order n. Thus $x^{-1}Hx \subseteq H$ as required.

THEOREM A.10.4 *Let G be a group and let H be a normal subgroup. Then* $G = \langle X \rangle$ *implies* $G/H = \langle Hx \mid x \in X \rangle$.

Proof: Let $Hg \in G/H$. Then $g \in G$, so $g = x_1{}^{e_1}x_2{}^{e_2} \ldots x_n{}^{e_n}$ where each $x_i \in X$. Consequently $Hg = Hx_1{}^{e_1}x_2{}^{e_2} \ldots x_n{}^{e_n} = (Hx_1)^{e_1}(Hx_2)^{e_2} \ldots (Hx_n)^{e_n}$.

THEOREM A.10.5 *Let G be a group such that* $G/Z(G)$ *is cyclic. Then* $G = Z(G)$.

Proof: Suppose $Z(G)g$ generates $G/Z(G)$. To prove the theorem, it suffices to show any two elements x and y of G commute. Since $Z(G)x, Z(G)y \in G/Z(G)$, $Z(G)x = (Z(G)g)^n$ and $Z(G)y = (Z(G)g)^m$ for some integers n and m. Hence $x = zg^n$ and $y = z^*g^m$ where $z, z^* \in Z(G)$. Then $xy = zg^nz^*g^m = zz^*g^{n+m}$ because $z^* \in Z(G)$, while $yx = z^*g^mzg^n = zz^*g^{n+m}$ because $z \in Z(G)$. Thus $xy = yx$.

THEOREM A.10.6 *If G is a group, H is a normal subgroup of G, and B is a characteristic subgroup of H, then* $B \lhd G$.

Proof: Let $b \in B$ and $x \in G$, and seek to show $x^{-1}bx \in B$. Define a homomorphism d on H by $d(h) = x^{-1}hx$ for all $h \in H$. Then d is onto H because $H \lhd G$, that is, $d:H \to H$. Note d is injective because $d(h) = 1$ implies $x^{-1}hx = 1$ and hence $h = 1$. Also d is surjective because if $y \in H$, then $xyx^{-1} \in H$ and $d(xyx^{-1}) = y$. Conclude $d \in \text{Aut}(H)$. Consequently $d[B] \subseteq B$ by hypothesis on B hence $x^{-1}bx \in B$ as required.

THEOREM A.10.7 *Let G be a group with subgroups H and K.*

(a) *If HK is also a subgroup, then* $[HK:H] = [K:H \cap K]$.

(b) *If* $H \lhd G$, *then* $[HK:H] = [K:H \cap K]$.

Proof: Suppose HK is a subgroup. Let $W = H \cap K$, let $R = \{Hx \mid x \in HK\}$, let $S = \{Wx \mid x \in K\}$, and seek to shown card(R) = card(S). Define $d:S \to R$ by $d(Wx) = Hx$. To show d is a function, suppose $Wx = Wy$. Then $xy^{-1} \in H \cap K$ so $xy^{-1} \in H$, hence $Hx = Hy$. Next it will be shown that d is injective. Suppose $d(Wx) = d(Wy)$. Then $Hx = Hy$ so $xy^{-1} \in H$. But $Wx, Wy \in S$ so $x, y \in K$, hence $xy^{-1} \in K$. Thus $xy^{-1} \in H \cap K$ so $Wx = Wy$ showing d is injective. Finally surjectivity of d will be demonstrated. Let $Hx \in R$. Then $x \in HK$ so $x = hk$ where $h \in H$ and $k \in K$. Thus $Wk \in S$ and $d(Wk) = Hk = Hhk = Hx$. Consequently d is surjective and hence a bijection showing card(R) = card(S) and hence $[HK:H] = [K:H \cap K]$ proving Part (a).

Part (b) follows immediately from Part (a) because $K \lhd G$ implies HK is a subgroup of G.

THEOREM A.10.8 *For natural numbers* $n \leqslant m$, S_n *can be embedded in* S_m.

Proof: Define $d:S_n \to S_m$ by

$$d(f)(i) = \begin{cases} f(i) & \text{if } 1 \leqslant i \leqslant n, \\ i & \text{if } n + 1 \leqslant i \leqslant m. \end{cases}$$

First it must be shown that $d(f)$ is a permutation whenever f is. If $1 \leqslant t \leqslant m$, then $t = d(f)(f^{-1}(t))$ for $t \leqslant n$ while $t = d(f)(t)$ for $n + 1 \leqslant t \leqslant m$. Therefore $d(f): \{1,2, \ldots ,m\} \to \{1,2, \ldots ,m\}$ is surjective. Application of the pigeonhole principle (or an easily supplied straightforward proof) shows $d(f)$ is also injective, hence a permutation.

Next it must be shown that d is a homomorphism. Suppose $f,g \in S_n$. For $i \leqslant n$, $(d(f) \circ d(g))(i) = d(f)(g(i)) = f \circ g(i) = d(f \circ g)(i)$, while for $n + 1 \leqslant i \leqslant m$, $(d(f) \circ d(g))(i) = d(f)(i) = i = d(f \circ g)(i)$. Thus $d(f) \circ d(g)$ and $d(f \circ g)$ agree on all i, hence are equal.

Finally it must be shown that d is injective. Suppose $d(f) = 1$ where 1 denotes the identity element of S_m. Then for $1 \leqslant i \leqslant n$, $d(f)(i) = i$ hence $f(i) = i$. Conclude f is the identity element of S_n proving d is injective.

THEOREM A.10.9 *If G is a finitely generated group and n is a positive integer, then G has only a finite number of subgroups of index n.*

Proof: Let $\mathcal{R} = \{J \mid J = \mathrm{Core}_G(H)$ for some subgroup H of G of index $n\}$. For all $J \in \mathcal{R}$ there is, by Theorem A.1.3, an embedding $e_J: G/J \to S_n$. Let p_J denote the natural map from G to G/J. Hence for each J, $f_J: G \to S_n$ defined by $f_J = e_J \circ p_J$ is a homomorphism. Also $\ker f_J = J$ because $\ker p_J = J$ and e_J is injective.

Let $\mathcal{E} = \{f_J \mid J \in \mathcal{R}\}$. Define the function $\phi: \mathcal{R} \to \mathcal{E}$ by $\phi(J) = f_J$. Then ϕ is injective because $\phi(J) = \phi(J^*)$ implies $f_J = f_{J^*}$ and hence $J = \ker f_J = \ker f_{J^*} = J^*$.

Since G is finitely generated and since S_n is finite, there can be only a finite number of homomorphisms from G into S_n because each homomorphism is determined by its action on the generators of G and can assume only a finite number of values on any given generator. Thus $\mathcal{E}$ is finite. But then existence of the injection ϕ shows $\mathcal{R}$ is finite. To complete the proof, it suffices to eliminate the possibility that some $J \in \mathcal{R}$ is the core of an infinite number of subgroups of G of index n. So suppose $\mathcal{L}$ is an infinite collection of subgroups of G of index n each having J for a core. Then $J \subseteq D$ for all $D \in \mathcal{L}$ hence $\{D/J \mid D \in \mathcal{L}\}$ is an infinite collection of subgroups of G/J. This contradicts the fact that G/J is finite (it can be embedded in S_n) and completes the proof.

THEOREM A.10.10 *Subdirectly irreducible groups are indecomposable.*

Proof: Let A be a subdirectly irreducible group and suppose $A = B \oplus C$. Suppose both B and C are nontrivial and seek a contradiction. Then $B, C \lhd A$ imply $\mathrm{H}(A) \subseteq B$ and $\mathrm{H}(A) \subseteq C$. Thus $\mathrm{H}(A) \subseteq B \cap C = \{1\}$ in contradiction to subdirect irreducibility.

The next three theorems are concerned with maximal subgroups. These are defined as follows.

DEFINITION Let G be a group. Then M is a **maximal subgroup** of G if and only if

(a) M is a subgroup of G.
(b) $M \neq G$.
(c) If H is a subgroup of G and $M \subseteq H \subseteq G$, then either $H = M$ or $H = G$.

CAUTION A maximal subgroup of G is *not* a maximal element (as defined in Appendix A.3) of the partially ordered set $(\mathrm{L}(G),\subseteq)$, but rather is a maximal element of the partially ordered set $(\mathrm{L}(G) \setminus \{G\},\subseteq)$. Thus the phrase *maximal subgroup* is slightly misleading but popular usage prefers it to the phrase *maximal proper subgroup.*

DEFINITION The group G is **coatomic** if and only if every proper subgroup of G is contained in a maximal subgroup.

THEOREM A.10.11 *Finitely generated groups are coatomic.*

Proof: Let $G = \langle x_1,x_2,\ldots,x_n\rangle$ be a finitely generated group and suppose H is a proper subgroup. Let $S = \{B \mid B$ is a proper subgroup of G and $H \subseteq B\}$. Then $(S,\subseteq)$ is a partially ordered set and S is nonempty because $H \in S$. Let C be a subchain of S and seek an upper bound for C. Clearly if C is empty, then any element of S, H say, will serve as an upper bound for C. Assume then that C is nonempty, so some B_0 belongs to C. Let $J = \cup C$. It is easily seen that J is a subgroup of G. Consequently J will meet the conditions of being an upper bound for C provided it is shown that $J \in S$. To this end first note $H \subseteq J$ because $B_0 \in S$ implies $H \subseteq B_0$ and $B_0 \in C$ implies $B_0 \subseteq J$. What remains to be shown is that $J \neq G$.

Assume that $J = G$ and seek a contradiction. Then $x_i \in J$ for all i. For each i then there exists B_i such that $x_i \in B_i \in C$. But C is a subchain of $(S,\subseteq)$. Hence there exists some k such that $B_i \subseteq B_k$ for all i. Thus $x_1,x_2,\ldots,x_n \in B_k$ implying $G = B_k$ since $G = \langle x_1,x_2,\ldots,x_n\rangle$. However, $B_k \in C \subseteq S$ implies $B_k \neq G$ so a contradiction has been reached, proving that $J \neq G$.

The hypothesis of Zorn's lemma has thus been shown to hold. Application of Zorn's lemma yields a maximal element of $(S,\subseteq)$, call it M. Then M is clearly a maximal subgroup of G with $H \subseteq M$.

THEOREM A.10.12 *Nilpotent groups are* MN *groups.*

Proof: This will be shown by induction on the nilpotency class. If G is class-1 nilpotent (abelian), then maximal subgroups are normal because all subgroups are normal. Hence induction begins.

Assume now the induction hypothesis, class-$(n-1)$ nilpotent groups are MN. Let G be a class-n nilpotent group, let M be a maximal subgroup of G, and seek to show $M \lhd G$. By Theorem A.5.5, $G/\mathrm{Z}(G)$ is class-$(n-1)$ nilpotent. If $\mathrm{Z}(G) \subseteq M$, then $M/\mathrm{Z}(G)$ is a maximal subgroup of $G/\mathrm{Z}(G)$, hence is normal in $G/\mathrm{Z}(G)$ by the induction hypothesis, thus implying $M \lhd G$.

Assume, on the other hand, that $\mathrm{Z}(G) \not\subseteq M$. Then $G = \mathrm{Z}(G)M$ by maximality of M. To show $M \lhd G$, let $m \in M$ and $x \in G$. Then $x = zm^*$ for some $z \in \mathrm{Z}(G)$ and $m^* \in M$, so $x^{-1}mx = m^{*-1}z^{-1}mzm^* = m^{*-1}mz^{-1}zm^*$ (since $z \in \mathrm{Z}(G)$) $= m^{*-1}mm^* \in M$. Hence $M \lhd G$.

DEFINITION AND NOTATION Let G be a group. Let $\Phi(G) = G$ if G has no maximal subgroups and let $\Phi(G) = \cap\{M \mid M$ is a maximal subgroup of $G\}$ if G does have maximal subgroups. Then $\Phi(G)$ is called the **Frattini subgroup** of G.

THEOREM A.10.13 *If G is a group, then $\Phi(G) \lhd G$.*

Proof: It is immediate from the definition that $\Phi(G)$ is a subgroup. It is also immediate that $\Phi(G) \lhd G$ when G has no maximal subgroups. Assume then G has maximal subgroups, let $x \in \Phi(G)$, and seek to show $g^{-1}xg \in \Phi(G)$ for any $g \in G$.

Thus it must be shown that $g^{-1}xg \in M$ whenever M is a maximal subgroup of G. For such M, gMg^{-1} is also a maximal subgroup of G (if B is some subgroup such that $gMg^{-1} \subseteq B \subseteq G$, then $M \subseteq g^{-1}Bg \subseteq G$ so $g^{-1}Bg = M$ or G and hence $B = gMg^{-1}$ or G) so $x \in \Phi(G)$ implies $x \in gMg^{-1}$ and hence $g^{-1}xg \in M$ as required.

DEFINITION Let G be a group and let H be a subgroup of G. Then T is a **transversal** of H in G if and only if $T \subseteq G$ such that

(a) For all $x \in G$, $Hx = H\theta$ for some $\theta \in T$.

(b) If $\theta, \theta^* \in T$ such that $H\theta = H\theta^*$, then $\theta = \theta^*$.

(c) $1 \in T$.

It is clear that transversals exist in finite groups. Their existence in arbitrary groups follows from the axiom of choice.

THEOREM A.10.14 *Let G be a finite group, let H be a subgroup, and let T be a transversal for H in T. Then*

(a) *For all $x \in G$, there exists a unique $h \in H$ and a unique $\theta \in T$ such that $x = h\theta$.*

(b) *If G is finite, then* $\text{card}(T) = \text{card}(G)/\text{card}(H)$.

Proof: Suppose $x = h\theta = h^*\theta^*$ where $h, h^* \in H$ and $\theta, \theta^* \in T$. Then $H\theta = H\theta^*$ implying $\theta = \theta^*$ by definition of transversal. This and $h\theta = h^*\theta^*$ imply $h = h^*$ proving Part (a).

By Part (a), each $x \in G$ has a unique decomposition $x = h_x\theta_x$ where $h_x \in H$ and $\theta_x \in T$. Define $\phi: G \to H \times T$ by $\phi(x) = (h_x, \theta_x)$. Then ϕ is injective because $\phi(x) = \phi(y)$ implies $(h_x, \theta_x) = (h_y, \theta_y)$ hence $x = h_x\theta_x = h_y\theta_y = y$. Furthermore ϕ is surjective because if $(h, \theta) \in H \times T$, then $\phi(h\theta) = (h, \theta)$. Conclude ϕ is a bijective function so $\text{card}(G) = \text{card}(H \times T) = \text{card}(H) \cdot \text{card}(T)$. The result follows.

THEOREM A.10.15 *Let p be a prime, G be a finite group, and S be a p-Sylow subgroup of G. Then*

(a) $[G:N_G(S)] = n_p(G)$.

(b) *If S is the only p-Sylow subgroup of G, then $S \lhd G$.*

Proof: By Theorem A.1.4, $[G:N_G(S)]$ is exactly the number of subgroups of G that are conjugate to S in G. Clearly each conjugate of S is also a p-Sylow subgroup of G, and by Sylow's theorem (Theorem A.1.6) each p-Sylow subgroup of G is conjugate to S in G. Thus there are exactly $n_p(G)$ subgroups of G conjugate to S in G. Conclude $[G:N_G(S)] = n_p(G)$ proving Part (a).

Now suppose S is the only p-Sylow subgroup of G. Then $[G:N_G(S)] = 1$ by Part (a). Hence $G = N_G(S)$ so S is normal in G.

References

[1] Baer, R., Supersolvable groups, *Proceedings of the American Mathematical Society* (1955) 16–32.

[2] Baumslag, B., and Chandler, B., *Theory and Problems of Group Theory,* McGraw-Hill, New York, 1968.

[3] Brauer, R., Investigations of group characters, *Annals of Mathematics* 2, 42 (1941) 936–958.

[4] Burnside, W., *Theory of Groups of Finite Order,* Dover, New York, 1955.

[5] Carmichael, R., *Introduction to the Theory of Groups of Finite Order,* Dover, New York, 1956.

[6] Coxeter, H., and Moser, W., *Generators and Relations for Discrete Groups,* Springer-Verlag, Berlin, 1972.

[7] Deskins, W.E., A characterization of finite supersolvable groups, *The American Mathematical Monthly,* vol 75, number 2, (1968), 180–182.

[8] Dickson, L.E., *Linear Groups with an Exposition of the Galois Field Theory,* Dover, New York, 1958.

[9] Dugundji, J., *Topology,* Allyn and Bacon, Boston, 1966.

[10] Feit, W., and Thompson, J., Finite groups which contain a self-centralizing subgroup of order three, *Nagoya Mathematics Journal,* 21, (1962) 185–197.

[11] Fuchs, L., *Infinite Abelian Groups,* Academic Press, New York, 1970 (vol I), 1973 (vol II).

[12] Gorenstein, D., *Finite Groups,* Harper and Row, New York, 1968.

[13] Greenberg, P.J., *Mathieu Groups,* Courant Institute of Mathematical Sciences, New York, 1973.

[14] Hall, M., *The Theory of Groups,* Macmillan, New York, 1959.

[15] Hall, M., and Senior, J.K., *The Groups of Order* 2^n $(n \leqslant 6)$, Macmillan, New York, 1964.

[16] Hall, P., Some constructions for locally finite groups, *Journal of the London Mathematical Society,* 34, (1959) 305–319.

[17] Hall, P., Finiteness conditions for solvable groups, *Proceedings of the London Mathematical Society* (3) vol 4, (1954) 419–436.

[18] Herstein, I.N., *Topics in Algebra,* Blaisdell, New York, 1964.

[19] Hewitt, E., and Stromberg, K., *Real and Abstract Analysis,* Springer-Verlag, New York, 1965.

[20] Higman, G., On infinite simple permutation groups, *Publ Math Debrecen,* 3, (1954) 221–226.

[21] Huppert, B., *Endliche Gruppen,* Springer-Verlag, Berlin, 1967.

[22] Kegel, O.H., and Wehrfritz, B.A.F., *Locally Finite Groups,* North-Holland, Amsterdam, 1973.

[23] Lewis, F.A., Note on the defining relations for the simple group of order 660, *Bulletin of the American Mathematical Society* 44, (1938) p 456.

[24] Mac Lane, S., *Homology,* Springer-Verlag, Berlin, 1967.

[25] Magnus, W., Karass, A., and Solitar, D., *Combinatorial Group Theory: Presentations of Groups in Terms of Generators and Relations,* John Wiley and Sons, New York, 1966.

[26] McCarthy, D., Infinite groups whose proper quotient groups are finite, *Comm Pure and Applied Math* 21, (1968) 545–562.

[27] Miller, G.A., Abstract definitions of all the substitution groups whose degrees do not exceed 7, *American Journal of Mathematics* 33, (1911) 363–372.

[28] Newman, M., *Integral Matrices,* Academic Press, New York, 1972.

[29] Rankin, R.A., *The Modular Group and its Subgroups,* The Ramanujan Institute, University of Madras, 1969.

[30] Robinson, D., *Finiteness Conditions and Generalized Solvable Groups,* Springer-Verlag, Berlin, 1972.

[31] Rotman, J., *The Theory of Groups: An Introduction,* Allyn and Bacon, Boston, 1973.

[32] Schenkman, E., *Group Theory,* D. Van Nostrand, Princeton, 1965.

[33] Scott, W.R., *Group Theory,* Prentice-Hall, Englewood Cliffs, 1964.

[34] Suzuki, M., The nonexistence of a certain type of simple group of odd order, *Proceedings of the American Mathematical Society* 8, (1957) 686–695.

[35] Todd, J.A., and Coxeter, H.S.M., A practical method for enumerating cosets of a finite abstract group, *Proceedings of the Edinburgh Mathematical Society* (2), 5, (1936) 26–34.

[36] Zassenhaus, H.J., *The Theory of Groups,* Chelsea Publishing Company, New York, 1958.

Hints to Some of the Exercises

Some of these "hints" are complete solutions, some are outlines of proofs, some are answers that require proving and some are no more than suggestions of how to proceed. It should be kept in mind that the hint to a given exercise may not be the only way to solve that exercise.

Section 1.4, Exercise 15: Let $G = \Sigma\{A_i\}_{i\in I}$. Suppose G/H is nontrivial. Then $\mathrm{D}_G(\{i\}) \not\subseteq H$ for some $i \in I$. Hence $\mathrm{D}_G(\{i\})/(\mathrm{D}_G(\{i\}) \cap H)$ is nontrivial. Continue. **Exercise 24**: Let B denote $\Sigma\{G_p\}_{p\in PR}$. Suppose $(Bf)^n = (Bg)^n$. Show

$S(fg^{-1}) \subseteq S(f^n g^{-n}) \cup \{q \mid q$ is a prime divisor of $n\}$. **Section 1.5, Exercise 15:** Use the Lemma to Theorem 1.6.10. **Section 1.7, Exercise 6:** Let $X = \times\{A_i\}_{i \in I}$. Then $\pi_i : X \to A_i$ induces $\pi_i^{**} : X$ wr $B \to A_i$ wr B. Show $\cap\{\ker \pi_i^{**} \mid i \in I\}$ is trivial. **Exercise 7:** Use Exercise 6. **Section 1.8, Exercise 5:** Denote $\Sigma\{A_i\}_{i \in I}$ by P, $\mathrm{Hom}(\Sigma\{A_i\}_{i \in I}, B)$ by H, and $\times\{\mathrm{Hom}(A_i, B)\}_{i \in I}$ by X. By Exercise 3, there are homomorphisms $\lambda_i^* : H \to \mathrm{Hom}(A_i, B)$. Use these and Exercise 14 of Section 1.3 to get $k : H \to X$ such that $\pi_i \circ k = \lambda_i^*$ for all i. To show k is onto, let $\phi \in X$. Then $\phi(i) \in \mathrm{Hom}(A_i, B)$ for all i. Use Theorem 1.4.13 to get $\beta : P \to B$ such that $\beta \circ \lambda_i = \phi(i)$. Show $k(\beta) = \phi$. Then show k is injective. **Exercise 6:** Let H denote $\mathrm{Hom}(A, \times\{B_i\}_{i \in I})$ and X denote $\times\{\mathrm{Hom}(A, B_i)\}_{i \in I}$. Let $\pi_i : \times\{B_i\}_{i \in I} \to B_i$ and $\bar{\pi}_i : X \to \mathrm{Hom}(A, B_i)$ be the coordinate projections. Use Exercise 14 of Section 1.3 to get $k : H \to X$ such that $\bar{\pi}_i \circ k = (\pi_i)_*$ for all i. To show k is surjective let $\phi \in X$. Then $\phi(i) \in \mathrm{Hom}(A, B_i)$ for all i. Use Exercise 14 again to get $h : A \to \times\{B_i\}_{i \in I}$ such that $\pi_i \circ h = \phi(i)$. Show $k(h) = \phi$. Then show k is injective. **Chapter 1, Miscellaneous Exercise I.10:** For Part (b) show the class of groups satisfying the ascending chain condition on normal subgroups is extension closed.

Section 3.2, Exercise I.6: Define $\phi : Z_i(UT(n,R)) \to R \times R \times \ldots \times R$ (i times) by $\phi(A) = (A_{n-i+1,1}, A_{n-i+2,2}, \ldots, A_{n,i})$. Show ϕ is a surjective homomorphism with kernel $Z_{i-1}(UT(n,R))$. **Exercises I.7 and I.8:** Use Exercise I.6. **Exercise IV.1:** By Theorem 3.2.18 it suffices to show every transvection is an nth power. **Exercises IV.6, IV.8, and IV.9:** See the proof of Theorem 3.2.26.

Section 4.2, Exercise 8: First prove the lemma: $B/Z(B)$ is isomorphic to the group of inner automorphisms of B, for any group B. **Section 4.3, Exercise 2:** Use the fact (prove if necessary) that if B is any group and $b \in B$, then b has $[B : C_B(b)]$ conjugates. The class equation is $8 = 2 + 2 + 2 + 2$. **Exercise 3:** For Part (a), define $f : \{X, Y\} \to \{x, xy\}$ by $f(X) = x$ and $f(Y) = xy$. Use Theorem 2.2.6. **Exercise 8:** First prove the lemma: If H is abelian with subgroups I, J, K such that $I \subseteq K$, then $I \vee (J \wedge K) = (I \vee J) \wedge K$. Use this lemma to show that A is nonabelian. Also use Theorems A.9.4 and A.10.3. **Section 4.4, Exercise 2:** See hint to Section 4.3, Exercise 2. **Exercise 3:** Use Theorem 2.2.6. **Exercise 4:** Use Exercise 3. **Exercise 7:** Use Theorems A.9.4 and A.10.13 and Exercise 5. **Section 4.5, Exercise 6:** Use Theorem 2.2.6. **Exercise 8:** Use the fact stated in the hint to Section 4.3, Exercise 2. The class equation is $12 = 2 + 2 + 2 + 3 + 3$. **Section 4.6, Exercise 6:** Use Theorem 2.2.6. **Exercise 7:** Let T be a 3-Sylow subgroup of A, show $\mathrm{Core}_A(T)$ is trivial, then use Theorem A.1.3 to get A embedded in S_4. Next show S_4 has a unique subgroup of cardinality 12. **Exercise 11:** Using Theorem A.9.4, note that A has a unique subgroup B isomorphic to $Z(2) \times Z(2)$. Therefore $B \lhd A$. Show $\mathrm{card}(A) = 4p$ for some prime p. Eliminate the case $p = 2$ by using Note 2 to Example 4.3. Show A is nonsupersolvable. Finally apply Exercise 8. **Section 4.7, Exercise 5:** Use the Lemma to Result 4.8.7. **Exercise 6:** Use the fact stated in the hint to Section 4.3, Exercise 2. **Ex-**

ercise 7: First show that for any $d \in \mathrm{Aut}(G)$, $d(y) = y$ or x^4y and there are 8 choices for $d(x)$. Next show any d satisfying these conditions is an automorphism by using Theorem 2.2.6. Consider $f,g,h \in \mathrm{Aut}(G)$ such that $f(x) = x^3y$, $f(y) = y$; $g(x) = x^3$, $g(y) = y$; and $h(x) = x^3$, $h(y) = x^4y$. Show $\mathrm{Aut}(G) = \langle f,g,h\rangle = \langle f,g\rangle \oplus \langle h\rangle$. **Exercise 8**: Use f defined in the previous hint to show $\langle x\rangle$ and $\langle xy\rangle$ are not characteristic. **Section 4.8, Exercise 4**: There exists $g \in \mathrm{Aut}(G)$ such that $g(x) = xy^2$ and $g(y) = y$. This shows neither $\langle x\rangle$ nor $\langle xy^2\rangle$ is characteristic. There exists $h \in \mathrm{Aut}(G)$ such that $h(x) = xy^2$ and $h(y) = xy$. This shows neither $\mathrm{Z}(G)\langle y\rangle$ nor $\mathrm{Z}(G)\langle xy\rangle$ are characteristic. **Section 4.9, Exercise 9**: By the hint to Section 4.2, Exercise 8, it suffices to show $\mathrm{card}(\mathrm{Aut}(G)) = 24$. First show $\mathrm{card}(\mathrm{Aut}(G)) \leqslant 36$ by showing there are at most 6 choices for $g(y)$ and 6 choices for $g(xy)$ for any automorphism g, and that these choices determine g. **Exercise 13**: First show $n_3(A) = 4$. Let S be a 3-Sylow subgroup of A and let $C = \mathrm{Core}_A(S)$. Show C is trivial then use Theorem A.1.3. **Exercise 15**: Let $T = \langle h,d\rangle$ where h and d are automorphisms of Q as in Section 4.4, Exercise 3. Show $T \cong G$. Then show $T = \mathrm{Aut}(Q)$ by showing $\mathrm{card}(\mathrm{Aut}(Q)) \leqslant 24$. **Section 4.10, Exercise 4**: Cyclic subgroups generated by elements of order 6 are Carter subgroups. **Exercise 5**: First show that if B is a nontrivial subgroup of G such that $\mathrm{Z}(G) \not\subseteq B$, then $\mathrm{card}(B) = 3$. Next show $n_3(G) = 4$. Thirdly show $G/\mathrm{Z}(G) \cong PSL(2,3)$, hence there are exactly 10 subgroups of G that contain $\mathrm{Z}(G)$. **Exercise 7**: 2 classes of elements of order 3, 2 classes of elements of order 6, 1 class of elements of order 4, and the center. **Exercise 8**: Show A has a normal subgroup H such that A/H has cardinality 12 and is nonsupersolvable. Conclude $A/H \cong PSL(2,3)$ by Section 4.6, Exercise 7. Thus $A/H = \langle Hx,Hy\rangle$ where $x^3, y^3, (xy)^2 \in H$ and $H = \langle h\rangle$ where $o(h) = 2$. Divide into cases; $x^3 = y^3 = (xy)^2 = 1$; $x^3 = 1$, $y^3 = h$, $(xy)^2 = 1$; etcetera. **Exercise 9**: Identify $\Phi(A)$, note that $\mathrm{L}(A/\Phi(A)) \cong \mathrm{L}(PSL(2,3))$, and conclude $A/\Phi(A) \cong PSL(2,3)$ by Section 4.6, Exercise 11. Next note A has a subgroup B such that $\mathrm{L}(B) \cong \mathrm{L}(Q)$ so conclude $B \cong Q$ by Section 4.4, Exercise 7. Show $\mathrm{card}(A) = 24$. Show A is nonsupersolvable and $\mathrm{Z}(A)$ is nontrivial. Then use Exercise 8. **Section 4.11, Exercise 3**: Let J,K be distinct 2-Sylow subgroups, suppose $\mathrm{card}(J \cap K) = 2$, and show $\mathrm{N}_G(J \cap K) = G$. **Exercise 5**: Suppose f is inner, determined by t. Then $x = f(y) = t^{-1}yt = t^{-1}(t^{-1}xt)t$ implies $t^2 \in \mathrm{C}_G(\langle x\rangle)$. Show $\mathrm{C}_G(\langle x\rangle) = \langle x\rangle$. Thus $t^2 \in \langle x\rangle$ and similarly $t^2 \in \langle y\rangle$. Hence $t^2 = 1$. Show $(ty)^4 = 1$ and conclude $(ty)^2 = 1$. Therefore $xy = (ty)^2 = 1$, the required contradiction. **Exercise 6**: 15 of cardinality 2, 10 of cardinality 3, 5 of cardinality 4, 6 of cardinality 5, 10 of cardinality 6, 6 of cardinality 10, 5 of cardinality 12. **Exercise 7**: 2 classes of elements of order 5. **Exercise 8**: It suffices to show $n_2(A) = 5$ for then, letting S be a 2-Sylow subgroup of A, $A \cong A/\mathrm{Core}_A(\mathrm{N}_A(S))$ which can be embedded in S_5 by Theorem A.1.3. Also G can be embedded in S_5. But S_5 has only 1 subgroup of cardinality 60. Hence $A \cong G$. **Exercise 10**: Let $G^* = \mathrm{G}(R,S \mid R^2 = 1, S^3 = 1, (RS)^5 = 1)$. Define $a,b \in G^*$ by $a = (s^2rs^2)^2$ and $b = rs$. Show $a^5 = b^5 = (ab)^2 = (b^{-1}a)^3 = 1$. Show $G^* = \langle a,b\rangle$. **Section 4.12, Exercise 4**: Use Section 4.9, Exercise 14. **Exercise 9**: There are two classes of elements of order 7 and the class equation is $168 = 1 + 21 + 56 + 42 + 24 + 24$. **Exercise 10**: Show subgroups of cardinalities 1, 2, 3, 4, 6, 7, 8, or 12 are nonmaximal. To do this show subgroups of cardinalities 6, 8, or 12 are contained in subgroups of cardinality 24. **Section 4.13, Exercise 1**: Use Theorem A.1.3 to embed both A and G in S_6 then show S_6 has a unique subgroup of cardinality 360. **Exercise 7**: Show H is nonsupersolvable and indecomposable. Therefore (Section 4.9, Exercise 13 and Section 4.10, Exercise 8) $H \cong S_4$ or $SL(2,3)$. Eliminate the second case by consideration of the 2-Sylow subgroup of H. **Exercise 8**: Reduce to showing $n_2(G) = 15$ is false. Let S be a 2-Sylow subgroup. Show $Z(S) \lhd N_G(S)$. Show $n_3(N_G(S)) = 1$ leads to a contradiction, hence $n_3(N_G(S)) = 4$. Let T be a 3-Sylow subgroup of $N_G(S)$. Show $N_G(S)/Z(S) \cong PSL(2,3)$. Conclude $N_G(S)$ is nonsupersolvable. Consider separately the cases $N_G(S)$ is decomposable/indecomposable. **Section 4.14, Exercise 8**: Eliminate the possibility $n_2(G) = 11$ as follows. Let S be a 2-Sylow subgroup of G, let $H = N_G(S)$, and let T be a 3-Sylow subgroup of H. Show $\text{card}(N_H(T)) = 6$. Show $\text{card}(S \cap N_H(T)) = 2$. Let $B = SN_H(T)$. Divide into cases, $n_3(B) = 4$ or 1. Find contradictions in either case. **Exercise 9**: It suffices to show that H is simple. Suppose $T \lhd H$. Divide into cases according to the possible cardinalities of T. Consider $n_5(T)$ and/or $n_3(T)$ in each case. One case ($\text{card}(T) = 30$) as follows. If $n_5(T) = 1$, T has a unique 5-Sylow subgroup S. Then $S \lhd H$ which leads to a contradiction. If $n_3(T) = 1$, then T has a unique 3-Sylow subgroup W. Then $W \lhd H$, again a contradiction. Therefore $n_5(T) = 6$ and $n_3(T) = 10$. Conclude that there are $4 \cdot 6 + 2 \cdot 10 = 44$ elements of order 5 or 3 in T. **Exercise 10**: Divide into cases, A is/is not supersolvable. **Chapter 4, Miscellaneous Exercises II.1**: There are 16 subgroups. **II.2**: Let $G = Z(4) \times Z(2)$, $A = \langle(1,0)\rangle$, $B = \langle(1,1)\rangle$. **III.4**: First show $G_{4,4}$, $D_4 \times Z(2)$, and $Q \times Z(2)$ have 7, 11, and 3 elements of order 2 respectively. Arrange the nine groups into 6 classes determined by the number of elements of order 2 then separate the groups in each class. **III.11**: Show $Z(K) = \{A \mid A \in K \text{ and } A_{31} = A_{32} = A_{43} = 0\}$. Thus $g: Z(K) \to Z(2) \times Z(2)$ defined by $g(A) = (A_{41}, A_{42})$ is an isomorphism. To prove indecomposability, eliminate the possibilities $K = C \oplus D$ where $\text{card}(C) = 2$, $\text{card}(D) = 16$; or where $\text{card}(C) = 4$, $\text{card}(D) = 8$. **IV.3**: By the hint to Section 4.2, Exercise 8, it suffices to show $\text{card}(\text{Aut}(G)) = 20$. Let $x = (1,1)$ and $y = (0,\beta)$ where β is defined by $\beta(w) = 2w$. Show $G = \langle x,y \rangle$. For $\lambda \in \text{Aut}(G)$, there are at most 4 choices for $\lambda(x)$ and 10 choices for $\lambda(y)$. Therefore $\text{card}(\text{Aut}(G)) \leqslant 40$. Show however that the choice $\lambda(x) = x$, $\lambda(y) = y^{-1}$ can be eliminated. This and $20 \mid \text{card}(\text{Aut}(G))$ imply $20 = \text{card}(\text{Aut}(G))$. **V.1 and V.3**: Let $G = PSL(2,3) \times Z(2)$. **V.4**: Exhibit all (four) transvections of $SL(2,3)$ as commutators in G. **VI.1 Part (b)**: By Miscellaneous Exercise 6 of Chapter 3, G is an extension of $P \cong PSL(2,7)$ by $Z(2)$. Suppose $\text{card}(H) = 48$. Then $G = PH$ and consequently $\text{card}(P \cap H) = 24$. Therefore by Section 4.12, Exercise 4, $P \cap H \cong S_4$. Apply Section 4.9, Exercise 10 to get $H \cong S_4 \times Z(2)$. Therefore the 2-Sylow subgroups of H, and hence of G, are isomorphic to $D_4 \times Z(2)$. **VI.4**: $1 = x^3yx^3yx^3y$ implies $x^{-3}y^{-1}x^3 = yx^3y$ so $yx^{-3}y^{-1}x^3 = y^2x^3y$ and hence $y^2x^3y \in G'$. Also

$x^{-3}y^{-1}x^3y = yx^3y^2$ so $yx^3y^2 \in G'$. Hence $(yx^3y^2)(y^2x^3y) = yx^3yx^3y = x^{-3} \in G'$. Continue. **VI.5:** $G = SL(2,7)$. Show the class of finite groups whose maximal subgroups have prime-power index is closed under central extensions.

Section 5.1, Exercise 2: Cf. Theorem A.10.5. **Exercises 6 and 8:** Use Result 5.1.14. **Exercise 9:** Let G denote the group with the given presentation. Show Q is isomorphic to a quotient group of G. Show G is torsion-free, divisible, and indecomposable. Then apply Exercise 8. **Section 5.2, Exercise 3:** Use Result 5.1.14. **Section 5.3, Exercise 4:** Use Result 5.1.14. **Exercise 5:** Let G denote the group with the given presentation. Show G is torsion-free. Show G can be embedded in Q and hence that G is pure-simple. Show $G/K \cong Q^p$ for some K. **Section 5.4, Exercise 3:** Let A be a nontrivial divisible abelian p-group. Let $a_1 \in A$ have order p. By divisibility there exists $a_{i+1} \in A$ such that $pa_{i+1} = a_i$ for all $i \in N$. Show $o(a_i) = p^i$ for all i. Define $g: Q^p \to A$ by $g(t/p^n) = ta_n$. Show $g[Z]$ is trivial and use the Corollary to Theorem A.10.1. **Exercises 5 and 7:** Use Exercise 4. **Section 5.5, Exercises 1 and 2:** Use Result 5.1.14. **Exercise 8:** For Part (a) use Theorem A.8.7. **Exercise 10:** Use Section 5.4, Exercise 2 to show there are no zero divisors. **Section 5.6, Exercise 5:** Use Exercise 1 to show $A/(B + C) \cong Z(r)$. Show $B + C$ is a hopfian fully invariant subgroup of A. Use the Lemma to Result 5.12.6. **Section 5.7, Exercise 1:** First show that $Z(n) \times Z(m)$ is cyclic whenever n and m are relatively prime. **Exercise 6:** Use Theorem 1.4.13. **Section 5.8, Exercise 1:** Show $Z(p) \times Z(p)$ can be embedded in G. **Section 5.10, Exercise 3:** First show that for any abelian group A and any relatively prime numbers n and m, $nA \cap mA \subseteq nmA$. Then use Result 5.10.16. **Exercise 6:** First show $\mathrm{Aut}(Z) \cong Z(2)$. Then show any automorphism of SF is determined by its action on Z. **Section 5.11, Exercise 3:** Suppose B is an isomorphic subgroup of G. Use the fact that $t(G)$ is cohopfian to show $t(B) = t(G)$ and hence that $t(G) \subseteq B$. Show $B/t(B)$ is divisible, then use Theorem A.8.4. Show B is a pure subgroup of G, then use algebraic compactness. **Section 5.13, Exercise 2:** Consider G as a subgroup of Example 5.12 and use Example 5.12, Exercise 3. **Section 5.14, Exercises 3 and 4:** First show that direct factors of hopfian (respectively cohopfian) groups are hopfian (cohopfian). Then use Section 1.4, Exercise 17. **Section 5.15, Exercise 4:** First show that there exists a surjective homomorphism from $\Sigma\{Z_i\}_{i\in N}$ onto Q. Next embed $\Sigma\{Z_i\}_{i\in N}$ in A and use Theorem A.8.4. **Chapter 5, Miscellaneous Exercises I.6:** For Part (b), let H be a proper subgroup so there exists $f \in B\setminus H$. Use Zorn's lemma to produce a subgroup M maximal with respect to $H \subseteq M$ and $f \notin M$. Show B/M is subdirectly irreducible. Then use the Lemma to Result 6.4.12. **I.7:** Use Zorn's lemma and the Lemma to Result 6.4.12. **II.2:** For Part (b), apply the Lemma to Result 5.5.7 to $\{p^kA \mid p \in PR$ and $k \in N\}$. **II.7:** Let $G = J \oplus K$ where $J \cong Q \cong K$. Suppose $G/H \cong G$. First show $J \cap H$ is trivial, hence $(H + J)/H$ is divisible. Use Theorem A.8.4. Also use Result 5.1.7. **II.9:** Let H be a characteristic subgroup. First show H is a pure subgroup. Next let $A = Q \times \{0\}$ and $B = \{0\} \times Q$. Show $H \cap A$ is a characteristic subgroup of A, and $H \cap B$ is a characteristic subgroup of B. Then divide into cases, $H \cap A = H \cap B = \{0\}$, $H \cap A = A$, and $H \cap B = B$. **III.3**: $G \times Q$ where G is Example 5.12. **III.6**: Let $A = \times\{G_p\}_{p \in PR}$ where G_p denotes Example 5.9.

Section 6.1, Exercise 4: Note that $Z(G) = G'$ and use the Lemma to Result 5.12.6. **Section 6.2, Exercise 1**: For Part (b) first show that the class of groups satisfying the descending chain condition on normal subgroups is quotient closed. **Exercise 2**: Use Theorems A.10.7 and A.8.2. **Section 6.3, Exercise 1**: Use Section 1.2, Exercise 11. **Exercise 4**: Define $d{:}M \to M$ by $d(x,y,i) = (2x, y, 2i)$. **Exercise 5**: Use the Lemma to Result 5.12.6. **Section 6.5, Exercise 3**: Let G be the group with the given presentation. Show $\langle a \rangle \cap \langle b \rangle$ is trivial. Show $\langle a \rangle \lhd G$. Show G is just-infinite. **Exercise 4**: Use Exercise 3. **Exercise 6**: First show that the class of groups satisfying the ascending chain condition is extension closed. **Section 6.6, Exercise 2**: First show that if A is any group, B is a fully invariant subgroup of A and C is a fully invariant subgroup of B, then C is fully invariant as a subgroup of A. **Section 6.7, Exercise 1**: Use Result 6.7.2. **Exercise 8**: Use the Lemma to Result 5.12.7. **Exercise 11**: Let G be the group with the given presentation and let R be the subgroup $\langle x_i \mid i \in N, i \neq 0 \rangle$. Show $R \cong Q$. Show $G/K \cong H$ for some K. Divide into cases $RK = G$ or $RK = R$. **Exercise 12**: Let C be a characteristic subgroup, assume $C \neq H$, and show C is a pure subgroup of Q^*. **Section 6.8, Exercise 1**: Use the Lemma to Result 6.8.2. **Exercise 6**: First show $Q^2 \times Q^2$ is hopfian. Using this and the Lemma to Result 5.12.6, show U is hopfian. Then use this lemma again to show G is hopfian. **Exercise 9**: By Result 6.8.3 there is a maximal subgroup M containing $\langle A \rangle$. **Section 6.9, Exercise 6**: If $K/Z(K)$ were hypercyclic it would have a nontrivial normal cyclic subgroup $R/Z(K)$ contained in $U/Z(K)$. Assume this and derive a contradiction. **Section 6.11, Exercise 2**: The center is trivial because of Lemma 2 to Theorem A.7.4. **Section 6.13, Exercise 6**: Eliminate the possibility $G = G^2$ by Exercise 5, Part (a) and the fact that $D_3 \neq \{w^2 \mid w \in D_3\}$. **Exercise 7**: Eliminate the possibility $G = G^3$ by Exercise 5, Part (b) and the fact that $PSL(2,3) \neq \{w^3 \mid w \in PSL(2,3)\}$. **Section 6.15, Exercise 1**: Since B is finite there exists an embedding $k{:}B \to G$. Let h be the restriction of k to A. Note that $g^{-1}{:}g[A] \to A$ is an isomorphism and let $\phi = h \circ g^{-1}$. Apply Result 6.15.3, Part (b) to get $x \in G$ such that $\phi(t) = x^{-1}tx$ for all $t \in g[A]$. Continue. **Exercise 2**: Show $G \times G$ can be embedded in, but is not isomorphic to G. **Chapter 6, Miscellaneous Exercises I.7**: To see that $Z(p)$ is a Q_p-module, note that if $a/b \in Q_p$, then b and p are relatively prime so there are integers satisfying $ib + jp = 1$. Then for all $x \in Z(p)$ define scalar multiplication by $(a/b)x = iax$. **III.6**: Generators are $(0,1)$ and $(f,0)$ where f is defined by $f(1) = 1$ and $f(i) = 0$ for $i \neq 1$. **IV.3**: To show infinite index, note that if F_n has finite index, then so does its core. **V.1**: Use the hint to Section 4.3, Exercise 2. Also use Theorem A.1.3.

Glossary and Index of Definitions

The number following a term is the page number its definition appears on, if the term has been defined in text. Otherwise a definition is given.

The abbreviation *iff* stands for if and only if.

abelian. The group A is abelian iff $xy = yx$ for all $x,y \in A$.

algebraically compact. 177

ascending central series. 265

ascending chain condition (respectively ascending chain condition on normal subgroups). The group G satisfies this condition iff whenever $H_0, H_1, H_2, \ldots$ are subgroups (respectively normal subgroups) of G such that $H_0 \subseteq H_1 \subseteq H_2 \subseteq \ldots$, there exists some $i \in N$ such that $H_i = H_m$ for all $m \geqslant i$.

bonding maps (in direct systems). 32

bonding maps (in inverse systems). 51

boolean algebra. 184

bounded. The group G is bounded iff $\exp(G) = n$ for some $n \in N$.

Carter subgroup. If G is a group, then the subgroup H is a Carter subgroup iff H is nilpotent and $N_G(H) = H$.

central extension. The group G is a central extension of H by K iff $H \lhd G$, $H \subseteq Z(G)$, and $G/H \cong K$.

central pushout (of a central pushout system). 2

central pushout (of a countable central pushout system). 39

central pushout system. 2

chain. 263

characteristically simple. The group G is characteristically simple iff $\{1\}$ and G are the only characteristic subgroups of G.

characteristic subgroup. The subgroup H of the group G is characteristic iff $g[H] \subseteq H$ for all $g \in \mathrm{Aut}(G)$.

Chernikov. 223

coatomic. 284

cohopfian. The group G is cohopfian iff G has no proper isomorphic subgroups. This condition is equivalent to the requirement that all injective homomorphisms $f: G \to G$ also be surjective.

commutes (diagram commutativity). While the following definition is unsuitable for arbitrary diagrams (it cannot, for instance, be applied to a diagram in the shape of a pentagon) it is sufficiently general for all instances of its use in this book.

A triangular diagram

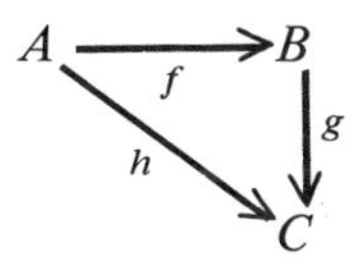

commutes iff $h = g \circ f$.

A rectangular diagram

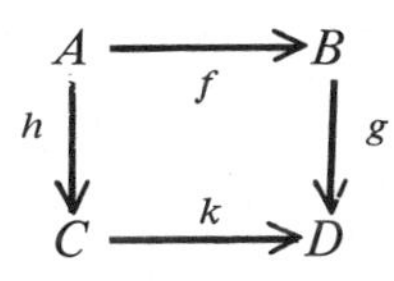

commutes iff $g \circ f = k \circ h$.

A diagram commutes iff all of its triangular and rectangular subdiagrams commute.

complemented lattice. 184

complete. The group G is complete iff $Z(G) = \{1\}$ and all automorphisms of G are inner automorphisms.

converse lagrange. The finite group G

is converse lagrange iff G has a subgroup of cardinality n for all n dividing card(G).

coordinate injection function. 24

coordinate projection function. 19

core (of a subgroup). If S is a subgroup of the group G, then $\text{Core}_G(S) = \bigcap \{g^{-1}Sg \mid g \in G\}$ is the core of S.

countable central pushout system. 38

derived series. 270

descending chain condition (respectively descending chain condition on normal subgroups). The group G satisfies this condition iff whenever H_0, H_1, H_2 . . . are subgroups (respectively normal subgroups) of G such that . . . $H_2 \subseteq H_1 \subseteq H_0$, there exists some $i \in N$ such that $H_i = H_m$ for all $m \geqslant i$.

determinant (of a 2 by 2 matrix). 70

diagonal. 44

directed set. 32

direct factor. 260

direct limit. 33

direct product. 18

direct system. 32

distributive lattice. 278

divisible. The group G is divisible iff for all $x \in G$ and all positive integers n, there exists $y \in G$ such that $x = y^n$.

embedded. The group A can be embedded in the group B iff there exists an injective homomorphism $j:A \to B$.

exponent. The group G has exponent n where n is a positive integer iff (a) $x^n = 1$ for all $x \in G$, and (b) if m is a positive integer such that $x^m = 1$ for all $x \in G$, then $n \mid m$. An arbitrary group may not have an exponent. Groups with exponents are called bounded.

extended Sylow properties. Let π be a set of primes and let G be a finite group. Then G satisfies extended Sylow proper $E(\pi)$ iff (a) G has (one or more) π-Hall subgroups; G satisfies extended Sylow property $C(\pi)$ iff (a) and (b) any two π-Hall subgroups of G are conjugate in G; G satisfies extended Sylow property $D(\pi)$ iff (a), (b), and (c) if H is any subgroup of G of cardinality divisible by only those primes in π, then $H \subseteq K$ for some π-Hall subgroup K of G.

extension. The group G is an extension of H by K iff $H \lhd G$ and $G/H \cong K$.

extension-closed class. The class of groups $\mathcal{H}$ is extension closed iff $H,K \in \mathcal{H}$ and G is an extension of H by K imply $G \in \mathcal{H}$.

FC. The group G is an FC group iff every element of G has only finitely many conjugates.

finitely generated. The group G is finitely generated iff $G = \langle S \rangle$ for some finite subset S of G.

Frattini subgroup. 284

free abelian group. 57

free group. 55

fully invariant subgroup. The subgroup H of the group G is fully invariant iff $f[H] \subseteq H$ for all homomorphisms $f:G \to G$.

generalized nilpotent class. 207

generalized solvable class. 207

greatest lower bound. 276

Hall subgroup. Let π be a set of primes let G be a finite group and let H be a subgroup of G. Then H is a π-Hall subgroup iff (a) all primes dividing card(H) belong to π, and (b) $[G:H]$ and card(H) are relatively prime.

hamiltonian. The group G is hamiltonian iff G is nonabelian and every subgroup of G is normal.

holomorph. 16

hopfian. The group G is hopfian iff G has no proper isomorphic quotient groups. This condition is equivalent to the requirement that all surjective homomorphisms $f:G \to G$ also be injective.

hypercentral. The group G is hypercentral iff every nontrivial quotient group has a nontrivial center.

hyper-$\mathcal{H}$. If $\mathcal{H}$ is a class of groups, then the group G is hyper-$\mathcal{H}$ iff every nontrivial quotient group of G has a nontrivial normal subgroup belonging to $\mathcal{H}$.

identity map. If A is any set, then $i:A \to A$ is the identity map iff $i(a) = a$ for all $a \in A$.

indecomposable. The group G is indecomposable iff whenever $G = B \oplus C$ either B or C is trivial.

indexed set. 18

inner automorphism. Let G be a group and let $x \in G$. Then $f:G \to G$ is the inner automorphism determined by x iff $f(y) = x^{-1}yx$ for all $y \in G$.

inverse limit. 51

inverse system. 50

isomorphism (of lattices). 278

just-infinite. The group G is just-infinite iff G is infinite and all proper quotient groups of G are finite.

lattice. 276

least element (of a directed set). 36

least upper bound. 276

length (of a word). 54

linear. 239

locally $\mathcal{H}$. If $\mathcal{H}$ is a class of groups, then the group G is locally $\mathcal{H}$ iff all finitely generated subgroups of G belong to $\mathcal{H}$.

lower bound. 276

$\mathbf{M}_1$. 160

$\mathbf{M}_2$. 160

$\mathbf{M}_3$. 160

main diagonal. The main diagonal of the n by n matrix A is the set of entries $\{A_{ii} \mid 1 \leqslant i \leqslant n\}$.

matrix. 64

maximal element. 263

maximal subgroup. 283

minimal normal subgroup. 100

minimal simple. The simple group G is minimal simple iff all proper subgroups of G are solvable.

mixed. 159

MN. The group G is an MN group iff its maximal subgroups are normal.

nilpotent. 265

partially ordered set. 262

p-divisible. When p is a prime, the group G is p-divisible iff for all $x \in G$, there exists $y \in G$ such that $x = y^p$.

perfect. The group G is perfect iff $G = G'$.

π-Hall subgroup. See Hall subgroup.

p-group. When p is a prime, the group G is a p-group iff the order of every element of G is a power of p.

presentation. 58

principal congruence subgroup. 81

proper quotient group. Let G be a group. Then G/H is a proper quotient group of G iff $H \neq \{1\}$.

proper subgroup. H is a proper subgroup of the group G iff H is a subgroup of G and $H \neq G$.

pullback. 50

pullback system. 50

pure-simple. The abelian group A is pure-simple iff $\{0\}$ and A are the only pure subgroups of A.

pure subgroup. Let A be an abelian

group. Then H is a pure subgroup of A iff H is a subgroup of A and $nA \cap H \subseteq nH$ for all positive integers n.

quotient-closed class. The class $\mathcal{H}$ is quotient closed iff $G \in \mathcal{H}$ and $H \lhd G$ imply $G/H \in \mathcal{H}$.

rank (of a torsion-free abelian group). 180

reduced. An abelian group is reduced iff it has no nontrivial divisible subgroups.

reduced word. 54

residually $\mathcal{H}$. 21

restricted direct product. 24

restricted wreath product. 44

R-group. The group G is an R-group iff the equation $x^n = y^n$ where $x,y \in G$ and n is a positive integer implies $x = y$.

"satisfies the relations." 59

self-dual lattice. 278

semidirect product. 8

semidirect system. 7

semidivisible. The group G is semidivisible iff for all $x \in G$ and all $n \in N$, $n \neq 0$, there exist $y_1, y_2, \ldots, y_k \in G$ (the number depends on x and n) such that $x = y_1{}^n y_2{}^n \ldots y_k{}^n$.

separates. If $\mathcal{H}$ and $\mathcal{E}$ are classes of groups such that $\mathcal{H} \subseteq \mathcal{E}$, then the group G separates $\mathcal{H}$ and $\mathcal{E}$ iff $G \in \mathcal{E} \setminus \mathcal{H}$.

simple. The group G is simple iff $\{1\}$ and G are the only normal subgroups of G.

solvable. 271

splitting. 159

subchain. 263

subdirectly irreducible. The group G is subdirectly irreducible iff $\mathrm{H}(G) \neq \{1\}$.

subgroup-closed class. The class of groups $\mathcal{H}$ is subgroup closed iff $A \in \mathcal{H}$ and B is a subgroup of A imply $B \in \mathcal{H}$.

subnormal subgroup. If G is a group, then the subgroup H is subnormal iff there are subgroups $B_1, B_2, \ldots, B_n$ of G such that $H \lhd B_1$, $B_1 \lhd B_2$, $B_2 \lhd B_3, \ldots, B_n \lhd G$.

supersolvable. 99

support. 24

torsion. The group G is torsion iff every element of G has finite order.

torsion-free. The group G is torsion-free iff 1 is the only element of G having finite order.

trace. 85

transpose. 94

transvection. 66

transversal. 285

triangular matrix. 67

trivial element. This is another name for the identity element of a group.

trivial group. A group G is trivial iff $\mathrm{card}(G) = 1$.

Ulm subgroup. If A is an abelian group then $\mathrm{U}(A) = \cap \{nA \mid n \in N, n \neq 0\}$ is the Ulm subgroup of A.

unit. If R is a ring with identity, then $x \in R$ is a unit of R iff there is some $y \in R$ such that $xy = yx = 1$.

unitriangular matrix. 67

upper bound. 263

word. 54

wreath product. 52 (See also restricted wreath product.)

Notation

Additive notation is used throughout for abelian groups and multiplicative notation for nonabelian and arbitrary groups. Identity elements in all groups are denoted by 1 (multiplicatively) or 0 (additively). In particular, this holds for quotient groups. Thus the notation $Hx = 1$ (or $H + x = 0$ additively) is used if and only if $x \in H$.

Alphabetic, English

$\mathrm{Aut}(G)$	automorphism group of G
$C(J)$	principal congruence subgroup of $SL(2,R)$ determined by the ideal J
$\mathrm{C}_G(H)$	$= \{x \mid x \in G$ and $xh = hx$ for all $h \in H\}$; centralizer of H
$\mathrm{card}(G)$	cardinality of G
$\mathrm{char}(F)$	characteristic of the field F: if $0 = 1 + 1 + \ldots + 1$ (p times), then $\mathrm{char}(F) = p$; otherwise $\mathrm{char}(F) = 0$
$\mathrm{Core}_G(H)$	$= \cap \{g^{-1}Hg \mid g \in G\}$; core of the subgroup H
D	if (A, H, K, h, k) is a central pushout system, then D denotes $\{(h(x), k(x^{-1})) \mid x \in A\}$
D_n	nth dihedral group, Example 4.1
$\mathrm{D}_G(J)$	when $G = \Sigma\{A_i\}_{i \in I}$, $\mathrm{D}_G(J)$ denotes $\{f \mid f \in G$ and $f(i) = 1$ for all $i \in I \backslash J\}$
$\bar{\mathrm{D}}_G(J)$	when $G = \mathsf{X}\{A_i\}_{i \in I}$, $\bar{\mathrm{D}}_G(J)$ denotes $\{f \mid f \in G$ and $f(i) = 1$ for all $i \in I \backslash J\}$
$\det(A)$	determinant of the matrix A
$\mathrm{Diag}(\mathrm{F}[A,B])$	$= \{f \mid f \in \mathrm{F}[A,B]$ and $f(p) = f(q)$ for all $p,q \in B\}$; the diagonal of $\mathrm{F}[A,B]$
$Dih(B)$	the generalized dihedral group determined by the abelian group B; see Section 1.2
$\exp(G)$	exponent of G
$\mathrm{F}[A,B]$	denotes $\Sigma\{A_b\}_{b \in B}$ where $A_b = A$ for all $b \in B$
$\bar{\mathrm{F}}[A,B]$	denotes $\mathsf{X}\{A_b\}_{b \in B}$ where $A_b = A$ for all $b \in B$
$\mathrm{G}(X \mid \mathcal{R})$	group of the presentation $(X \mid \mathcal{R})$
$G_{n,m}$	see Chapter 4, Miscellaneous Exercise I.2
$GL(2,R)$	group of 2 by 2 matrices whose entries are in R and whose determinants are units of R
h_d	dth associated map for direct limits
$\mathrm{H}(G)$	$= \cap \{B \mid B \lhd G$ and B is nontrivial$\}$
$\mathrm{Hol}(B)$	holomorph of B
$\mathrm{Hom}(A,B)$	$= \{f \mid f: A \to B$ is a homomorphism$\}$
I	identity matrix
$-I$	$\begin{pmatrix} -1 & 0 \\ 0 & -1 \end{pmatrix}$

Im f	$= \{x \mid x = f(y)$ for some $y\}$; image of the function f
J_p	p-adic integers, Example 5.5
ker f	$= \{x \mid f(x) = 1\}$; kernel of f
L(G)	lattice of subgroups of G
len(T)	length of the word T
$M(B)$	semidirect product of $(Z, g, B \times B)$, see Section 1.2
N	set of all natural numbers
N(G)	lattice of normal subgroups of G
$\mathrm{N}_G(H)$	$= \{x \mid x \in G$ and $x^{-1}Hx \subseteq H\}$; normalizer of H
$n_p(G)$	number of p-Sylow subgroups of G
o(x)	order of x
P(A, x)	$= \{a \mid a \in A$ and $ma \in \langle x \rangle$ for some positive integer $m\}$
$PGL(2,F)$	$= GL(2,F)/\mathrm{Z}(GL(2,F))$
PR	set of all prime numbers
$PSL(2,F)$	$= SL(2,F)/\mathrm{Z}(SL(2,F))$
Q	(1) group of rational numbers, Example 5.1 (2) quaternion group, Example 4.4 (this usage in Chapter 4 only)
Q^p	$= \{a/p^n \mid a/p^n \in Q\}$
Q_p	$= \{a/b \mid a/b \in Q$ and b is relatively prime to $p\}$
R(G)	(1) if (A, H, K, h, k) is a central pushout system, then R(G) denotes $h^*[H] \cap k^*[K]$ (2) for countable central pushouts, $\mathrm{R}(L) = h_0 \circ f_0 [A]$
S_A	symmetric group on the set A
S_n	symmetric group on the set $\{1, 2, \ldots, n\}$
S(A, x)	$= \{a \mid a \in A$ and $ma \in \langle x \rangle$ for some square-free positive integer $m\}$
S(f)	support of f
SD_n	nth semidihedral group, see Chapter 4, Miscellaneous Exercise I.1
SF	$= \{a/b \mid a/b \in Q$ and b is square-free$\}$
$SL(2,R)$	group of 2 by 2 matrices of determinant 1 with entries in R
t(A)	$= \{x \mid x \in A$ and o(x) is finite$\}$; torsion subgroup of A
$T(n,F)$	group of n by n triangular matrices M with entries in F such that $1 \leqslant i \leqslant n$ implies $M_{ii} \neq 0$
tr(M)	trace of the matrix M
U(A)	Ulm subgroup of A
$UT(n,R)$	group of n by n unitriangular matrices with entries in R
$UT(\infty,R)$	direct limit of $(\{UT(k_n,R)\}_{n \in N}, \{\pi_n^m \mid n \leqslant m\})$; see Section 3.2
W(X)	set of words formed by X
$\mathrm{W}_r(X)$	set of reduced words formed by X; equivalently the free group generated by X

wr	A wr B denotes the restricted wreath product of A by B
WR	A WR B denotes the (unrestricted) wreath product of A by B
Z	group of integers (under addition)
$Z(n)$	cyclic group of integers mod n (under addition); $Z(n)$ is the set of equivalence classes of integers under the equivalence relation $a \equiv b \bmod n$ if and only if $n \mid a-b$. By an abuse of notation, an equivalence class will be denoted by any integer belonging to it. Thus $a = b$ in $Z(n)$ if and only if $n \mid a-b$
$Z_n(G)$	nth term of the ascending central series for G
$[Z]_m(UT(n,R))$	$= \{A \mid A \in UT(n,R)$ and $0 < i-j < n-m$ implies $A_{ij} = 0\}$
$Z(p^\infty)$	Example 5.4

Alphabetic, Greek

γ_b	wreath-product notation; $\gamma_b(f) = f_b$
θ	the automorphism $\theta(x) = -x$ of an abelian group
$\Theta_n(i,j,x)$	transvection whose i,j entry is x
λ_j	jth coordinate injection $\lambda_j: A_j \to \Sigma\{A_i\}_{i \in I}$
π	any set of prime numbers
π_j	jth coordinate projection $\pi_j: \mathsf{X}\{A_i\}_{i \in I} \to A_j$
π_A	when $G = A \oplus B$, $\pi_A: G \to A$ is defined by $\pi_A(x) = a$ if and only if $x = ab$ where $a \in A$ and $b \in B$
Σ	(1) restricted direct product notation (2) summation symbol
$\Phi(G)$	Frattini subgroup of G

Nonalphabetic

$\langle\ \rangle$	$\langle S \rangle$ denotes the subgroup generated by the subset S; when S is finite, say $S = \{x_1, x_2, \ldots, x_n\}$, $\langle x_1, x_2, \ldots, x_n \rangle$ will be used instead to $\langle \{x_1, x_2, \ldots, x_n\} \rangle$
*	(1) if (A,g,B) is a semidirect system and S is a subgroup of A, then $*S$ denotes $\{(1,s) \mid s \in S\}$ (2) if (A,g,B) is a semidirect system and T is a subgroup of B, then $T*$ denotes $\{(t,1) \mid t \in T\}$ (3) if (A,H,K,h,k) is a central pushout system (respectively a pullback system), then h^* and k^* denote the associated homomorphisms, see page 3 (respectively page 50) (4) f^* denotes an induced homomorphism, see Theorem 1.2.8; Theorem 1.2.20; Section 1.7, Exercise 4; Section

		1.8, Exercises 1 and 3
$\{\ \}$		$\{A_i\}_{i\in I}$ denotes an indexed set of groups
$[\ ]$	(1)	in direct limits, $[(x,d)]$ denotes the equivalence class containing (x,d)
	(2)	if $M \in SL(2,F)$, then $[M]$ denotes the coset of M relative to the center, $Z(SL(2,F))$; thus $[M] \in PSL(2,F)$
	(3)	$[G{:}H]$ denotes the index of the subgroup H in G
	(4)	if A is an abelian group and n is a positive integer, then $A[n]$ denotes $\{a \mid a \in A \text{ and } na = 0\}$
	(5)	see also $\mathrm{F}[A,B]$ and $\overline{\mathrm{F}}[A,B]$ in the alphabetic list
$\oplus$		$G = A \oplus B$ if and only if $A,B \lhd G$, $A \cap B = \{1\}$, and $G = AB$
$\mid$		$n \mid m$ means n divides m
$!$		$n! = n(n-1)(n-2)\ldots 3\cdot 2\cdot 1$
$\wedge$		$x \wedge y$ is the greatest lower bound of $\{x,y\}$ in a lattice
$\cap$	(1)	$A \cap B = \{x \mid x \in A \text{ and } x \in B\}$
	(2)	$\cap\mathcal{E} = \{x \mid x \in E \text{ for all } E \in \mathcal{E}\}$
$\cong$		isomorphism (groups or lattices)
$\vee$		$x \vee y$ is the least upper bound of $\{x,y\}$ in a lattice
(lowercase)		x is the element of $G(X \mid \mathcal{R})$ associated with $x \in X$; that is, $x = N(x)$; see page 59
$\times_g$		semidirect-product notation
$\times$		direct-product notation
$\lhd$		$H \lhd G$ means H is a normal subgroup of G
$(\ \mid\ )$	(1)	$(X \mid \mathcal{R})$ denotes a presentation
	(2)	$A(h \mid k)$ denotes the matrix formed from the matrix A by deletion of row h and column k
	(1)	X' and a', see Section 2.1
	(2)	G' denotes the commutator subgroup of G
	(3)	induced homomorphisms, see Theorem A.10.1
$\setminus$		$A \setminus B = \{x \mid x \in A \text{ and } x \notin B\}$
(subscript)	(1)	if (A,g,B) is a semidirect system, then g_a is used instead of $g(a)$
	(2)	M_{ij} is the entry in row i and column j of the matrix M
$\subseteq$		$A \subseteq B$ means A is a subset of B
$\subset$		$A \subset B$ means A is a proper subset of B ($A \subseteq B$ and $A \neq B$)
(superscript)	(1)	in wreath-product notation, f^b denotes the function $f^b(\beta) = f(\beta b)$
	(2)	$G^{(n)}$ is the nth term of the derived series of G
$\cup$	(1)	$A \cup B = \{x \mid x \in A \text{ or } x \in B\}$
	(2)	$\cup\mathcal{E} = \{x \mid x \in E \text{ for some } E \in \mathcal{E}\}$

Index of Examples Arranged by Group Properties

WARNING: The failure of a group to be listed corresponding to a property below should not be interpreted to mean that the group does not have this property. It may mean only that the group was not shown (either in text or in the exercises) to have the property.

Index of Some Interesting Examples

I Abelian Groups

II Finiteness Conditions

III Generalized Nilpotent Groups

IV Generalized Solvable Groups

V Nilpotent Groups

Index of Counterexamples

WARNING: Failure of a class of groups to be listed in categories I, II, or III below should not be interpreted to mean that the class is subgroup closed, quotient closed, or extension closed respectively. It may mean only that no suitable counterexample was given in the book to show that the class is not subgroup, quotient, or extension closed.

I Classes that are not subgroup closed

II Classes that are not quotient closed

III Classes that are not extension closed

IV Separations

---

*This example, being finite, also shows all generalized nilpotent classes fail to be extension closed.